图灵教育

站在巨人的肩上
Standing on the Shoulders of Giants

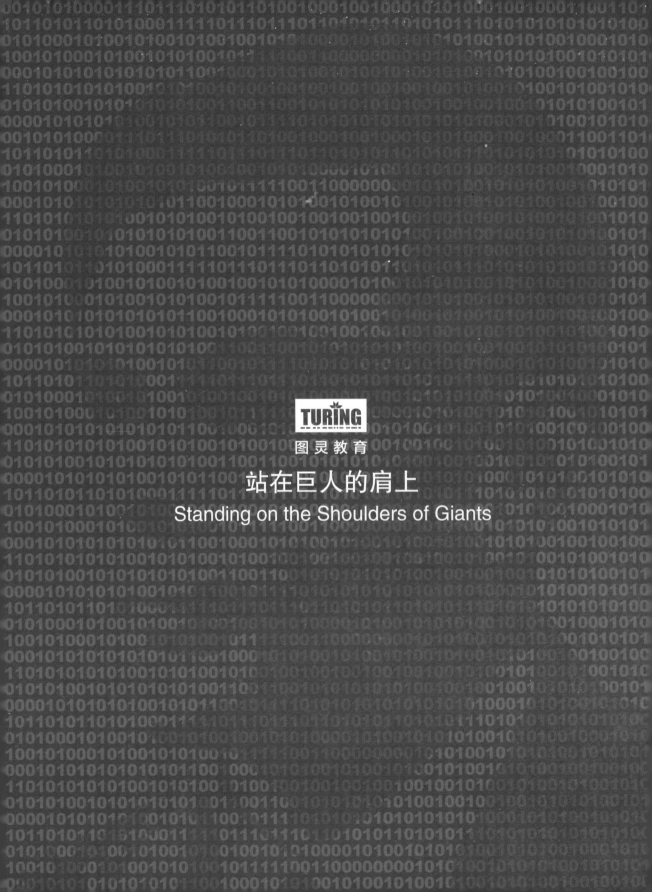

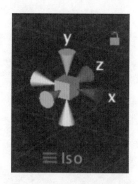

图 2-22 坐标系控制器

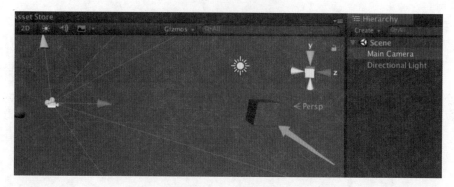

图 3-17 辅助元素

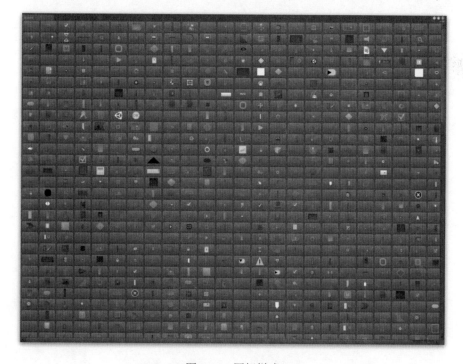

图 3-36 图标样式

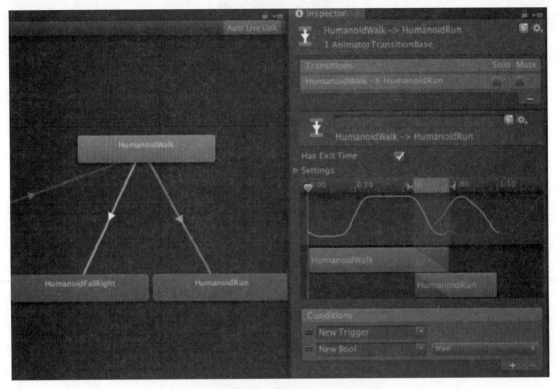

图 7-16 配置条件

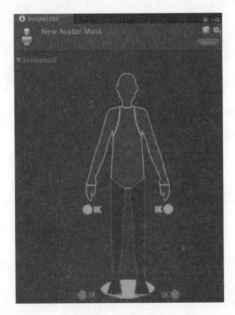

图 7-21 人形遮罩

1	2	3	
4	我	2	3
6	4	我	6
	7	8	9

图 9-17 移动

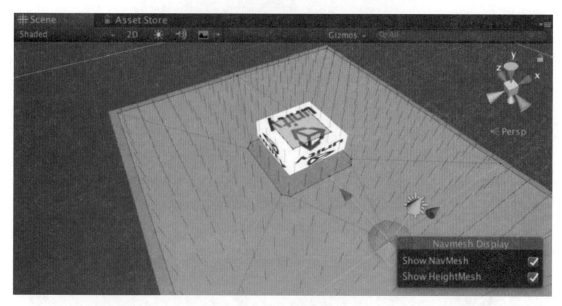

图 9-26 计算阻挡

图 9-30 反射效果

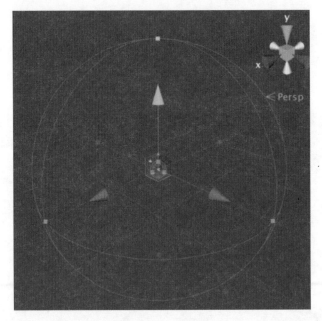

图 10-4 3D 距离

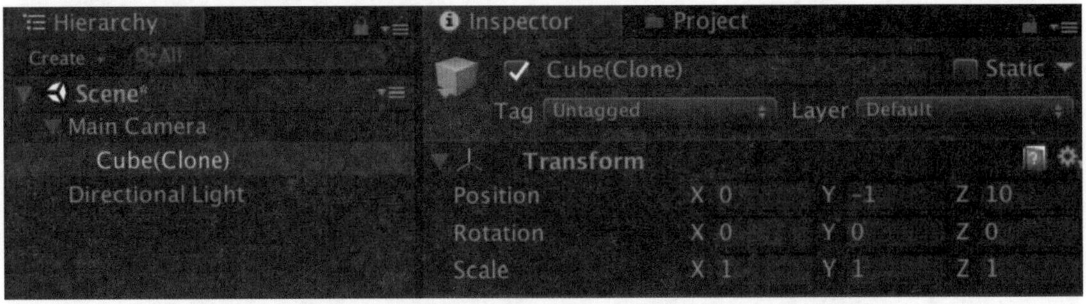

图 11-2 实例化游戏对象

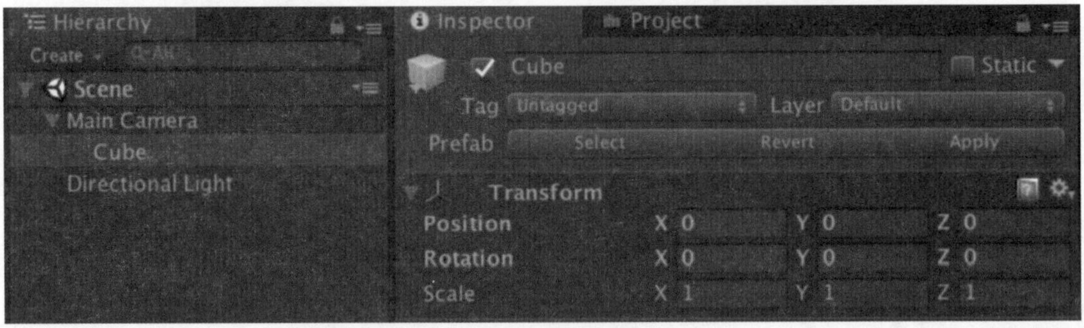

图 11-3 实例化 Prefab

Unity 3D 游戏开发

（第2版）

宣雨松◎编著

图书在版编目（CIP）数据

Unity 3D游戏开发 / 宣雨松编著. -- 2版. -- 北京：人民邮电出版社, 2018.9（2022.1重印）
（图灵原创）
ISBN 978-7-115-49294-4

Ⅰ．①U… Ⅱ．①宣… Ⅲ．①游戏程序－程序设计 Ⅳ．①TP317.6

中国版本图书馆CIP数据核字(2018)第204995号

内 容 提 要

　　Unity是一款市场占有率非常高的商业游戏引擎，横跨25个主流游戏平台。本书基于Unity 2018，结合2D游戏开发和3D游戏开发的案例，详细介绍了它的方方面面，内容涉及编辑器、游戏脚本、UGUI游戏界面、动画系统、持久化数据、静态对象、多媒体、资源加载与优化、自动化与打包等。

　　本书适合初学者或者有一定基础的开发者阅读。

◆ 编　著　宣雨松
责任编辑　王军花
责任印制　周昇亮

◆ 人民邮电出版社出版发行　　北京市丰台区成寿寺路11号
邮编　100164　电子邮件　315@ptpress.com.cn
网址　http://www.ptpress.com.cn
固安县铭成印刷有限公司印刷

◆ 开本：800×1000　1/16
印张：27.25　　彩插：2
字数：644千字　　2018年9月第2版
　　　　　　　　2022年1月河北第19次印刷

定价：89.00元

读者服务热线：(010)84084456-6009　印装质量热线：(010)81055316
反盗版热线：(010)81055315
广告经营许可证：京东市监广登字 20170147 号

序 一

第一次读雨松的书，也就是这本书的上一版，应该是五六年前的事情了。当时，我刚刚加入Unity，而Unity在中国也只能算是刚刚起步，相关的技术书很少，雨松的书应该算是市面上最早的相关书之一了。在我个人来看，雨松的第一本书对于Unity引擎在中国的普及起到了很大的推动作用。可以毫不夸张地说，当时看雨松的书入门Unity的人，要远远多于看Unity用户手册的人。所以，当雨松问我是否可以为他的第二本书作序时，我的回答是：非常荣幸！

在读这本书时，我已经离开Unity三年了，创办了自己的公司，每天解答着大量来自开发者的技术问题，其中有很大一部分是Unity相关的。原本以为这些烂熟于心的知识点，读起来不会再有什么味道，但是我发现我错了。原来哪怕是相同的问题，不同的人关注的点不同、所处的环境不同，那么解决方式也会完全不同。同时，相较于第一本书，这本书多了大量的内容锤炼和经验沉淀。书中关于UGUI的讲解、自动化打包的推荐以及3D游戏开发中的实战技巧，都是雨松这几年来的一线研发经验，同时也是目前大量游戏团队所需要了解和掌握的知识和技巧。我相信这本书的问世，可以帮助更多游戏团队解决更多棘手的问题。

我非常钦佩能够持续更新自己技术博客的人，因为这相当于在每天或每周进行大量工作的同时，还能将自己的经验和教训向外分享，从而去帮助更多的后来者解决问题。雨松就是这样一个人，他的博客是我闲暇时经常光顾的网站，通过阅读他的技术随笔来不断补充自己的知识点。我和雨松每次交谈时，都会非常欣喜地谈到，最近几年来游戏圈愿意分享经验和解答问题的人越来越多，分享的内容也越来越深入，而这些正在潜移默化地推动整个行业快速发展。如果你正在阅读此书，那么我希望你读完之后，不仅可以对Unity游戏开发有更为深入的了解，同时还可以将以往的经验进行总结、分享，并尝试去帮助身边的朋友解决问题。一个人的财富并不在于他拥有了多少金钱，而在于他影响了多少人。

张鑫，侑虎科技CEO

序　二

移动互联网和手机智能化浪潮带来了全新的手机游戏模式。而随着手机游戏开发逐渐成熟，手机游戏开发门槛的降低，越来越多的开发者希望加入到这一行业中。Unity作为一款优秀的游戏引擎，为广大游戏开发者提供了高效、简洁的开发流程，使游戏开发充满了乐趣。

Unity最早诞生于丹麦，目前总部位于美国旧金山，是全球最大的游戏引擎开发商之一。

Unity引擎的强大编辑器，使游戏开发过程简单、快捷。组件式开发和可视化开发，使过去艰涩难懂的开发情景变得清楚明了。

Unity引擎具有优秀的默认渲染管线，即使是不了解计算机图形学的开发者，也可以渲染出炫丽的游戏画面。而最新的Unity版本也允许开发者自由定制渲染管线，给有丰富图形学经验的开发者更高的自由度，从而做出更加优秀的差异化产品。

Unity引擎可以对接目前市面上几乎所有主流的开发工具，开发者可以根据自身需求，建立灵活的制作流水线。整个开发过程更加自由、可控。

总而言之，Unity引擎是一款伟大的游戏引擎，它真正使游戏开发大众化，给更多有才华、有创意的开发者实现自己梦想的机会。

雨松是中国Unity技术普及的先行者。很多开发者（包括我）都是通过雨松的分享，开始学习和使用Unity的。可以说，雨松为Unity在中国的普及做了很大贡献，是Unity技术布道的先行者之一。

本书是雨松的第二本书，对Unity引擎进行了全面而翔实的讲解。书中的内容依然是雨松一贯的深入浅出风格。无论是对于初学者还是有经验的开发者，本书都会非常有帮助。

基于作者自己多年的经验和扎实的技术功底，雨松的这本书，可以说是Unity技术书中不可多得的佳作。

我与雨松相识近10年，能为他的这部大作写序，倍感荣幸。同时也希望通过此书，能有更多优秀的开发者参与到Unity开发中来。

祝你开卷有益。

高川，Unity资深开发者

推　荐　语

雨松从 2012 年就开始写 Unity 相关的技术博客，而游戏蛮牛于 2013 年成立，我们一起见证了游戏行业的兴衰荣辱，也一起帮助很多游戏行业的从业者成长。雨松作为 Unity 领域的前辈，拥有丰富的知识沉淀和实战经验。这是他的第二本 Unity 图书，其内容系统、全面，对于想系统学习的开发者们，这无疑是一个不错的福利。

崇慕，游戏蛮牛创始人

Unity 在中国发展多年，以绝对的优势引领 3D 引擎类内容开发，俘获大量粉丝，目前仍是 3D 内容开发的首选引擎。本书不仅介绍了作者在工作中积累的经验，还涵盖了实际应用中的各类功能，很适合新手系统学习 Unity，并全方面了解其运行原理。

杨博，Unity 圣典社区创始人，威爱教育生态公司 CTO

本书条理清晰，覆盖了 Unity 游戏开发过程中所需的绝大部分知识点，稍有基础的新手按顺序阅读，即可快速了解游戏制作的整个过程。而本书的精华在于作者大量商业项目经验的详细呈现，即使是老司机，也会受益匪浅。

陈小飞，游族网络技术总监

本书比较全面地介绍了在 Unity 项目开发过程中经常用到的一些知识以及作者的一些经验总结。此书非常适合 Unity 初学者，同时书中的一些知识点对中高级开发人员也有一定的借鉴意义。

窦玉波，天马时空技术总监，游戏技术全能专家

本书是面向 Unity 游戏开发者的进阶指南，更宝贵的是作者结合过去几年的商业游戏实战经验，对游戏框架、工作流、解决方案和性能调优等做了很详细的描述。我相信通过阅读本书，能让读者的游戏开发经验和技巧更上一层楼。

张宏亮，蓝港互动技术 VP

雨松结合自身经验撰写了此书。书中全方位覆盖 Unity 引擎技术，比较适合 Unity 初级开发者阅读。同时作为游戏开发从业人员，你可以通过阅读本书，点亮自身技能树，拓展眼界，作出更深入的思考。

郭智，腾讯游戏《穿越火线》手游主程

自从 2011 年正式在中国推广 Unity 不久，我就认识了雨松。作为一位游戏开发专家，雨松对技术（特别是对 Unity 技术）的执着让我印象极为深刻。雨松不仅自己专注探讨，而且愿意和他人分享，此书为证。书中既有雨松对 Unity 的深刻理解，又有他的丰富经验总结。本书必定是广大从事游戏开发朋友们的福音。

郭振平，前 Unity 布道师，华宸互动 CTO

雨松这本书条理清晰、循序渐进、内容描述非常详细、面面俱到，会让 Unity 初学者快速、扎实地掌握 Unity 开发技能。同时书中包含了大量实战中积累的技巧和知识点，即使对 Unity 有一定使用经验的人，也会受益匪浅。

胡志波，网元圣堂技术总监

Unity 的从业者或多或少都知道雨松，或是看过他的上一本书作为入门，或是关注他的博客来获取新的开发经验。而雨松的这本新书则兼具两者的特点，既适合初学者作为入门读物，快速掌握 Unity 引擎的基本概念和操作，同时也包含了他在 Unity 项目实战中积累的经验，可供有一定基础的开发者用来提升自己的能力。我相信通过阅读这本书，读者一定会对游戏开发和 Unity 的用法更加了解与娴熟。

陈嘉栋，Unity 中国 Field Engineer，《Unity 3D 脚本编程》作者

雨松多年来始终在 Unity 一线开发，积累和分享了大量经验和知识，非常值得大家学习。书中的内容有深有浅，不管是新手上路还是资深开发，都可以学到很多技巧。本书最大的亮点在于结合项目开发过程中遇到的各种情况，我相信作为 Unity 开发者的你一定能从中收获颇丰。

钱康来，北京 UUG 社区讲师，UVP 价值专家

宣雨松是中国 Unity 引擎发展的布道者。Unity 在国内发展的蛮荒时代里，是他的《Unity 3D 游戏开发》第一版以及技术站点"雨松 MOMO 程序研究院"引领广大工程师走入 Unity 的殿堂。今天，这本姗姗来迟的《Unity 3D 游戏开发（第 2 版）》更是结合了 2D 和 3D 游戏开发的实战案例，由浅入深地讲解了基础入门到引擎最新的进阶操作。无论是作为 Unity 基础开发者的入门书，还是作为熟练工程师的速查手册，你都应该拥有这本书。

王亮，Unity 资深专家，钛核互动联合创始人

初学 Unity 最好的学习资料无疑是官方的手册和最佳实践，之后需要去学习各种游戏开发的领域知识。雨松的这本书除了介绍 Unity 的知识外，还加入了很多实际开发过程中的经验。基于这种综合，推荐初学者通过此书快速建立对 Unity 游戏开发的认知。

程小山，完美世界游戏开发工程师，《武林外传》手游主程

时隔 6 年，终于盼到雨松的第二本书了。在新书中，系统阐述了 Unity 引擎的方方面面，如 UGUI、Atlas、Shader、Timeline 等。另外，更为难能可贵的是，雨松把自己最近几年实战一线的游戏经验技巧等也融会贯通到本书中，如游戏优化、团队源码管控、Excel 持久化、AssetBundle 动态加载等。我相信摆在你面前的这本书，一定会令你大快朵颐的。

王文刚，微软 MVP，UVP 价值专家

前　　言

　　Unity 是一款市场占有率非常高的商业游戏引擎，横跨 25 个主流游戏平台，在市面上遥遥领先。尤其是在手机游戏平台，Unity 已经处于无可撼动的霸主地位。在 App Store 和 Google Market 上，已经有太多用 Unity 开发的游戏了。另外，Unity 不仅仅是一款游戏引擎，它已经渗透到了各个传统行业，在建筑、医疗、工业、娱乐、虚拟现实、动画、电影和艺术等行业中，都可以看到 Unity 的身影。为了兼容各平台特性以及复杂的需求，Unity 的版本更新也非常及时，最新最酷的功能会第一时间给到开发者手中。

　　近些年，硬件设备处于飞速发展中，这促使游戏开发得到进一步发展。尤其是现在的移动设备，移动硬件几乎每年都会得到一次提升，这意味着将能做出效果更加酷炫的手机游戏来。这无疑也是对开发者的一种挑战，意味着每年都需要学习和应用新技术来使游戏效果更加逼真。

　　Unity 底层是由 C++ 编写的。为了让开发者上手更加容易，它利用 Mono 跨平台的特性使用 C# 作为游戏脚本语言。为了与 C++ 底层接口交互，引擎还封装了 UnityEngine.dll 和 UnityEditor.dll，这样开发者只需要用 C# 就可以开发自己的游戏了。另外，在编辑模式下，Unity 也非常灵活，开发者可以任意拓展专属自己的编辑器。此外，它还提供了资源商店，从中能找到全球顶尖开发者上传的插件，其中大多数都是源码形式的，这给学习和工作提供了非常大的便利。另外，Unity 还提供了完善的开发文档，甚至每一条 API 都有详细的解释，尤其是用户手册中还有详细的例子以及功能介绍。此外，官网中还有很多视频教程供开发者学习。

　　Unity 不仅仅可以做 3D 游戏，也全面支持 2D 游戏的开发，尤其是最新版本的 Unity 2018，它增加了 2D 的骨骼动画，对 2D 物理和地形也有更好的支持。在 UI 方面，UGUI 作为官方支持的界面工具，太多游戏都采用它来制作游戏界面，3D 方面就更不用说了，全新的动画系统以及时间线编辑工具让开发者可以更好地驾驭它。

　　本书主要面向初学者或者有一定基础的开发者，无论你是转行学习还是入门学习，都可以阅读本书。通过阅读本书，读者可以快速学习并入门 3D 游戏开发。此外，本书还包含了很多实战经验，其中丰富精彩的案例同样适用于晋级的读者。书中每一章都包含丰富的实例和源码，尤其是拓展编辑部分以及自动化部分，都是非常宝贵的实战经验，读者可以直接将其应用在实际开发中。

阅读本书

书中所有例子的源代码都可以在图灵社区（www.ituring.com.cn）本书主页免费注册下载，下载的源码按章编号，如图 0-1 所示。查看源码前，请确保 Unity 已经无误地安装在了本机当中。按照图中所示找到章节源码对应的游戏场景文件，双击该场景文件即可打开游戏工程，继而查看阅读。本书是在 macOS 下讲解 Unity 开发，若需在 Windows 下查看源码，查看方式与 macOS 下完全一致。

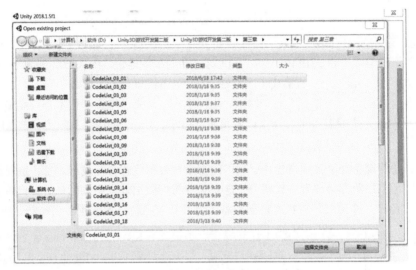

图 0-1　打开游戏工程

致谢

时至今日，我已经从事游戏开发十余年了。这本书是第 2 版，距离上一版出版已经 6 年了。其实这些年我也看到了上一版中的很多问题，包括一些读者给我的反馈，鉴于此，这本书我决定完全重写。这本书的写作也是一波三折，它本该在 2014 年就写完，可是一直拖到了 2018 年才完成。这几年我的身边发生了很多事情，我也做了好几款游戏，遗憾的是没有特别成功的作品。对于游戏开发，我的内心永远是狂热的，我将不忘初心，永远坚持钻研技术，励志走在研发的第一梯队中。

我最想感谢的人就是这些年努力的自己，我对技术非常痴迷，我非常享受解决一个复杂难题的瞬间。由于这份痴迷，我经常会写代码到凌晨，我最应该感谢的就是我的妻子，早在我创作上一版书的时候，她就在我身旁，并一直支持我到今天。我必须要感谢图灵公司的编辑王军花，她教会我很多写作技巧并且审阅了本书的书稿，保证这本迟到了 4 年的书顺利出版。最后，祝图灵公司越做越好，为祖国的 IT 人才培养贡献出伟大的力量。

2018 年 7 月 19 日

宣雨松

目　　录

第1章　基础知识 ································· 1
　1.1　Unity 简介 ······························· 1
　1.2　跨平台与多工种协作 ················· 1
　1.3　Unity 版本 ······························· 2
　1.4　Unity 内置资源或拓展资源 ········ 3
　1.5　示例项目打包与发布 ················· 5
　1.6　Unity 服务 ······························· 7
　1.7　小结 ······································· 8

第2章　编辑器的结构 ·························· 9
　2.1　游戏项目 ································· 9
　　2.1.1　创建项目 ························· 9
　　2.1.2　打开项目 ······················· 10
　　2.1.3　版本管理 ······················· 10
　　2.1.4　安装多个版本的 Unity ···· 11
　2.2　Project 视图 ··························· 12
　　2.2.1　创建资源 ······················· 12
　　2.2.2　搜索资源 ······················· 13
　　2.2.3　搜索标签 ······················· 14
　2.3　Hierarchy 视图 ······················· 16
　　2.3.1　创建游戏对象 ················ 16
　　2.3.2　搜索游戏对象 ················ 16
　2.4　Inspector 视图 ························ 17
　　2.4.1　标题栏 ··························· 17
　　2.4.2　组件栏 ··························· 18
　2.5　Scene 视图 ····························· 19
　　2.5.1　导航栏 ··························· 19
　　2.5.2　标题栏 ··························· 20
　　2.5.3　坐标系控制器 ················ 21
　　2.5.4　视图预览 ······················· 21
　2.6　Game 视图 ···························· 22

　　2.6.1　标题栏 ··························· 22
　　2.6.2　视图预览 ······················· 23
　2.7　导航栏视图 ···························· 23
　2.8　其他功能 ······························· 25
　　2.8.1　小锁头 ··························· 25
　　2.8.2　窗口菜单 ······················· 25
　　2.8.3　保存组件参数 ················ 26
　　2.8.4　Package Manager ············ 26
　2.9　小结 ······································ 27

第3章　拓展编辑器 ···························· 28
　3.1　拓展 Project 视图 ··················· 28
　　3.1.1　拓展右键菜单 ················ 28
　　3.1.2　创建菜单 ······················· 30
　　3.1.3　拓展布局 ······················· 31
　　3.1.4　监听事件 ······················· 32
　3.2　拓展 Hierarchy 视图 ··············· 33
　　3.2.1　拓展菜单 ······················· 33
　　3.2.2　拓展布局 ······················· 34
　　3.2.3　重写菜单 ······················· 35
　3.3　拓展 Inspector 视图 ················ 37
　　3.3.1　拓展源生组件 ················ 38
　　3.3.2　拓展继承组件 ················ 38
　　3.3.3　组件不可编辑 ················ 40
　　3.3.4　Context 菜单 ·················· 42
　3.4　拓展 Scene 视图 ····················· 44
　　3.4.1　辅助元素 ······················· 44
　　3.4.2　辅助 UI ·························· 45
　　3.4.3　常驻辅助 UI ··················· 46
　　3.4.4　禁用选中对象 ················ 47
　3.5　拓展 Game 视图 ····················· 48

3.6　MenuItem 菜单 ································· 49
　　　　3.6.1　覆盖系统菜单 ························· 50
　　　　3.6.2　自定义菜单 ···························· 50
　　　　3.6.3　源生自定义菜单 ····················· 51
　　　　3.6.4　拓展全局自定义快捷键 ············ 52
　　3.7　面板拓展 ·· 53
　　　　3.7.1　Inspector 面板 ························· 53
　　　　3.7.2　EditorWindows 窗口 ················ 55
　　　　3.7.3　EditorWindows 下拉菜单 ········· 56
　　　　3.7.4　预览窗口 ································· 57
　　　　3.7.5　获取预览信息 ·························· 59
　　3.8　Unity 编辑器的源码 ··························· 60
　　　　3.8.1　查看 DLL ································ 60
　　　　3.8.2　清空控制台日志 ····················· 61
　　　　3.8.3　获取 EditorStyles 样式 ············· 63
　　　　3.8.4　获取内置图标样式 ··················· 64
　　　　3.8.5　拓展默认面板 ·························· 66
　　　　3.8.6　例子：查找 ManagedStatic-
　　　　　　　References() 静态引用 ············ 67
　　　　3.8.7　UIElements ······························ 69
　　　　3.8.8　查询系统窗口 ·························· 71
　　　　3.8.9　自定义资源导入类型 ··············· 73
　　3.9　小结 ··· 74

第 4 章　游戏脚本 ·· 75
　　4.1　创建脚本 ·· 75
　　　　4.1.1　脚本模板 ································· 75
　　　　4.1.2　拓展脚本模板 ·························· 76
　　4.2　脚本的生命周期 ·································· 78
　　　　4.2.1　脚本绑定事件 ·························· 80
　　　　4.2.2　脚本初始化和销毁 ··················· 80
　　　　4.2.3　脚本更新与协程任务 ··············· 82
　　　　4.2.4　停止协程任务 ·························· 83
　　　　4.2.5　使用 OnGUI 显示 FPS ·············· 84
　　4.3　多脚本管理 ··· 85
　　　　4.3.1　脚本的执行顺序 ······················· 85
　　　　4.3.2　多脚本优化 ······························ 86
　　4.4　脚本序列化 ··· 86
　　　　4.4.1　查看数据 ································· 87
　　　　4.4.2　私有序列化数据 ······················· 88

　　　　4.4.3　serializedObject ······················· 90
　　　　4.4.4　监听部分元素修改事件 ············ 91
　　　　4.4.5　序列化/反序列化监听 ·············· 92
　　　　4.4.6　ScriptableObject ······················ 95
　　　　4.4.7　脚本的 Attributes 特性 ·············· 97
　　　　4.4.8　单例脚本 ································· 99
　　　　4.4.9　定时器 ···································· 100
　　　　4.4.10　CustomYieldInstruction ········· 101
　　　　4.4.11　工作线程 ······························ 103
　　4.5　脚本编译 ·· 104
　　　　4.5.1　编译规则 ································ 105
　　　　4.5.2　优化编译 ································ 105
　　　　4.5.3　编译 DLL ······························· 105
　　　　4.5.4　脚本跨平台 ···························· 106
　　　　4.5.5　程序集定义 ···························· 106
　　　　4.5.6　日志 ······································· 107
　　4.6　脚本调试 ·· 109
　　4.7　小结 ··· 111

第 5 章　UGUI 游戏界面 ······························ 112
　　5.1　基础元素 ·· 112
　　　　5.1.1　Text ·· 112
　　　　5.1.2　描边和阴影 ···························· 113
　　　　5.1.3　动态字体 ································ 113
　　　　5.1.4　字体花屏 ································ 114
　　　　5.1.5　Image 组件 ····························· 115
　　　　5.1.6　Raw Image 组件 ····················· 116
　　　　5.1.7　Button 组件 ···························· 116
　　　　5.1.8　Toggle 组件 ··························· 117
　　　　5.1.9　Slider 组件 ····························· 118
　　　　5.1.10　Scrollbar & ScrollView 组件 ··· 119
　　　　5.1.11　使用 ScrollRect 组件制作
　　　　　　　　游戏摇杆 ································ 120
　　5.2　事件系统 ·· 121
　　　　5.2.1　UI 事件 ··································· 122
　　　　5.2.2　UI 事件管理 ···························· 124
　　　　5.2.3　UnityAction 和 UnityEvent ······ 126
　　　　5.2.4　RaycastTarget 优化 ················· 128
　　　　5.2.5　渗透 UI 事件 ·························· 129
　　　　5.2.6　例子——新手引导聚合动画 ····· 131

目录

- 5.3 Canvas 组件 ·········· 135
  - 5.3.1 自适应屏幕 ·········· 135
  - 5.3.2 锚点对齐方式 ·········· 137
  - 5.3.3 背景图全屏 ·········· 138
  - 5.3.4 布局组件 ·········· 139
  - 5.3.5 Canvas 优化 ·········· 139
- 5.4 Atlas ·········· 139
  - 5.4.1 创建 Atlas ·········· 139
  - 5.4.2 读取 Atlas ·········· 140
  - 5.4.3 Variant ·········· 141
  - 5.4.4 多图集管理 ·········· 142
- 5.5 UGUI 实例 ·········· 142
  - 5.5.1 置灰 ·········· 142
  - 5.5.2 粒子特效与 UI 的排序 ·········· 145
  - 5.5.3 Mask & RectMask2D 裁切 ·········· 147
  - 5.5.4 粒子的裁切 ·········· 148
  - 5.5.5 粒子自适应 ·········· 152
  - 5.5.6 滑动列表嵌套 ·········· 154
  - 5.5.7 UI 模板嵌套 ·········· 156
  - 5.5.8 UI 特效与界面分离 ·········· 157
  - 5.5.9 输入事件 ·········· 161
  - 5.5.10 按钮不规则点击区域 ·········· 162
  - 5.5.11 更换默认 Shader ·········· 164
  - 5.5.12 小地图优化 ·········· 164
  - 5.5.13 查看 UGUI 源码 ·········· 165
- 5.6 小结 ·········· 165

第 6 章 2D 游戏开发 ·········· 166
- 6.1 Sprite ·········· 166
  - 6.1.1 2D 摄像机与分辨率自适应 ·········· 167
  - 6.1.2 Sprite Renderer 排序 ·········· 169
  - 6.1.3 裁切 ·········· 170
- 6.2 Sprite 动画 ·········· 171
  - 6.2.1 创建 2D 动画 ·········· 171
  - 6.2.2 2D 动画控制器 ·········· 173
- 6.3 Tile 地图 ·········· 176
  - 6.3.1 创建 Tile ·········· 176
  - 6.3.2 Tile Palette ·········· 178
  - 6.3.3 编辑 Tile ·········· 178
  - 6.3.4 多 Tile 编辑与排序 ·········· 179
  - 6.3.5 拓展 Tile Palette ·········· 180
  - 6.3.6 拓展 Tile ·········· 183
  - 6.3.7 更新 Tile ·········· 184
- 6.4 2D 碰撞检测 ·········· 186
  - 6.4.1 Collider 2D ·········· 186
  - 6.4.2 Rigidbody 2D ·········· 187
  - 6.4.3 碰撞事件 ·········· 189
  - 6.4.4 碰撞方向 ·········· 191
  - 6.4.5 触发器监听 ·········· 193
  - 6.4.6 Effectors 2D ·········· 193
  - 6.4.7 优化 ·········· 194
  - 6.4.8 计算区域 ·········· 194
- 6.5 小结 ·········· 195

第 7 章 动画系统 ·········· 196
- 7.1 模型 ·········· 196
  - 7.1.1 Mesh Filter ·········· 196
  - 7.1.2 Mesh Renderer ·········· 197
  - 7.1.3 Prefab ·········· 197
- 7.2 动画编辑器 ·········· 198
  - 7.2.1 编辑器面板 ·········· 198
  - 7.2.2 在编辑器中添加事件 ·········· 199
- 7.3 导入类动画 ·········· 200
  - 7.3.1 人形重定向动画 ·········· 200
  - 7.3.2 通用动画 ·········· 202
  - 7.3.3 老版动画 ·········· 203
  - 7.3.4 导入类动画事件 ·········· 203
- 7.4 动画控制器 ·········· 203
  - 7.4.1 系统状态 ·········· 204
  - 7.4.2 切换条件 ·········· 205
  - 7.4.3 状态机脚本 ·········· 206
  - 7.4.4 IK 动画 ·········· 207
  - 7.4.5 Root Motion ·········· 209
  - 7.4.6 Avatar Mask ·········· 210
  - 7.4.7 层 ·········· 211
  - 7.4.8 Blend Tree ·········· 212
  - 7.4.9 非运行播放动画 ·········· 213
  - 7.4.10 Animator Override Controller ·········· 214
  - 7.4.11 RuntimeAnimator- Controller ·········· 215

7.5 TimeLine 编辑器 215
　7.5.1 创建 Timeline 216
　7.5.2 Activation Track 217
　7.5.3 Animation Track 217
　7.5.4 Audio Track 218
　7.5.5 Control Track 218
　7.5.6 Playable Track 219
　7.5.7 自定义 Track 221
7.6 Playables 222
　7.6.1 播放动画 222
　7.6.2 动画混合 224
　7.6.3 音频混合 225
　7.6.4 自定义脚本 226
7.7 Constraint 227
　7.7.1 Aim Constraint 227
　7.7.2 Parent Constraint 228
　7.7.3 脚本控制约束 229
7.8 小结 229

第8章 持久化数据 230

8.1 Excel 230
　8.1.1 EPPlus 230
　8.1.2 读取 Excel 230
　8.1.3 写入 Excel 232
　8.1.4 JSON 233
　8.1.5 JSON 支持字典 235
8.2 文件读取与写入 236
　8.2.1 `PlayerPrefs` 237
　8.2.2 `EditorPrefs` 237
　8.2.3 `PlayerPrefs` 保存复杂结构 238
　8.2.4 `TextAsset` 239
　8.2.5 读写文本 240
　8.2.6 运行期读写文本 240
　8.2.7 PersistentDataPath 目录 242
　8.2.8 游戏存档 243
8.3 XML 247
　8.3.1 创建 XML 247
　8.3.2 读取与修改 248
　8.3.3 XML 文件 249
8.4 YAML 251

8.4.1 YamlDotNet 252
8.4.2 序列化和反序列化 252
8.4.3 读取配置 253
8.5 小结 255

第9章 静态对象 256

9.1 Lightmap 256
　9.1.1 设置烘焙贴图 256
　9.1.2 实时光和焙贴光共存 258
　9.1.3 灯光管理 258
　9.1.4 运行时更换烘焙贴图 258
　9.1.5 动态更换游戏对象 260
　9.1.6 复制游戏对象 262
9.2 遮挡剔除 263
　9.2.1 遮挡与被遮挡 264
　9.2.2 遮挡与被遮挡事件 265
　9.2.3 动态剔除 267
　9.2.4 自定义遮挡剔除 267
9.3 Batching（静态合批） 268
　9.3.1 设置静态合批 268
　9.3.2 脚本静态合批 269
　9.3.3 动态合批 269
　9.3.4 静态合批的隐患 270
9.4 寻路网格 270
　9.4.1 设置寻路 270
　9.4.2 连接两点 272
　9.4.3 获取寻路路径 272
　9.4.4 动态阻挡 274
　9.4.5 导出寻路网格信息 274
9.5 反射探头 277
9.6 小结 278

第10章 多媒体 279

10.1 音频 279
　10.1.1 音频文件 279
　10.1.2 Audio Source 280
　10.1.3 3D 音频 280
　10.1.4 代码控制播放 281
　10.1.5 混音区 283
　10.1.6 Audio Mixer 284

10.1.7	录音	284
10.1.8	声音进度	285
10.2	视频	287
10.2.1	视频文件	287
10.2.2	视频渲染模式	287
10.2.3	视频自适应	288
10.2.4	UI 盖在视频之上	289
10.2.5	视频渲染在材质上	290
10.2.6	视频 Render Texture	291
10.2.7	播放工程外视频	291
10.2.8	自定义视频显示	293
10.2.9	视频进度	293
10.3	小结	294

第 11 章 资源加载与优化 … 295

11.1	编辑模式	295
11.1.1	加载资源	295
11.1.2	实例化 Prefab	296
11.1.3	创建 Prefab	297
11.1.4	更新 Prefab	298
11.1.5	卸载资源	299
11.1.6	游戏对象与资源的关系	299
11.2	版本管理	300
11.2.1	.meta 文件	300
11.2.2	多工程	301
11.2.3	同步文件	302
11.2.4	SVN 外链	303
11.3	运行模式	304
11.3.1	引用资源	304
11.3.2	Resources	305
11.3.3	删除对象	306
11.3.4	删除资源	306
11.3.5	GC	307
11.4	AssetBundle	308
11.4.1	设置 AssetBundle	308
11.4.2	依赖关系	310
11.4.3	通过脚本设置依赖关系	311
11.4.4	压缩格式	311
11.4.5	加载包体内的 AssetBundle	312
11.4.6	下载 AssetBundle	313
11.4.7	加载场景	314
11.4.8	卸载 AssetBundle	315
11.5	游戏对象	315
11.5.1	创建游戏对象	316
11.5.2	Transform 设置排序	316
11.5.3	删除节点	317
11.5.4	获取游戏对象	317
11.5.5	管理游戏组件	318
11.6	优化工具	319
11.6.1	重复无用资源	319
11.6.2	查看内存	321
11.6.3	查看 CPU 效率	321
11.6.4	自定义观察区间	322
11.6.5	Profiler 信息的导出与导入	323
11.6.6	Frame Debugger	324
11.7	资源管理实例	325
11.7.1	特殊内置的文件夹	325
11.7.2	隐藏文件夹	326
11.7.3	Resources 与 AssetBundle 无缝切换	326
11.7.4	资源加载策略	329
11.7.5	资源更新	329
11.7.6	资源引用关系	329
11.7.7	系统资源修改	332
11.7.8	AssetBundle 里的脚本	333
11.7.9	热更新代码	334
11.8	小结	335

第 12 章 自动化与打包 … 336

12.1	资源导入	336
12.1.1	监听导入事件	336
12.1.2	自动设置贴图压缩格式	337
12.1.3	自动设置模型	339
12.1.4	禁止模型生成材质文件	340
12.1.5	删除移动资源事件	341
12.1.6	选择性自动设置	342
12.1.7	主动设置	342
12.1.8	待保存状态	343
12.1.9	自动执行 MenuItem	344
12.2	配置错误	345

12.2.1 主动检查工具 345
12.2.2 被动检查工具 346
12.2.3 导出类检查工具 346
12.2.4 导出 UI 346
12.2.5 生成 UI 代码 347
12.2.6 导出模型 351
12.2.7 不参与保存对象 352
12.2.8 导出场景 353
12.2.9 过滤无用场景 355
12.3 自动打包 356
12.3.1 打包前后事件 356
12.3.2 打包后自动压缩 357
12.3.3 用 C# 调用 shell 脚本 359
12.3.4 等待 shell 结束 359
12.3.5 shell 脚本调用 C# 360
12.3.6 shell 脚本自动打包 362
12.3.7 彻底解放程序员的双手 .. 365
12.4 自动构建图集与压缩 366
12.4.1 UI 图集的压缩 366
12.4.2 非 UI 图集压缩 370
12.5 小结 371

第 13 章 3D 游戏开发 372

13.1 Shader 372
13.1.1 固定渲染管线 372
13.1.2 可编程渲染管线 375
13.1.3 可编程渲染管线的表面着色器 378
13.1.4 深度排序 379
13.1.5 透明 381
13.1.6 裁切 382
13.1.7 着色器变体采集 383
13.2 摄像机 384
13.2.1 3D 坐标转换屏幕坐标 384
13.2.2 3D 坐标转换 UI 坐标 385
13.2.3 主摄像机 387
13.2.4 在 UI 上显示模型 387
13.2.5 Render Texture 389
13.2.6 在 Render Texture 上显示带特效的模型 390
13.2.7 LOD Group 392
13.3 场景管理 393
13.3.1 切换场景 393
13.3.2 `DontDestroyOnLoad` ... 393
13.3.3 异步加载场景以及进度 .. 393
13.3.4 多场景 395
13.3.5 多场景游戏对象管理 396
13.3.6 场景切换事件 397
13.3.7 多场景烘焙 398
13.4 输入事件 398
13.4.1 全局事件 398
13.4.2 射线 399
13.4.3 点选 3D 模型 400
13.4.4 通过点击控制人物移动 .. 402
13.5 物理碰撞 403
13.5.1 碰撞器 404
13.5.2 角色控制器 404
13.5.3 碰撞区域 405
13.5.4 碰撞检测 406
13.5.5 碰撞触发器 407
13.5.6 物理调试器 408
13.6 实战技巧 409
13.6.1 3D 头顶文字 409
13.6.2 图文混排 410
13.6.3 图片数字 410
13.6.4 游戏对象缓存池 412
13.6.5 游戏资源缓存池 414
13.6.6 运行时合并网格 414
13.6.7 运行时合并贴图 416
13.7 小结 419

第 1 章

基础知识

Unity 是一款 3D 跨平台的次世代游戏引擎，"Unity"一词的释义为"团结"，好比集合所有人的力量一起来完成这件伟大的巨作。Unity 于 2005 年发布了第一个版本，至今已经 13 年了。起初，该公司致力于游戏引擎的研发，由于 Unity 具有强大的跨平台开发能力与绚丽的 3D 渲染效果，因而很快被广大游戏开发厂商以及开发者所信赖，尤其是在手游领域已经处于无可撼动的地位。近年来，该公司处于飞速发展中，它涉及的层面已经不仅仅局限于游戏引擎，还在 AI、VR、教育、医疗、工业、动画片等领域大放异彩。

1.1 Unity 简介

Unity 是一款标准的商业游戏引擎，商业引擎的主要特点就是：收费、封闭源码和功能强大。关于收费情况，Unity 目前分为 3 个版本：个人版、加强版和专业版。个人版是完全免费的，只能使用引擎核心的基础功能，适用于刚接触 Unity 的初学者或者学生。加强版和专业版可用于发布正式的游戏，当然也是收费的。专业版更加昂贵，但是它可使用的 Unity 服务是最全面的，主要包括 Unity 的 Analytics 分析服务等。如果你的公司或独立工作室的启动资金或年度收入大于 20 万美元的话，可以购买 Unity 专业版，反之可以购买 Unity 加强版。

1.2 跨平台与多工种协作

Unity 是一个跨平台游戏引擎，其中跨平台分为开发跨平台和发布跨平台。开发跨平台就是开发者可以在不同的操作系统下开发 Unity 游戏，目前可以跨 Windows 和 macOS 这两种操作系统。而发布跨平台则表示使用 Unity 开发出来的游戏能在多平台下运行。如图 1-1 所示，目前 Unity 已经可以横跨 25 个主流游戏平台，是不是很强大呢？不过每个平台可能有些自己特有的需求，详细的跨平台信息可以在如下网址中查阅到：https://unity3d.com/cn/unity/features/multiplatform。

传统的游戏引擎更多的是程序员在使用，而 Unity 则是多工种同时配合使用。Unity 提供了丰富的编辑工具，可以辅助美术人员以及策划人员在引擎中进行编辑工作，比如 TimeLine 时间线工具，程序员可以在不需要编写一行代码的情况下编辑各种复杂的剧情动画或者动画片。此外，Unity 还提供了拓展编辑器的接口，程序员可以开发自定义的辅助编辑工具供其他工种人员使用。

图 1-1 支持平台

Unity 秉承"所见即所得"的开发理念，任何人点击播放后即可预览游戏，实时查看游戏的效果。此外，它还具有强大的图形渲染以及引擎性能。无论是策划人员、美术人员，还是程序员和测试人员等，都可以很好地使用 Unity 开始创作。

1.3 Unity 版本

在 Unity 5.X 版本以后，取消了以数字命名的版本规则，而采用年份来表示。此外，Unity 公司宣布了 Unity 全新的版本发布计划，包括技术前瞻版本（简称：Unity TECH 版）和 Unity LTS 稳定支持版本（简称：Unity LTS 版）。其中，Unity TECH 版每年都会有 3 个大版本更新，例如 Unity 2017.1.X、Unity 2017.2.X 和 Unity 2017.3.X。而 Unity LTS 版则从 Unity TECH 版的最后一个版本开始，持续支持两年的时间，例如 Unity 2017.4.X。所以实际开发项目时，最好使用 LTS 版。目前只有 Unity 2017 才有 LTS 版，最新版本为 LTS Release 2017.4.5f1，结尾以 f1 表示，可以在如下链接中下载到：https://unity3d.com/cn/unity/qa/lts-releases。

Unity 会为每个发布过的版本提供对应的补丁版本，补丁版本的数量一般是 1 至 5 个不等。补丁版本一般以修复 bug 为主，直接覆盖原版本的安装目录即可。例如，Unity 2017.2.3p1 中结尾以 p1 表示对应的补丁。补丁版本可以在如下链接中下载到：https://unity3d.com/cn/unity/qa/patch-releases。

此外，Unity 还提供了测试版本，也就是 Beta 版本。此版本仅用于测试新功能而已，不排除会有其他严重的 bug，所以平常开发中就不要使用 Beta 版本了。每年发布 TECH 版本之前，Unity 公司都会提前发布测试版本。例如，Unity 2018.2.0b1 中结尾以 b1 表示对应的测试版本号。测试版可以在如下链接中下载到：https://unity3d.com/cn/unity/beta-download。

另外，由于开发周期比较长，开发版本很有可能已经和最新版本相差很远了，所以有时还需要快速找到 Unity 的旧版本，此时可以通过如下链接找到：https://unity3d.com/cn/get-unity/download/archive。

通过以上这些版本的介绍，我们可以发现 Unity 的版本其实是很多的。通常，在实际开发中，为了测试多个版本之间的差异性，很有可能需要同时安装好几个版本，如何管理就成了一个难题。还好，Unity 提供了新工具 Unity Hub 来专门管理多版本（目前还是预览版）。如图 1-2 所示，打开 Unity Hub 后，可以安装与管理多个不同的 Unity 版本，并且可以很方便地用指定版本打开不同的游戏工程。有关 Unity Hub 的详细介绍，可以参考这里：https://blogs.unity3d.com/cn/2018/01/24/streamline-your-workflow-introducing-unity-hub-beta/。

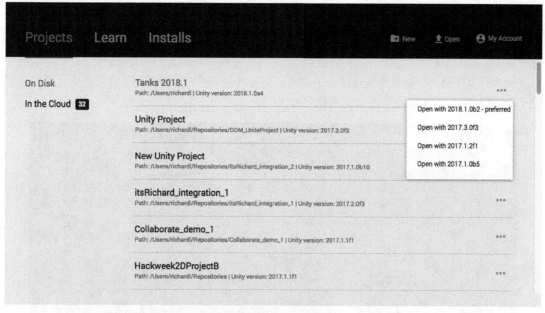

图 1-2　Unity Hub

1.4　Unity 内置资源或拓展资源

Unity 有很多资源内置在引擎中，开发者是无法看到的。此外，还有一部分资源作为拓展资源，需要开发者自行下载将其放入工程中。如图 1-3 所示，打开 Unity 旧版本的网址后，选择对应查看的版本并点击下拉菜单，即可下载该版本对应的资源，其中包括 Unity 编辑器、Cache Server、内置着色器、标准的资源和示例项目。

图 1-3　下载资源

打开 Unity 引擎后,在导航菜单栏中选择 Window→Package Manager 菜单项,即可打开资源包管理器。如图 1-4 所示,Unity 会将比较重要的一些包放在 Package Manager 中,这样可以极大限度地为引擎瘦身。开发者只需要选择下载自己需要的包即可。可以看到,目前这里大部分都标记了 preview 字样,未来的功能会越来越完善。

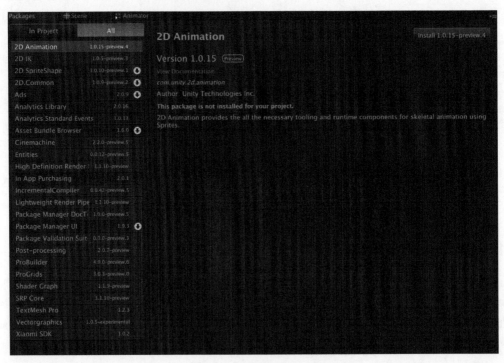

图 1-4　Package Manager

此外，Unity 还提供了资源商店（Asset Store），在导航菜单栏中选择 Window→Asset Store 即可打开。这里面有很多好用的资源以及代码插件，当然有些是需要支付一定费用的。Unity 自己也提供了大量的插件以及资源，并且都是免费的，非常适合新手来学习。如图 1-5 所示，在搜索栏中搜索你感兴趣的内容，即可得到相关的插件。很炫酷吧！

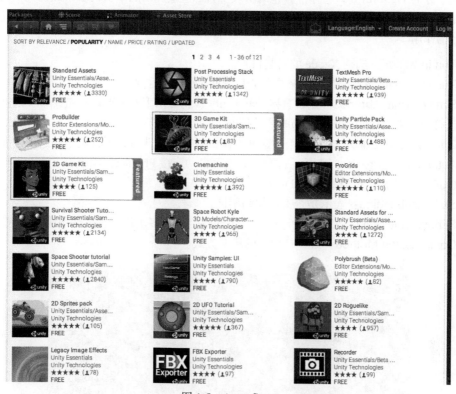

图 1-5　Asset Store

1.5　示例项目打包与发布

导入示例项目，这里是在 Asset Store 中下载并且导入 Unity 最经典的 angryBots 项目。首先，需要在导航菜单栏中选择 Edit→Project Settings→Player。默认情况下，Unity 会选择 Device SDK（它表示只能导出到真机上运行），这里我们选择 Simulator SDK（表示可以在模拟器上运行），如图 1-6 所示。

图 1-6　目标设备

接着，在导航菜单栏中选择 File→Build Settings 菜单项，此时会弹出构建窗口。如图 1-7 所示，在 Scenes In Build 中添加待打包的场景，只需打开需要打包的场景后，点击 Add Open Scenes 按钮即可。这里提供一个技巧，如果不需要打包某些场景，可以在该窗口中删除它或者取消勾选。

在左下方的 Platform 处，选择待打包的游戏平台。这里列出的平台需要下载对应的支持，这里选择 iOS 平台，表示可以打包 iPhone 或者 iPad。右侧是用于设置打包的参数，下面简要介绍各个参数的作用。

- Run in Xcode：选择 Xcode 的安装目录。
- Run in Xcode as：设置 Xcode 中是否以 Release 方式运行。
- Symlink Unity libraries：是否直接关联 Unity 安装目录下的 iOS 动态链接库。勾选后，调试打包会更快一些，正式发布时要关掉它。
- Development Build：表示是否构建开发调试版本。勾选后，下方两个勾选框会亮起来。
- Autoconnect Profiler：表示运行游戏后是否自动连接 Profiler，用于查看游戏性能。关于 Profiler 的用法，可以参见第 11 章。
- Script Debugging：表示是否支持代码调试。
- Scripts Only Build：表示只构建脚本，不构建资源，多次构建将大幅度降低打包时间。
- Compression Method：选择打包时的压缩方式。

图 1-7　Build Settings 窗口

参数设置完毕后，点击 Build 或者 Build And Run 按钮即可。由于 iOS 平台比较特殊，需要预先生成 Xcode 工程，所以这里点击的构建并非真正构建成 IPA 安装文件。运行后，经典的 angryBots 项目已经在模拟器中打开了，如图 1-8 所示。

图 1-8　angryBots

1.6　Unity 服务

Unity 预制了很多服务，使用这些服务不需要接入第三方 SDK，直接就可以设置它们。在导航菜单栏中选择 Window→Services 菜单项，即可弹出服务窗口，如图 1-9 所示。更多服务的用法，读者也可以自行拓展学习。

- Cloud Build：远程云构建，云打包。
- Ads：内置广告，开发者变现平台。
- Analytics：挖掘玩家行为数据，添加标准以及自定义事件，提供分析面板等。
- Collaborate：项目版本管理服务，多人开发可以很方便地保存并且同步项目。
- Performance Reporting：捕获玩家在游戏中产生的错误日志，可以后台中分析查询。
- In-App Purchasing：跨平台充值接口。
- Multiplayer：点对点创建实时多人联网游戏框架。

图 1-9　服务窗口

1.7　小结

本章主要向读者介绍了学习 Unity 的基础知识，做好开发 Unity 3D 游戏之前的一切准备工作。首先介绍了 Unity 这款商业游戏引擎的特点，之后讲述了 Unity Hub 多版本管理工具以及内置或者拓展资源的用法。在学习中，我们可以通过强大的 Asset Store 下载适合自己的游戏插件。此外，还介绍了 Unity 的拓展服务。作为 Unity 学习基础中的基础，希望读者们认真学习本章内容，以便为后续章节的学习做好铺垫。

第 2 章

编辑器的结构

经过多年的发展，Unity 编辑器已经越来越完善，使用起来也相当方便、快捷。Unity 秉承"所见即所得"的开发原理，将编辑器与游戏引擎融合在了一起。传统游戏引擎几乎没有任何游戏界面，它们提供给开发者的往往都是赤裸裸的源代码，以至于想实现任何功能，都需要编写代码才可以完成。而 Unity 的理念是为开发者节省时间，让他们爱上 Unity 并成为它的粉丝。它以可视化方式提供大部分开发工作，开发者无须编写任何代码，只需在界面中执行一些赋值操作就可以了。

2.1 游戏项目

Unity 游戏项目有若干游戏场景组件，游戏场景由若干游戏对象组成，游戏对象又由若干游戏资源组成。比如，在 Hierarchy 视图窗口中可以创建所有游戏对象，在 Project 视图中可以管理所有游戏资源，在游戏界面场景中可以实现游戏的所有业务逻辑等。游戏工程需要在 Unity 编辑器中打开，可视化的编辑器界面无疑让开发者使用起来更加有条理，使开发者可以更清晰地看到整个游戏工程的层次与概念。使用 Unity，我们可以快速敏捷地制作 3D 游戏。

2.1.1 创建项目

打开 Unity，点击右上角的 New 按钮即可创建游戏项目，如图 2-1 所示。根据游戏类型，可以选择 3D 或 2D，它将影响后面 SceneView 视图的样式。Add Asset Package 按钮可以给即将创建的工程添加预制游戏包。如果需要启动 Unity 分析器，请打开 Enable Unity Analytics（该选项建议打开）。接着输入正确的 Project name（项目名称）和 Location（项目路径），确认无误后点击 Create project（创建项目）按钮，Unity 就会自动打开这个项目了。

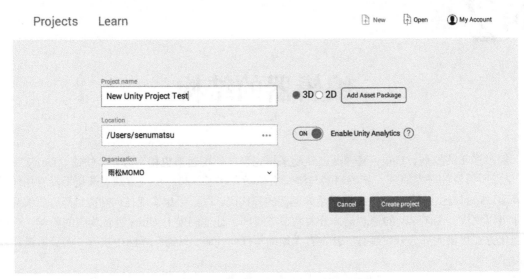

图 2-1 创建工程

2.1.2 打开项目

打开 Unity 后，如图 2-1 所示，点击右上角的 Open 按钮即可打开一个已有的 Unity 项目。如果打开项目时需要打开指定的场景，则在 Assets 文件夹中选择一个 Unity 场景，双击场景即可，如图 2-2 所示。

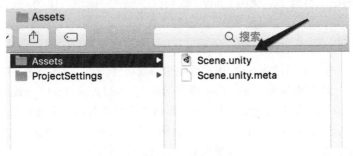

图 2-2 打开项目

2.1.3 版本管理

项目一般需要 SVN 或者 Git 一类的工具来进行版本管理。但是，打开 Unity 项目后会生成一些中间文件，这些不需要进行版本管理。如图 2-2 所示，只需要把 Assets 目录以及 ProjectSettings 目录上传 SVN 或者 Git 即可。

打开项目后，Unity 会自动生成 3 个文件夹，如图 2-3 所示。其中，Library 文件夹就是根据 Assets 目录下的游戏资源所生成的中间文件，Temp 文件夹是 Library 生成过程中产生的临时文件。这两个目录**一定不要**上传到版本管理中。另外，UnityPackageManager 保存的是所要用到的 package，也需要上传到 SVN 中。

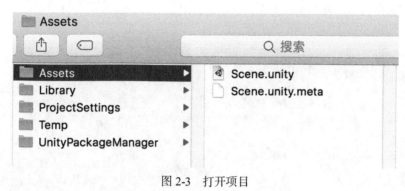

图 2-3　打开项目

2.1.4　安装多个版本的 Unity

有时候，想试试新版本的 Unity，但是又不想删除老版本，所以多个版本共存很有必要。然而 Unity 在安装的时候并不能选择目录，如果安装了新版本，就会自动把老版本覆盖。这里有一个小技巧，那就是可以把老版本的文件夹改个名字，比如 Unity2017，然后再安装新版本，如图 2-4 所示。如此一来，就可以多版本共存了。

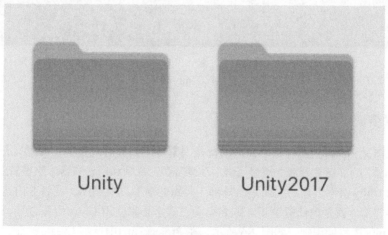

图 2-4　多版本共存

可能 Unity 也意识到项目中多版本共存的需求，所以在新版本中提供了 Unity Hub 的工具，可以专门处理多版本问题，不过目前还只是 Preview（预览版）。

2.2 Project 视图

新项目创建完毕后，Unity 的 5 大经典视图就映入眼帘了，如图 2-5 所示。接着，我们会详细介绍每个视图的功能以及工作方法。首先是 Project 视图，它是游戏资源的集合视图。换句话说，游戏内所能用到的资源，无论大小，都放在这里。

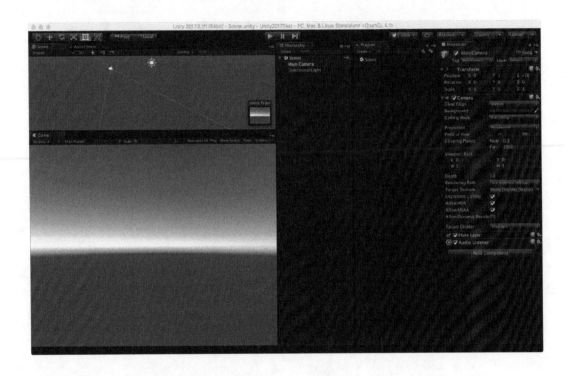

图 2-5 Unity 界面

2.2.1 创建资源

Project 视图又称资源视图，这里面放的都是引擎所用到的游戏资源。资源又分两部分，第一部分为外部资源，例如图片资源、模型资源、动画资源、视频和声音资源。这些资源的特点就是，它们并非使用 Unity 引擎创建的，而是由外部工具做的模型以及贴图，或者业内达成共识的通用格式资源。第二部分就是内部资源了。顾名思义，就是必须使用 Unity 引擎创建，并且这部分资源只有放在 Unity 中才能识别。如图 2-6 所示，在 Project 视图中点击 Create 按钮，会弹出创建资源面板，这里面列出了脚本、Shader、场景、预制、材质、精灵、动画控制器、角色遮罩、时间线、物理材质，等等（这些资源在游戏中都起到了非常重要的作用）。

2.2 Project 视图

图 2-6　创建资源

2.2.2　搜索资源

成型的游戏项目资源是海量的，开发者可能会通过多文件夹或者命名规则来规划它们。但是，如果数以万计的游戏资源都出现在 Project 视图中，快速搜索资源无疑是一个很必要的功能。在 Create 按钮右边的搜索栏中，可以通过字符串的方式模糊搜索资源，但是资源多了，很多不同类型的资源或者放在不同目录下的资源的重名概率就很大，这样搜索非常不便。因此，最佳的搜索应该可以指定文件夹以及资源类型，这样会更精准。

如图 2-7 所示，点击搜索栏右侧的第一个按钮，在弹出框中可以指定搜索类型。比如，我们选择 Scene，此时搜索栏中会自动出现 "t:Scene" 字样，表示在 Project 视图所有文件夹下搜索场景资源。如图 2-8 所示，在搜索栏中再添加一个需要搜索的字符 "a"，即 "a t:Scene"，这就表示在所有场景中搜索名字中包含 "a" 的场景了。聪明的朋友此时应该能联想到，如果想搜索其他类型的资源，也可以自行在输入框中输入 "标识 t:类型" 了。

14 第 2 章 编辑器的结构

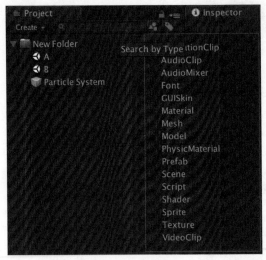

图 2-7 搜索资源

图 2-8 搜索类型

如果想进一步指定搜索的文件夹，点击 Search 下拉框，从中选择需要搜索的文件夹即可，如图 2-9 所示。然后直接在搜索框中输入"par t:Prefab"就表示搜索 New Folder 文件夹下，名字包含"par"字符串的所有 Prefab 类型的游戏资源。怎么样，Unity 的搜索功能很炫酷吧！

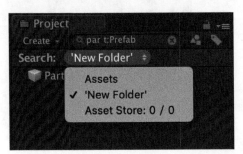

图 2-9 搜索目录

2.2.3 搜索标签

此外，Unity 编辑器还提供了搜索标签的功能。我们可以给特殊资源添加标签，这样即使它们的资源类型不同，也可以通过搜索标签的方式快速搜索出来。如图 2-10 所示，在 Project 中选择一个游戏资源，在右侧 Inspector 面板中的右下角点击添加标签按钮，然后就可以在弹出面板中添加一个或者多个特殊的搜索标签了。这里我们暂时给 Particle System 添加了"Effect"和"Ground"这两个标签。如果希望自定义一些别的标签，则可以在标签菜单栏上面的搜索栏中输入一个全新的自定义字符串标签，然后点击回车即可。

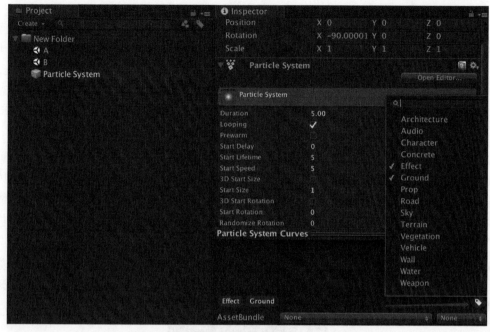

图 2-10　添加标签

要想搜索标签，可以按照图 2-11 所示，点击 Project 视图上面搜索栏右侧的第二个按钮，然后从弹出的标签菜单中选择需要搜索的标签，或者直接在搜索栏中输入"par l:Effect"即可。我相信聪明的读者此时定能明白，"par"表示资源名必须包含的字符串，"l:Effect"表示满足搜索的标签名为 Effect。

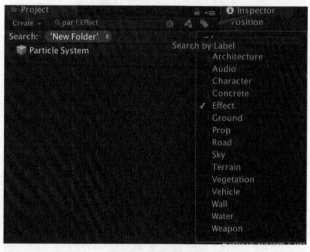

图 2-11　搜索标签

2.3 Hierarchy 视图

Project 视图中的游戏资源，如果需要出现在正式游戏中，那么就需要 Hierarchy 视图（层次视图）了。并不是所有资源都可以直接拖入 Hierarchy 视图中，通常只有游戏对象才能放进，比如 Prefab 资源（第 11 章会详细介绍这个特殊的资源）。游戏对象本身是一个很小的东西，它负责引用外部资源，比如模型、贴图和声音等，其原理其实就是通过程序来控制需要的游戏对象。

2.3.1 创建游戏对象

游戏对象分为两种，第一种是预先编辑的，第二种是运行时代码动态生成的。这里我们先讲预先编辑的游戏对象。如图 2-12 所示，在 Hierarchy 视图中点击 Create 按钮，从弹出的快捷菜单中选择需要创建的游戏对象，例如 3D 游戏对象、2D 游戏对象、特效、光、音频、视频、UI、摄像机等。

Unity 的游戏对象其实就是在一个空的 GameObject 上挂了对应属性的脚本。拿摄像机来说吧，就是在游戏对象上挂了个 Camera 组件（又称脚本），如果删除掉这个脚本，就和普通的游戏对象没有区别了。有些复杂的游戏对象甚至需要挂很多组件才行。由此可见，它非常灵活、方便。

图 2-12　创建游戏对象

2.3.2 搜索游戏对象

在 Hierarchy 视图中，可以同时编辑多场景，而游戏对象隶属于某个场景中。如图 2-13 所示，Create 按钮的右侧有个搜索栏，它和 Project 搜索栏不一样的是，Hierarchy 视图的搜索功能比较简单，只提供了按名称搜索，或者按类型搜索。按名称搜索就是模糊搜索匹配，按类型搜索可以搜索到某个游戏对象身上挂的组件。

Hierarchy 视图的搜索功能没有 Project 全面，其原因就是它并不需要复杂的搜索规则。因为 Hierarchy 视图中同时出现的游戏对象并没有那么多。从开发的角度讲，游戏要尽可能地剔除不需要的游戏对象，避免内存和渲染白白浪费，所以大部分情况下，搜索名称就可以很快定位了。

图 2-13　搜索游戏对象

2.4　Inspector 视图

Inspector 视图承载着所有游戏对象以及游戏资源组件参数的编辑工作。在 Project 视图或者 Hierarchy 视图中任意选择一个游戏对象或者游戏资源，右侧的 Inspector 面板都会列出它的详细属性。选择不同的对象，面板呈现的内容也不同，其原理就是键入一些数据并将其序列化在这个对象身上。如果是在 Hierarchy 中选择的对象，数据就会保存在这个对象所在的场景上。如果是在 Project 视图中选择的资源，数据就会保存在这个资源本身上。

2.4.1　标题栏

大部分对象都有标题栏，Inspector 面板也有，在最上方，如图 2-14 所示。上下共分两行，下面按照从左到右、从上到下的次序介绍。

- 立方体下拉按钮：用于给游戏对象选择一个特殊的标志，可以在 Scene 视图中快速定位到它。
- 第一个勾选框：设置激活状态。如果未勾选，则此游戏对象不会显示在 Scene 和 Game 视图中。
- 输入框：设置游戏对象的名称，可以起任意的字符串名。
- 第二个勾选框：设置对象是否为静态属性。静态属性包括烘焙、遮挡剔除、寻路、剔除等。如果勾选，则启动（第 9 章会详细介绍相关的内容）。
- Tag：给游戏对象设置一个特殊表示，可用于代码中动态获取，或者根据 tag 来判断逻辑。
- Layer：表示给游戏对象设置一个层。用于设置摄像机是否显示某个层，或者点击事件不可响应在某个层上。

以上内容其实都比较重要，后面会详细介绍它们。

图 2-14　标题栏

2.4.2 组件栏

组件又称脚本。习惯上，Unity 系统提供的脚本称为组件，我们自己创建的 C# 或者 JS 则称为脚本。如图 2-15 所示，Inspector 标题栏下方的都属于组件。可见，一个摄像机由 Transform、Camera、Flare Layer、Audio Listener 这 4 个组件组成。

每个组件上都有一些参数可以选择或者设置。拿 Transform 组件来说，所有游戏对象都有这个组件，并且无法删除。它用来记录游戏对象在 3D 世界中的坐标、旋转和缩放信息。在 Camera 组件上，也可以设置颜色、裁切、视口大小等。因为目前选择的这个游戏对象是在场景中的，所以修改后的属性也会记录在场景上。如果我们把这个 Camera 拖入 Project 视图让它变成 Prefab，那么这组信息就会记录在 Prefab 自身上，而与场景无关了。

总之，组件也好，脚本也罢，每个 Inspector 面板所显示的信息是完全不一样的，我们需要做的就是理解组件的含义，这样才能更好地控制它。另外，组件与组件的配合也是一门学问，比如摄像机的后处理效果，它必须依赖摄像机才能发挥作用。这样的例子还有很多，后面会慢慢介绍给大家的。

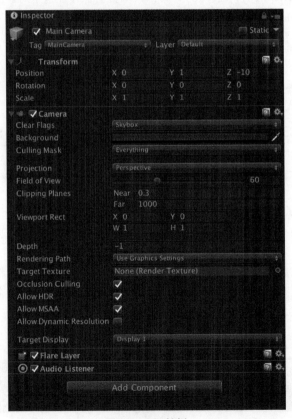

图 2-15　组件栏

2.5　Scene 视图

Scene 视图就是游戏最终画面的自由视角。在游戏开发过程中，首先需要在 Scene 视图中合理地摆放游戏对象，然后配合摄像机将最终画面输出给玩家。例如，游戏中一个主角在挥舞着自己的武器砍怪，我们可以通过 Scene 视图绕到他的身后，看看游戏画面看不到的位置。可见，Scene 视图默默承担着所有的幕后工作。

2.5.1　导航栏

Scene 视图的最上面是一组导航栏，如图 2-16 所示，左上的这 6 个按钮是操作栏按钮，分别是拖动、坐标、旋转、缩放、区域（常用 UI 区域）、整体（同时操作坐标、旋转和缩放），它们对应的快捷键分别是 Q、W、E、R、T 和 Y。在 Hierarchy 视图中任意选择一个游戏对象就可以操作了，读者可以自行在电脑上操作一下，很简单。

图 2-16　导航栏

在操作栏按钮的右边，第一组按钮用于设置游戏对象的操作原点。如图 2-17 和图 2-18 所示，两个 Cube 嵌套在了一起。Pivot 表示父对象的操作原点就是自身的坐标点，而 Center 则表示父对象的操作原点是所有子对象共同的中心点。

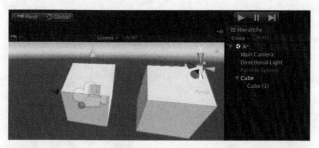

图 2-17　真实坐标

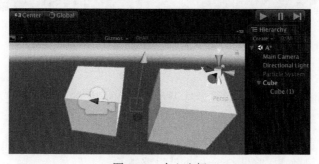

图 2-18　中心坐标

在坐标点设置按钮的右边还有一组按钮，用于设置旋转的方向，如图 2-19 和图 2-20 所示。Global 表示忽略游戏对象自身的旋转，Local 表示使用游戏对象自身的旋转角度。例如，在编辑模式下移动摄像机对象。由于摄像机自身是有旋转角度的，如果希望顺着它的角度来移动，需要设置成 Local。如果希望以水平方向来移动，需要设置成 Global。

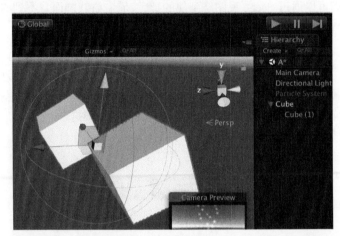

图 2-19　自身旋转坐标

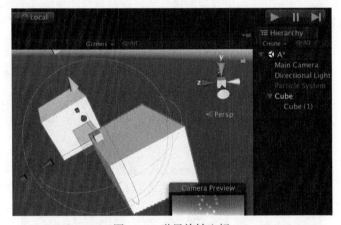

图 2-20　世界旋转坐标

2.5.2　标题栏

导航栏的下面是一组标题栏，如图 2-21 所示，Scene 视图的幕后工作离不开它。下面从左到右依次来介绍一下。

- Shaded：控制 Scene 视图中的显示，例如 GI、模型网格、渲染、shadow 以及 Overdraw（若要优化渲染效率，降低 Overdraw 就很重要了）。

- 2D：设置 Scene 视图是 2D 模式还是 3D 模式。如果是 2D 游戏，没有 Z 轴的概念，所以用不到旋转，这样查看会更方便一些。
- 太阳图标：是否需要光源。如果你发现 Scene 视图很暗，那么打开它即可（推荐打开）。
- 声音图标：如果 Hierarchy 中此时有音频（Audio）对象，非运行模式下点击这个按钮可以听到音频。
- 图片图标：设置与游戏视图相同的渲染信息，例如天空盒子、雾和动画等。这会带来一定开销（通常不建议打开）。
- Gizmos：Scene 视图启动或隐藏一些图标，例如摄像机图标。Unity 的脚本提供了强大的 Gizmos 功能，可以自定义 Gizmos（通过代码绘制）。
- 搜索框：根据名称模糊搜索游戏对象。

图 2-21　标题栏

2.5.3　坐标系控制器

在 Scene 视图的右上角，是坐标系控制器。如图 2-22 所示，点击 x、y、z 三个方向的标志后，会切换 Scene 视图的摄像机朝向；下面的 ISO 用来切换摄像机正交与透视；右上角有个小锁头，点击后即可锁定，此时坐标系控制器将无法操作。

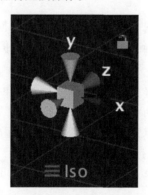

图 2-22　坐标系控制器（另见彩插）

2.5.4　视图预览

Hierarchy 视图以列表的形式列出所有的游戏对象，通过父子对象的关系来管理它们，而 Scene 视图展示了每个游戏对象的具体坐标、选择和缩放等，所以它们的关系是非常紧密的。如图 2-23 所示，在 Hierarchy 视图中选择任意游戏对象，即可在 Scene 视图中出现。结合 2.5.1 节介绍的视图导航栏操作按钮，可以编辑它们的位置。所见即所得，Unity 确实很强大啊！

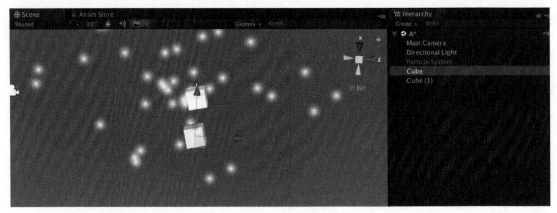

图 2-23 视图预览

2.6 Game 视图

Game 视图是游戏最终展示给玩家的一面,就是把游戏的主摄像机看到的内容显示到了这里。成品的游戏可能会有多个摄像机,比如场景摄像机和 UI 摄像机等,我们可以通过设置摄像机渲染的深度来决定渲染的先后顺序。我们先来看看 Game 视图的组成结构。

2.6.1 标题栏

和 Scene 视图一样,Game 视图也有标题栏,如图 2-24 所示。它的主要功能就是控制 Game 视图的显示,这里的设置并不能影响最终发布游戏的结果,但是可以让开发更方便一些。下面从左到右依次介绍一下。

- Display 1:摄像机可以设置当前的 Display 层,这里可以切换 Game 视图显示的是哪个 Display 层(这个属性在移动平台不支持)。
- Free Aspect:设置分辨率,这个属性很重要。一般游戏都要做自适应屏幕,可以在这里设置目标分辨率,从而验证自适应是否正确。
- Scale:用于整体调节 Game 视图的大小。放大后,可以更清楚地看到开发者关注的某一部分。
- Maximize On Play:全屏运行游戏。
- Mute Audio:是否关闭声音,这个属性很重要。游戏开发到后面,声音都已经植入了,但是程序员关注的可能并不是声音,可是戴上耳机总播放游戏背景音乐,那就太不方便了,所以适当的时候就关掉音频吧。
- Stats:当前游戏的性能面板。
- Gizmos:让 Scene 中的 Gizmos 都显示在 Game 视图中,打开后会影响性能(建议不要打开)。

图 2-24 标题栏

2.6.2 视图预览

如图 2-25 所示,游戏画面就出来了,打包发布到终端几乎就是这个效果了。为什么是几乎,其实还是会有一些差别的,比如分辨率自适应、优化、性能,等等。后面我们会慢慢向大家介绍。

图 2-25 游戏视图

2.7 导航栏视图

导航栏视图在 Unity 编辑器的最上方,主要是所有视图通用的一些功能以及设置信息。正上方的是 Unity 播放器,如图 2-26 所示,这三个按钮分别表示运行游戏、暂停游戏、逐帧播放游戏。

图 2-26 播放器

在调试游戏的时候,通常会用到暂停键,这里提供两个播放器的技巧。

- 点击暂停键以后再点击运行游戏按钮,游戏就会被暂停到第一帧。
- 有时候,调试时来不及点击鼠标暂停键,可以利用 Command+Shift+P 组合键暂停/恢复游戏。

在播放器右侧,有一组工具栏,如图 2-27 所示。

- **Collab**:可以用 Unity 自己提供的云服务来管理版本,包括它的云构建和云打包功能。缺点就是需要把整个工程都上传到 Unity 的服务器上,使用云构建查错的话,没有本地直观。
- **云朵图标**:登录 Unity 云服务,例如 AD 广告、闪退日志收集、性能分析等(有兴趣的朋友可以自行查阅)。
- **Account**:Unity 账号登录管理系统。

图 2-27　工具栏

- **Layers**:可以编辑层。如图 2-28 所示,菜单中每个层的右侧有个"小眼睛"按钮,它表示是否在 Scene 和 View 视图中显示属于这个层的游戏对象。如果你发现在 Game 视图中能看到某个游戏对象,但是在 Scene 视图中却看不到,多半就是这里给关闭了。Edit Layers 用于添加或删除新的层。

图 2-28　编辑层

- **Layout**:表示 Unity 编辑器的布局。如图 2-29 所示,默认提供了 5 种布局供开发者选择,下面还提供了保存和删除布局的功能。通常,Unity 默认提供的布局已经够用。但是 Unity 拓展编辑器的功能非常强大,如果在日后的开发中自己拓展了一些自定义窗口,就需要用到这个了。这可以保证每次打开 Unity,都是自定义的布局。

图 2-29　布局

2.8　其他功能

Unity 编辑器中还隐藏了一些容易忽略但是很有用处的小功能，这里将它们列出来，方便大家在日后的工作中使用。

2.8.1　小锁头

在 Unity 中，几乎每个系统窗口的右上角都有一个"小锁头"，不同窗口中它们的含义可能会有所不同。如图 2-30 所示，Inspector 面板的小锁头就是把这个面板彻底锁定起来。如果此时再选择别的 GameObject，这个面板的内容并不会切换，还是保留之前锁定的信息。

图 2-30　小锁头

2.8.2　窗口菜单

在 Unity 每个系统窗口或者自定义窗口的右上角，都有一个菜单窗口，如图 2-31 所示，下面介绍各个菜单的含义。

- **Normal**：表示正常面板。
- **Debug**：组件的私有属性默认是不显示在面板上，勾选该项后，所有私有属性将显示在面板中。
- **Lock**："小锁头"的功能。
- **Maximize**：最大化窗口。

- Close Tab：关闭窗口。
- Add Tab：添加一个新窗口，这个技巧非常有用，有时需要两个面板的信息，此时可以通过 Add Tab 添加多个面板。配合"小锁头"的功能，对比面板中的属性会更加方便。

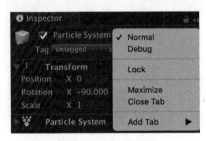

图 2-31　菜单窗口

2.8.3　保存组件参数

Unity 2018 提供了保存属性的功能，每个组件都可以保存。如图 2-32 所示，任意组件的右上方都有一个保存或选取的按钮，可以将参数保存到文件中，或者从文件中选取来设置新组件的参数。

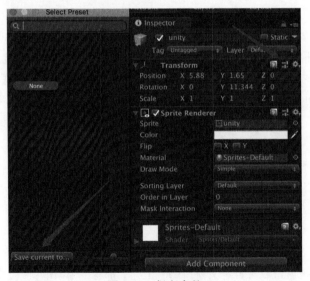

图 2-32　保存参数

2.8.4　Package Manager

Unity 内置了很多好用的开发插件，但是有些插件并不是所有项目都需要的。因此，它提供了 Package Manager 选择窗口（如图 2-33 所示），开发者可以自行安装自己需要的插件。它和 Asset Store 的区别就是，插件的代码以及资源并没有放在 Assets 目录下，这样就更像内置的功能了。

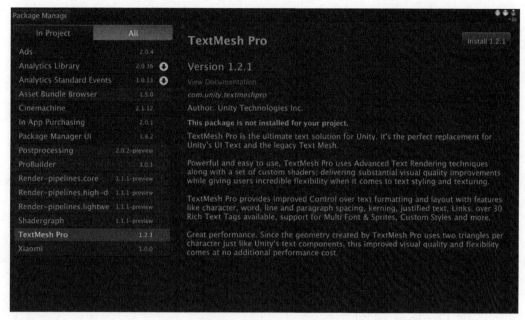

图 2-33　Package Manager

2.9　小结

本章中，我们介绍了 Unity 编辑器常用的 5 大布局，简要说明了布局之间的相互工作方法，以及 Unity 编辑器的一些小技巧。

- Project 视图：保存游戏资源。
- Hierarchy 视图：处理游戏对象的层级。
- Inspector 视图：游戏对象的详细信息展示。
- Scene 视图：编辑模式视图。
- Game 视图：游戏最终发布视图。

第3章

拓展编辑器

上一章中，我们学习了 Unity 的 5 大视图（Project 视图、Hierarchy 视图、Inspector 视图、Scene 视图和 Game 视图），它们都有一套自己的编辑布局，以及相互协作的工作方式。其实 Unity 还内置了很多工具视图，后面会陆续介绍。Unity 内置的编辑器做得再好，其实也满足不了多变的开发需求，不过 Unity 提供了灵活多变的编辑器拓展 API 接口，通过代码反射，可以修改一些系统自带的编辑器窗口。此外，丰富的 EditorGUI 接口也可以拓展出各式各样的编辑器窗口。学好本章，可以为日后开发专属的游戏编辑器打下良好的基础。

3.1 拓展 Project 视图

在 Project 视图下，存放着大量游戏资源，资源之间的依赖关系非常复杂，有效地管理这些资源尤其重要。默认的布局比较简单，按照文件夹为单位依次显示，文件夹下可以嵌套子文件夹，我们可以根据资源类型归类存放。右键弹出的菜单项的功能比较基础，但是可以满足绝大多数的需求。本章中，我们将学习如何拓展它，让菜单项更加丰富。

3.1.1 拓展右键菜单

在 Project 视图中，点击鼠标右键会弹出视图菜单，如图 3-1 所示。菜单的基础功能包括资源的创建、打开、删除、导入、导出、查找引用、刷新和重新导入等。其中，删除、打开和重新导入等操作需要在 Project 视图中选中一个或多个资源。选中某个资源时，该资源名称上将会出现蓝色的矩形框，此时右键弹出菜单即可针对选中的资源来处理。下面首先来学习如何拓展这个菜单。

编辑器使用的代码应该仅限于编辑模式下，也就是说正式的游戏包不应该包含这些代码。Unity 提供了一个规则：如果属于编辑模式下的代码，需要放在 Editor 文件夹下；如果属于运行时执行的代码，放在任意非 Editor 文件夹下即可。如图 3-2 所示，将编辑代码放在 Editor 文件夹下。这里需要说明的是，Editor 文件夹的位置比较灵活，它还可以作为多个目录的子文件夹存在，这样开发者就可以按功能来划分，将不同功能的编辑代码放在不同的 Editor 目录下。例如可以有多个 Editor 目录，它们各自处理各自的逻辑。

3.1 拓展 Project 视图

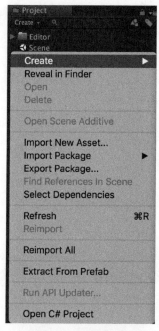

图 3-1 右键菜单

图 3-2 编辑模式

接着，看一下拓展鼠标右键的窗口菜单，如图 3-3 所示。先用鼠标左键选中 Scene 游戏资源，此时点击鼠标右键，此时将弹出系统菜单。这里出现了我们拓展的 My Tools 菜单集合，右侧出现了 Tools 2 和 Tools 1 菜单条，点击其中任意一个菜单条，程序将会自动在 Console 窗口中打印它的名称。本示例的代码如代码清单 3-1 所示。

图 3-3 拓展菜单

代码清单 3-1　Script_03_01.cs 文件

```
using UnityEngine;
using UnityEditor;

public class Script_03_01
{
    [MenuItem("Assets/My Tools/Tools 1",false,2)]
    static void MyTools1()
    {
        Debug.Log(Selection.activeObject.name);
    }
    [MenuItem("Assets/My Tools/Tools 2",false,1)]
    static void MyTools2()
    {
        Debug.Log(Selection.activeObject.name);
    }
}
```

自定义菜单的参数需要在 `MenuItem` 方法中写入显示的菜单路径。如果菜单条比较多，可以在第三个参数处输入表示排序的整数，数值越小，它的排序就越靠前。最后使用 `Debug.Log` 打印选择的游戏对象 `Selection.activeObject.name` 即可。

3.1.2　创建菜单

在 Project 视图中，点击左上方的 Create 按钮，可以弹出资源创建菜单，如图 3-4 所示。在最上面拓展了 My Create 菜单，点击 Sphere 和 Cube 菜单条，程序将创建球体和立方体游戏对象。

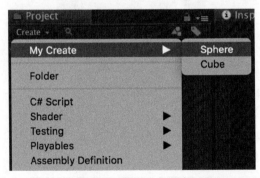

图 3-4　资源创建菜单

菜单条的排序方法和上一节介绍的完全一样，设置 `MenuItem` 方法的第三个参数即可。如代码清单 3-2 所示，重写 `MenuItem` 方法中的路径即可拓展菜单项。

代码清单 3-2　Script_03_02.cs 文件

```
using UnityEngine;
using UnityEditor;

public class Script_03_02
```

```
    {
        [MenuItem("Assets/Create/My Create/Cube",false,2)]
        static void CreateCube()
        {
            GameObject.CreatePrimitive(PrimitiveType.Cube);  //创建立方体
        }
        [MenuItem("Assets/Create/My Create/Sphere",false,1)]
        static void CreateSphere()
        {
            GameObject.CreatePrimitive(PrimitiveType.Sphere);//创建球体
        }
    }
```

通过观察可以发现,拓展菜单的关键就是找到正确的菜单路径,通过"/"符号将它们拼合而成即可。代码中的 `GameObject.CreatePrimitive()` 方法用于创建 Unity 基础模型体。

3.1.3 拓展布局

如图 3-5 所示,当用鼠标选中一个资源后,右边将出现拓展后的 click 按钮,点击这个按钮,程序会自动在 Console 窗口中打印选中的资源名。

图 3-5 拓展布局

如代码清单 3-3 所示,在 Project 视图代码的右侧拓展自定义按钮,在代码中既可设置拓展按钮的区域,也可监听按钮的点击事件。

代码清单 3-3　Script_03_03.cs 文件

```
using UnityEngine;
using UnityEditor;

public class Script_03_03
{
    [InitializeOnLoadMethod]
    static void InitializeOnLoadMethod()
    {
        EditorApplication.projectWindowItemOnGUI = delegate(string guid,
            Rect selectionRect) {
            //在 Project 视图中选择一个资源
            if(Selection.activeObject &&
                guid == AssetDatabase.AssetPathToGUID(AssetDatabase.GetAssetPath
                    (Selection.activeObject))){
                //设置拓展按钮区域
                float width=50f;
```

```
            selectionRect.x +=(selectionRect.width - width);
            selectionRect.y +=2f;
            selectionRect.width = width;
            GUI.color = Color.red;
            //点击事件
            if(GUI.Button(selectionRect,"click")){
                Debug.LogFormat("click : {0}",Selection.activeObject.name);
            }
            GUI.color = Color.white;
        }
    };
}
```

需要说明的是，在方法前面添加[InitializeOnLoadMethod]表示此方法会在C#代码每次编译完成后首先调用。监听EditorApplication.projectWindowItemOnGUI委托，即可使用GUI方法来绘制自定义的UI元素。这里我们添加了一个按钮。此外，GUI还提供了丰富的元素接口，可以用来添加文本、图片、滚动条和下拉框等复杂元素。

3.1.4 监听事件

Project视图中的资源比较多，如果不好好规划，资源就会很凌乱。有时候，我们可能需要借助程序来约束资源，这可以通过监听资源的创建、删除、移动和保存等事件来实现。例如，将某个文件移动到错误的目录下，此时就可以监听资源移动事件，程序判断资源的原始位置以及将要移动的位置是否合法，从而决定是否能阻止本次移动。Unity提供了监听的基类。

如代码清单3-4所示，首先需要继承UnityEditor.AssetModificationProcessor，接着重写监听资源创建、删除、移动和保存的方法，处理自己的特殊逻辑。

代码清单3-4　Script_03_04.cs文件

```csharp
using UnityEngine;
using UnityEditor;
using System.Collections.Generic;

public class Script_03_04 : UnityEditor.AssetModificationProcessor
{
    [InitializeOnLoadMethod]
    static void InitializeOnLoadMethod()
    {
        //全局监听Project视图下的资源是否发生变化（添加、删除和移动）
        EditorApplication.projectWindowChanged = delegate() {
            Debug.Log("change");
        };
    }
    //监听"双击鼠标左键，打开资源"事件
    public static bool IsOpenForEdit(string assetPath, out string message)
    {
        message = null;
        Debug.LogFormat("assetPath : {0} ", assetPath);
```

```csharp
        //true 表示该资源可以打开，false 表示不允许在 Unity 中打开该资源
        return true;
    }
    //监听"资源即将被创建"事件
    public static void OnWillCreateAsset(string path)
    {
        Debug.LogFormat("path : {0}", path);
    }
    //监听"资源即将被保存"事件
    public static string[] OnWillSaveAssets(string[] paths)
    {
        if(paths != null) {
            Debug.LogFormat("path : {0}",  string.Join(",",paths));
        }
        return paths;
    }
    //监听"资源即将被移动"事件
    public static AssetMoveResult OnWillMoveAsset(string oldPath, string newPath)
    {
        Debug.LogFormat("from : {0} to : {1}", oldPath,newPath);
        //AssetMoveResult.DidMove 表示该资源可以移动
        return AssetMoveResult.DidMove;
    }
    //监听"资源即将被删除"事件
    public static AssetDeleteResult OnWillDeleteAsset(string assetPath,
        RemoveAssetOptions option)
    {
        Debug.LogFormat("delete : {0}", assetPath);
        //AssetDeleteResult.DidNotDelete 表示该资源可以被删除
        return AssetDeleteResult.DidDelete;
    }
}
```

3.2 拓展 Hierarchy 视图

Hierarchy 视图中出现的都是游戏对象，这些对象之间同样具有一定的关联关系。我们可以用树状结构来表示游戏对象之间复杂的父子关系。Hierarchy 视图中的游戏对象会通过摄像机最终投影在发布的游戏中。Hierarchy 视图也比较简单，本章就来学习如何拓展它使其更加丰富。

3.2.1 拓展菜单

在 Hierarchy 视图中，也可以对 Create 菜单项进行拓展。如图 3-6 所示，在 Hierarchy 视图中点击 Create 按钮，弹出的菜单 My Create→Cube 就是自定义拓展的菜单，相关代码如代码清单 3-5 所示。

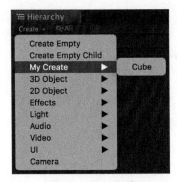

图 3-6 拓展菜单

代码清单 3-5　Script_03_05.cs 文件

```
using UnityEngine;
using UnityEditor;

public class Script_03_05
{
    [MenuItem("GameObject/My Create/Cube",false,0)]
    static void CreateCube()
    {
        GameObject.CreatePrimitive(PrimitiveType.Cube); //创建立方体
    }
}
```

菜单中已经包含了系统默认的一些菜单项，我们拓展的原理就是重写 `MenuItem` 的自定义路径。Create 按钮下的菜单项都在 `GameObject` 路径下面，所以只要开头是 `GameObject/xx/xx`，均可自由拓展。

3.2.2　拓展布局

在 Hierarchy 视图中，同样可以对布局进行拓展。如图 3-7 所示，选择不同的游戏对象后，在右侧可根据 EditorGUI 拓展出一组按钮，点击 Unity 图标按钮后，在 Console 窗口中输入这个游戏对象。它的工作原理就是监听 `EditorApplication.hierarchyWindowItemOnGUI` 渲染回调。相关代码如代码清单 3-6 所示。

图 3-7 拓展布局

代码清单 3-6　Script_03_06.cs 文件

```
using UnityEngine;
using UnityEditor;

public class Script_03_06
{
    [InitializeOnLoadMethod]
    static void InitializeOnLoadMethod()
    {
        EditorApplication.hierarchyWindowItemOnGUI = delegate(int instanceID,
            Rect selectionRect) {
            //在Hierarchy视图中选择一个资源
            if(Selection.activeObject &&
                instanceID ==Selection.activeObject.GetInstanceID()){
                //设置拓展按钮区域
                float width=50f;
                float height=20f;
                selectionRect.x +=(selectionRect.width - width);
                selectionRect.width = width;
                selectionRect.height= height;
                //点击事件
                if(GUI.Button(selectionRect,AssetDatabase.LoadAssetAtPath<Texture>
                    ("Assets/unity.png"))){
                    Debug.LogFormat("click : {0}",Selection.activeObject.name);
                }
            }
        };
    }
}
```

在代码中实现 `EditorApplication.hierarchyWindowItemOnGUI` 委托，就可以重写 Hierarchy 视图了。这里我们使用 `GUI.Button` 来绘制自定义按钮，点击按钮可监听事件。GUI 的种类比较丰富。此外，我们还可以拓展其他显示元素。

3.2.3　重写菜单

通过上面的学习，我们知道 Hierarchy 视图中的菜单可以在原有基础上拓展，那么如果想彻底抛弃它的菜单项，完全使用自己的菜单项是否可行呢？答案是可行的。如图 3-8 所示，先用鼠标选择一个游戏对象，点击右键即可弹出我们的重写菜单，这个菜单项已经和 Unity 自带的完全不一样了。它的工作原理就是监听点击的事件，打开一个新的菜单窗口。相关代码如代码清单 3-7 所示。

图 3-8　重写菜单

代码清单 3-7　Script_03_07.cs 文件

```
using UnityEngine;
using UnityEditor;

public class Script_03_07
{
    [MenuItem("Window/Test/yusong")]
    static void Test()
    {
    }

    [MenuItem("Window/Test/momo")]
    static void Test1()
    {
    }
    [MenuItem("Window/Test/雨松/MOMO")]
    static void Test2()
    {
    }


    [InitializeOnLoadMethod]
    static void StartInitializeOnLoadMethod()
    {
        EditorApplication.hierarchyWindowItemOnGUI += OnHierarchyGUI;
    }

    static void OnHierarchyGUI(int instanceID, Rect selectionRect)
    {
        if(Event.current != null && selectionRect.Contains(Event.current.mousePosition)
            && Event.current.button == 1 && Event.current.type <= EventType.MouseUp)
        {
            GameObject selectedGameObject = EditorUtility.InstanceIDToObject(instanceID)
                as GameObject;
            //这里可以判断 selectedGameObject 的条件
            if(selectedGameObject)
            {
                Vector2 mousePosition = Event.current.mousePosition;

                EditorUtility.DisplayPopupMenu(new Rect(mousePosition.x,
                    mousePosition.y, 0, 0), "Window/Test",null);
                Event.current.Use();
            }
        }
    }
}
```

在上述代码中，我们使用 Event.current 来获取当前的事件。当监听到鼠标抬起的事件后，并且满足游戏对象的选中状态，开始执行自定义事件。其中，EditorUtility.DisplayPopupMenu 用于弹出自定义菜单，Event.current.Use() 的含义是不再执行原有的操作，所以就实现了重写菜单。

此外，Hierarchy 视图还可以重写系统自带的菜单行为。例如，我觉得 Unity 创建的 Image 组件不好，可以复写它的行为，如图 3-9 所示。

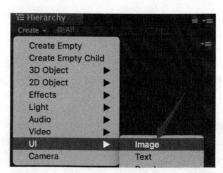

图 3-9　覆盖菜单

创建 Image 组件时，会自动勾选 RaycastTarget。如果图片不需要处理点击事件，这样会带来一些额外的开销。代码清单 3-8 就是复写了创建 Image 组件的逻辑，让 RaycastTarget 默认不勾选。

代码清单 3-8　Script_03_08.cs 文件

```
using UnityEngine;
using UnityEditor;
using UnityEngine.UI;

public class Script_03_08
{
    [MenuItem("GameObject/UI/Image")]
    static void CreatImage()
    {
        if(Selection.activeTransform)
        {
            if(Selection.activeTransform.GetComponentInParent<Canvas>())
            {
                Image image = new GameObject("image").AddComponent<Image>();
                image.raycastTarget = false;
                image.transform.SetParent(Selection.activeTransform,false);
                //设置选中状态
                Selection.activeTransform = image.transform;
            }
        }
    }
}
```

由于重写了菜单，所以需要通过脚本自行创建 Image 对象和组件。接着，获取到 image 组件对象，直接设置它的 raycastTarget 属性即可。

3.3　拓展 Inspector 视图

Inspector 视图可用来展示组件以及资源的详细信息面板，每个组件的面板信息是各不相同的。系统提供的大量组件通常可以满足开发需求,但是偶尔我们还是希望能在原有组件上去拓展,比如添加一些按钮或者添加一些逻辑等。

3.3.1 拓展源生组件

摄像机就是典型的源生组件。如图 3-10 所示,可以在摄像机组件的最上面添加一个按钮。它的局限性就是拓展组件只能加在源生组件的最上面或者最下面,不能插在中间,不过这样也就够了。相关代码如代码清单 3-9 所示。

图 3-10 源生组件

代码清单 3-9 Script_03_09.cs 文件

```
using UnityEngine;
using UnityEditor;

[CustomEditor(typeof(Camera))]
public class Script_03_09 : Editor
{
    public override void OnInspectorGUI(){
        if(GUILayout.Button("拓展按钮")) {
        }
        base.OnInspectorGUI();
    }
}
```

在代码清单 3-9 中,`CustomEditor()` 表示自定义哪个组件,`OnInspectorGUI()` 可以对它进行重新绘制,`base.OnInspectorGUI()` 表示是否绘制父类原有元素。

3.3.2 拓展继承组件

有些系统组件可能在 Unity 内部已经重写了绘制方法,但是外部是访问不了内部代码的,所以修改起来比较麻烦。不过,我们还是有办法的。如图 3-11 所示的 Transform 组件,按照前面介绍的方式直接拓展的话,它的面板就变得非常丑陋,最好拓展面板后,还能保留组件原有的绘制方式。

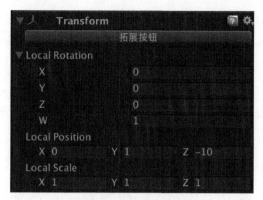

图 3-11　拓展组件

Unity 将大量的 Editor 绘制方法封装在内部的 DLL 文件里，开发者无法调用它的方法。如果想解决这个问题，可以使用 C# 反射的方式调用内部未公开的方法。如图 3-12 所示，通过拓展的 Transfom 组件，现在就可以保留原有的绘制方式了。

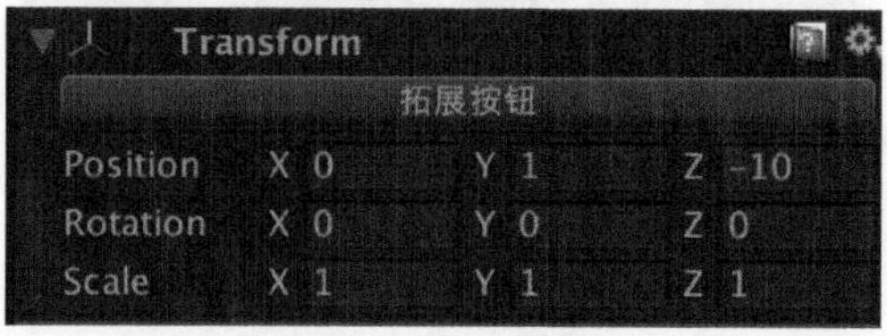

图 3-12　拓展组件

如代码清单 3-10 所示，通过反射先得到 UnityEditor.TransformInspector 对象，然后就可以调用它内部的 OnInspectorGUI() 方法了。

代码清单 3-10　Script_03_10.cs 文件

```
using UnityEngine;
using UnityEditor;
using System.Reflection;

[CustomEditor(typeof(Transform))]
public class Script_03_10 : Editor
{
    private Editor m_Editor;
    void OnEnable()
    {
        m_Editor = Editor.CreateEditor(target,
            Assembly.GetAssembly(typeof(Editor)).GetType("UnityEditor.
                TransformInspector",true));
    }

    public override void OnInspectorGUI(){
        if(GUILayout.Button("拓展按钮")) {
        }
```

```
            //调用系统绘制方法
            m_Editor.OnInspectorGUI();
            //base.OnInspectorGUI();
        }
    }
```

上述代码中,我们重写了 `OnInspectorGUI()` 方法。使用 `GUILayout.Button` 绘制了自定义的按钮元素,接着调用 `m_Editor.OnInspectorGUI()` 绘制 Transform 原有面板信息,这样我们拓展的按钮就会显示在 Transform 面板的上方。

3.3.3 组件不可编辑

在 Unity 中,我们可以给组件设置状态,这样它就无法编辑了。如图 3-13 所示,将 Transform 组件的原始功能禁掉(灰色表示不可编辑),而不影响我们上下拓展的两个按钮。相关代码如代码清单 3-11 所示。

图 3-13 禁用组件

代码清单 3-11   Script_03_11.cs 文件

```
using UnityEngine;
using UnityEditor;
using System.Reflection;

[CustomEditor(typeof(Transform))]
public class Script_03_11 : Editor
{
    private Editor m_Editor;
    void OnEnable()
    {
        m_Editor = Editor.CreateEditor(target,
            Assembly.GetAssembly(typeof(Editor)).GetType("UnityEditor.
                TransformInspector",true));
    }

    public override void OnInspectorGUI(){
        if(GUILayout.Button("拓展按钮上")) {
        }
        //开始禁止
        GUI.enabled = false;
        m_Editor.OnInspectorGUI();
        //结束禁止
        GUI.enabled = true;
```

```
            if(GUILayout.Button("拓展按钮下")) {
            }
        }
    }
```

如果想整体禁止组件，可以按照图 3-14 所示选择任意游戏对象，然后从右键菜单中选择 3D Object→Lock→Lock（锁定）或者 UnLock（解锁）。它的原理就是设置游戏对象的 hideFlags。需要说明的是，我们不一定非设置游戏对象的 hideFlags，也可以单独给某个组件设置 hideFlags，这样只会影响到某一个组件并非全部。相关代码如代码清单 3-12 所示。

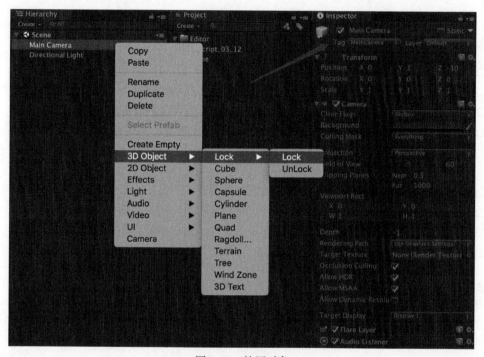

图 3-14 禁用对象

代码清单 3-12　Script_03_12.cs 文件

```
using UnityEngine;
using UnityEditor;

public class Script_03_12
{
    [MenuItem("GameObject/3D Object/Lock/Lock",false,0)]
    static void Lock()
    {
        if(Selection.gameObjects != null) {
            foreach(var gameObject in Selection.gameObjects) {
                gameObject.hideFlags = HideFlags.NotEditable;
            }
        }
    }
```

```
}
[MenuItem("GameObject/3D Object/Lock/UnLock",false,1)]
static void UnLock()
{
    if(Selection.gameObjects != null) {
        foreach(var gameObject in Selection.gameObjects) {
            gameObject.hideFlags = HideFlags.None;
        }
    }
}
```

HideFlags 可以使用按位或（|）同时保持多个属性，其含义都很好理解，大家可以自行输入代码调试一下。

- **HideFlags.None**：清除状态。
- **HideFlags.DontSave**：设置对象不会被保存（仅编辑模式下使用，运行时剔除掉）。
- **HideFlags.DontSaveInBuild**：设置对象构建后不会被保存。
- **HideFlags.DontSaveInEditor**：设置对象编辑模式下不会被保存。
- **HideFlags.DontUnloadUnusedAsset**：设置对象不会被 Resources.UnloadUnused-Assets() 卸载无用资源时卸掉。
- **HideFlags.HideAndDontSave**：设置对象隐藏，并且不会被保存。
- **HideFlags.HideInHierarchy**：设置对象在层次视图中隐藏。
- **HideFlags.HideInInspector**：设置对象在控制面板视图中隐藏。
- **HideFlags.NotEditable**：设置对象不可被编辑。

3.3.4 Context 菜单

点击组件中设置（鼠标右键），可以弹出 Context 菜单，如图 3-15 所示，我们可在原有菜单中拓展出新的菜单栏，相关代码如代码清单 3-13 所示。

图 3-15 Context 菜单

代码清单 3-13　Script_03_13.cs 文件

```
using UnityEngine;
using UnityEditor;

public class Script_03_13
{
    [MenuItem("CONTEXT/Transform/New Context 1")]
    public static void NewContext1(MenuCommand command)
    {
        //获取对象名
        Debug.Log(command.context.name);
    }
    [MenuItem("CONTEXT/Transform/New Context 2")]
    public static void NewContext2(MenuCommand command)
    {
        Debug.Log(command.context.name);
    }
}
```

其中，`[MenuItem("CONTEXT/Transform/New Context 1")]` 表示将新菜单扩展在 Transform 组件上。如果想拓展在别的组件上，例如摄像机组件，直接修改字符串中的 `Transform` 为 `Camera` 即可。如果想给所有组件都添加菜单栏，这里改成 `Compoment` 即可。

以上设置也可以应用在自己写的脚本中。如图 3-16 所示，Script_03_14.cs 是自己创建的脚本，在代码中可以通过 `MenuCommand` 来获取脚本对象，从而访问脚本中的变量，相关代码如代码清单 3-14 所示。

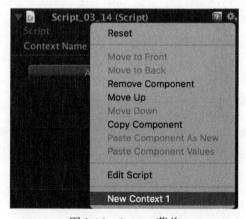

图 3-16　Context 菜单

代码清单 3-14　Script_03_14.cs 文件

```
using UnityEngine;
#if UNITY_EDITOR
using UnityEditor;
#endif
public class Script_03_14 : MonoBehaviour
```

```
{
    public string contextName;
#if UNITY_EDITOR
    [MenuItem("CONTEXT/Script_03_14/New Context 1")]
    public static void NewContext2(MenuCommand command)
    {
        Script_03_14 script = (command.context as Script_03_14);
        script.contextName = "hello world!!";
    }
#endif
}
```

在上述代码中，我们使用到了宏定义标签，其中 `UNITY_EDITOR` 表示这段代码只会在 Editor 模式下执行，发布后将被剔除掉。

当然，我们也可以在自己的脚本中这样写。如果和系统的菜单项名称一样，还可以覆盖它。比如，这里重写了删除组件的按钮，就可以执行一些自己的操作了：

```
[ContextMenu("Remove Component")]
void RemoveComponent()
{
    Debug.Log("RemoveComponent");
    //等一帧再删除自己
    UnityEditor.EditorApplication.delayCall = delegate() {
        DestroyImmediate(this);
    };
}
```

编辑模式下的代码同步时有可能会有问题，比如上述 `DestroyImmediate(this)` 删除自己的代码时，会触发引擎底层的一个错误。不过我们可以使用 `UnityEditor.EditorApplication.delayCall` 来延迟一帧调用。后续如果大家在开发编辑器代码时发现类似问题，也可以尝试等一帧再执行自己的代码。

3.4 拓展 Scene 视图

Scene 视图承担着游戏 "第三人称" 观察的工作。Unity 提供了强大的 Gizmos 工具 API，我们可以在 Scene 视图中绘制立方体、网格、贴图、射线和 UI 等，开发者可以自由地拓展显示组件。

3.4.1 辅助元素

场景在编辑的过程中，通常需要一些辅助元素，这样使用者可以更高效地完成编辑工作。如图 3-17 所示，选中 Main Camera 对象时，程序会给摄像机组件添加一条红色的辅助线，并且在线段终点处添加一个立方体辅助对象。请注意这里拓展的辅助元素只能用来编辑，并不会影响到最终发布的游戏。相关代码如代码清单 3-15 所示。

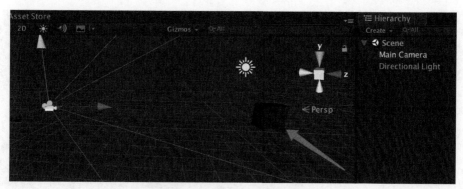

图 3-17 辅助元素（另见彩插）

Gizmo 的绘制原理就是在脚本中添加 OnDrawGizmosSelected()，此方法仅在编辑模式下生效。使用 Gizmos.cs 工具类，我们可以绘制出任意辅助元素。

代码清单 3-15 Script_03_15.cs 文件

```
using UnityEngine;

public class Script_03_15 : MonoBehaviour
{
    void OnDrawGizmosSelected()
    {
        Gizmos.color = Color.red;
        //画线
        Gizmos.DrawLine(transform.position, Vector3.one);
        //立方体
        Gizmos.DrawCube(Vector3.one, Vector3.one);
    }
}
```

如果希望辅助元素并不依赖选择对象出现，而是始终都出现在 Scene 视图中，可使用方法 OnDrawGizmos() 绘制元素：

```
void OnDrawGizmos()
{
    Gizmos.DrawSphere(transform.position, 1);
}
```

此外，Gizmos 工具类中还有很多常用绘制元素，读者也可以自行摸索。

3.4.2 辅助 UI

在 Scene 视图中，我们可以添加 EditorGUI，这样可以方便地在视图中处理一些操作事件。如图 3-18 所示，我们可以在 Scene 视图中绘制辅助 UI，EditorGUI 的代码需要在 Handles.BeginGUI() 和 Handles.EndGUI() 中间绘制完成。这里我们只设置摄像机辅助 UI，其实也可以修改成别的对象，比如游戏对象。相关代码如代码清单 3-16 所示。

第 3 章 拓展编辑器

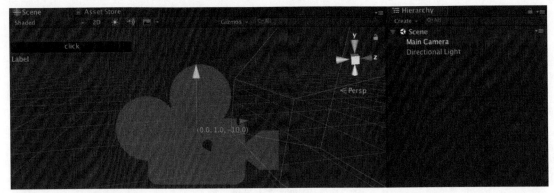

图 3-18 辅助 UI

代码清单 3-16　Script_03_16.cs 文件

```
using UnityEngine;
using UnityEditor;
[CustomEditor(typeof(Camera))]
public class Script_03_16 : Editor
{
    void OnSceneGUI()
    {
        Camera camera = target as Camera;
        if(camera!=null){
            Handles.color = Color.red;
            Handles.Label(camera.transform.position, camera.transform.position.
                ToString());

            Handles.BeginGUI();
            GUI.backgroundColor = Color.red;
            if(GUILayout.Button("click",GUILayout.Width(200f))) {
                Debug.LogFormat("click = {0}", camera.name);
            }
            GUILayout.Label("Label");
            Handles.EndGUI();
        }
    }
}
```

在上述代码中，我们继承了 Editor 类，这样重写 OnSceneGUI() 方法，就可以在 Scene 视图中拓展自定义元素了。

3.4.3　常驻辅助 UI

上一节中，我们介绍的辅助 UI 需要选中一个游戏对象。当然，我们也可以设置常驻辅助 UI。例如，无须选择游戏对象，EditorGUI 将常驻显示在 Scene 视图中，如图 3-19 所示。其原理就是要重写 SceneView.onSceneGUIDelegate，依然需要在 Handles.BeginGUI() 和 Handles.EndGUI() 中间绘制完成。相关代码如代码清单 3-17 所示。

图 3-19 常驻辅助 UI

代码清单 3-17   Script_03_17.cs 文件

```
using UnityEngine;
using UnityEditor;

public class Script_03_17
{
    [InitializeOnLoadMethod]
    static void InitializeOnLoadMethod(){
        SceneView.onSceneGUIDelegate = delegate(SceneView sceneView) {
            Handles.BeginGUI();

            GUI.Label(new Rect(0f,0f,50f,15f),"标题");
            GUI.Button(new Rect(0f,20f,50f,50f),
                AssetDatabase.LoadAssetAtPath<Texture>("Assets/unity.png"));

            Handles.EndGUI();
        };
    }
}
```

上述代码中，全局监听了 SceneView.onSceneGUIDelegate 委托，这样就可以使用 GUI 全局绘制元素了。

3.4.4　禁用选中对象

在 Scene 视图和 Hierarchy 视图中，都可以选择游戏对象。Scene 视图中因为东西很多，而且很可能大量重叠，很容易选错对象。在开发编辑器的时候，当操作某个对象时，如果不希望 Scene 视图中误操作别的对象，我们可以禁用选中对象的功能。相关代码如代码清单 3-18 所示。

代码清单 3-18   Script_03_18.cs 文件

```
using UnityEngine;
using UnityEditor;
```

```
public class Script_03_18
{
    [InitializeOnLoadMethod]
    static void InitializeOnLoadMethod(){
        SceneView.onSceneGUIDelegate = delegate(SceneView sceneView) {
            Event e = Event.current;
            if(e != null){
                int controlID = GUIUtility.GetControlID(FocusType.Passive);
                if(e.type == EventType.Layout)
                {
                    HandleUtility.AddDefaultControl(controlID);
                }
            }
        };
    }
}
```

在上述代码中，`FocusType.Passive`表示禁止接收控制焦点，获取它的`controlID`后，即可禁止将点击事件穿透下去。

此外，还有个办法可以禁止选中功能，即以层为单位设置某个层无法选中。如图3-20所示，右边有个"小锁头"的就无法选中了。

图3-20　禁止选中

直接在Scene视图中很容易选择到子节点，此时可以给它绑定一个`[SelectionBase]`标记，这样该脚本下的所有节点都会定位到绑定这个标记的对象身上：

```
[SelectionBase]
public class RootScript : MonoBehaviour {
}
```

3.5　拓展Game视图

Game视图输出的是最终的游戏画面，理论上是不需要拓展的，不过Unity也可以对其进行拓展。它的拓展主要分为两种：运行模式下以及非运行模式下。

脚本挂在游戏对象后，需要运行游戏才可以执行脚本的生命周期，不过非运行模式下其实也可以执行脚本。如图 3-21 所示，Game 视图在非运行模式下也可以绘制 GUI。其原理就是在脚本类名上方声明[ExecuteInEditMode]，表示此脚本可以在编辑模式中生效。此类脚本通常只是用来做编辑器，正式发布后是不需要的，此时可以使用 UNITY_EDITOR 条件编译发布后剥离掉。

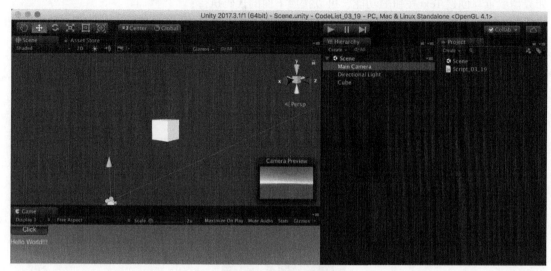

图 3-21　编辑执行脚本

如代码清单 3-19 所示，在类的上面标记[ExecuteInEditMode]，表示非运行模式下也会执行代码的生命周期，接着在 OnGUI() 方法中就可以绘制元素了。

代码清单 3-19　Script_03_19.cs 文件

```
using UnityEngine;
#if UNITY_EDITOR
[ExecuteInEditMode]
public class Script_03_19 : MonoBehaviour
{
    void OnGUI()
    {
        if(GUILayout.Button("Click")) {
            Debug.Log("click!!!");
        }
        GUILayout.Label("Hello World!!!");
    }
}
#endif
```

3.6　MenuItem 菜单

Unity 编辑器使用的拓展菜单是 MenuItem。当然，开发者也可以自由拓展，直接使用"/"符号区分开它的路径即可。系统上方自带的一排菜单也在大量使用这个功能。

3.6.1 覆盖系统菜单

Unity编辑器中自带了很多菜单，大多数情况下可以按照它的原有菜单路径拼合集合。例如，如果需要重写创建Text按钮的功能，可以按照下面的代码执行：

```
[MenuItem("GameObject/UI/Text")]
static void CreateNewText()
{
    Debug.Log("CreateNewText!");
}
```

3.6.2 自定义菜单

自定义菜单可以设置路径、排序、勾选框和禁止选中状态。如图3-22所示，菜单中下方有一个下划线。在代码中设置上一个菜单的 priority（优先级），一共预留了10个元素位置，只需要 priority+11 即可，就会自动增加这个下划线效果。相关代码如代码清单3-20所示。

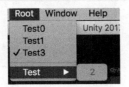

图3-22 自定义菜单

代码清单3-20　Script_03_20.cs 文件

```
using UnityEngine;
using UnityEditor;

public class Script_03_20
{
    [MenuItem("Root/Test1",false,1)]
    static void Test1()
    {
    }
    //菜单排序
    [MenuItem("Root/Test0",false,0)]
    static void Test0()
    {
    }
    [MenuItem("Root/Test/2")]
    static void Test2()
    {
    }
    [MenuItem("Root/Test/2", true,20)]
    static bool Test2Validation()
    {
        //false 表示Root/Test/2菜单将置灰，即不可点击
        return false;
    }
```

```
[MenuItem("Root/Test3",false,3)]
static void Test3()
{
    //勾选框中的菜单
    var menuPath = "Root/Test3";
    bool mchecked = Menu.GetChecked(menuPath);
    Menu.SetChecked(menuPath, !mchecked);
}
}
```

3.6.3 源生自定义菜单

MenuItem 是依托于 Unity 编辑器的菜单栏，换句话说就是无法设置它的位置。如果希望菜单的位置以及出现时机更加灵活的话，可以调用源生自定义菜单的方法。如图 3-23 所示，拓展后，可以在 Scene 视图中点击鼠标右键，此时会弹出一组源生自定义菜单。相关代码如代码清单 3-21 所示。

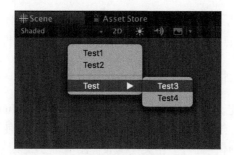

图 3-23　源生菜单

代码清单 3-21　Script_03_21.cs 文件

```
using UnityEngine;
using UnityEditor;

public class Script_03_21
{
    [InitializeOnLoadMethod]
    static void InitializeOnLoadMethod()
    {
        SceneView.onSceneGUIDelegate = delegate(SceneView sceneView) {
            Event e = Event.current;
            //鼠标右键抬起时
            if(e != null && e.button ==1 && e.type == EventType.MouseUp){
                Vector2 mousePosition = e.mousePosition;
                //设置菜单项
                var options = new GUIContent[]{
                    new GUIContent("Test1"),
                    new GUIContent("Test2"),
                    new GUIContent(""),
```

```
                new GUIContent("Test/Test3"),
                new GUIContent("Test/Test4"),
            };
            //设置菜单显示区域
            var selected= -1;
            var userData=Selection.activeGameObject;
            var width =100;
            var height =100;
            var position =new Rect(mousePosition.x,mousePosition.y - height,
                width,height);
            //显示菜单
            EditorUtility.DisplayCustomMenu(position,options,selected,
                delegate(object data, string[] opt, int select) {
                Debug.Log(opt[select]);
            },userData);
            e.Use();
        }
    };
}
```

在上述代码中,首先监听鼠标右键以获取鼠标位置,接着使用 `EditorUtility.DisplayCustomMenu()` 方法来弹出自定义菜单,以及监听菜单选择后的事件。

3.6.4 拓展全局自定义快捷键

Unity 没有提供全局自定义快捷键的拓展,不过可以利用 MenuItem 提供的快捷键来实现这个目的。如图 3-24 所示,我们自定义了快捷键 Command +Shift +D,使用者将需要执行的逻辑(即快捷键后的逻辑)写在方法体内即可。相关代码如代码清单 3-22 所示。

图 3-24 自定义快捷键

代码清单 3-22　Script_03_22.cs 文件

```
using UnityEngine;
using UnityEditor;

public class Script_03_22
{
    [MenuItem("Assets/HotKey %#d",false,-1)]
    private static void HotKey()
    {
        Debug.Log("Command Shift + D");
    }
}
```

热键可以相互组合，其中 %#d 就表示 Command+Shift+D。按照这个格式，我们也可以自由拓展热键组合。

其他热键如下。

- ❏ %：表示 Windows 下的 Ctrl 键和 macOS 下的 Command 键。
- ❏ #：表示 Shift 键。
- ❏ &：表示 Alt 键。
- ❏ LEFT/RIGHT/UP/DOWN：表示左、右、上、下 4 个方向键。
- ❏ F1...F12：表示 F1 至 F12 菜单键。
- ❏ HOME、END、PGUP 和 PGDN 键。

3.7 面板拓展

脚本挂在游戏对象上时，右侧会出现它的详细信息面板，这些信息是根据脚本中声明的 public 可序列化变量而来的。此外，也可以通过 EditorGUI 来对它进行绘制，让面板更具可操作性。

3.7.1 Inspector 面板

EditorGUI 和 GUI 的用法几乎完全一致，目前来说前者多用于编辑器开发，后者多用于发布后调试编辑器。总之，它们都是起辅助作用的。EditorGUI 提供的组件非常丰富，常用的绘制元素包括文本、按钮、图片和滚动框等。做一个好的编辑器，是离不开 EditorGUI 的。如图 3-25 所示，我们将 EditorGUI 拓展在 Inspector 面板上了，相关代码如代码清单 3-23 所示。

图 3-25　EditorGUI

代码清单 3-23　Script_03_23.cs 文件

```
using UnityEngine;
#if UNITY_EDITOR
using UnityEditor;
#endif

public class Script_03_23 : MonoBehaviour
{
    public Vector3 scrollPos;
    public int myId;
```

```csharp
    public string myName;
    public GameObject prefab;
    public MyEnum myEnum = MyEnum.One;
    public bool toogle1;
    public bool toogle2;

    public enum MyEnum
    {
        One=1,
        Two,
    }
}

#if UNITY_EDITOR
[CustomEditor(typeof(Script_03_23))]
public class ScriptEditor_03_23 : Editor
{
    private bool m_EnableToogle;

    public override void OnInspectorGUI()
    {
        //获取脚本对象
        Script_03_23 script = target as Script_03_23;
        //绘制滚动条
        script.scrollPos = 
            EditorGUILayout.BeginScrollView(script.scrollPos,false,true);

        script.myName = EditorGUILayout.TextField("text",script.myName);
        script.myId = EditorGUILayout.IntField("int", script.myId);
        script.prefab = EditorGUILayout.ObjectField("GameObject", script.prefab,
            typeof(GameObject),true)as GameObject;

        //绘制按钮
        EditorGUILayout.BeginHorizontal();
        GUILayout.Button("1");
        GUILayout.Button("2");
        script.myEnum =(Script_03_23.MyEnum)EditorGUILayout.EnumPopup("MyEnum:",
            script.myEnum);
        EditorGUILayout.EndHorizontal();
        //Toogle 组件
        m_EnableToogle = EditorGUILayout.BeginToggleGroup("EnableToogle",
            m_EnableToogle);
        script.toogle1 = EditorGUILayout.Toggle("toogle1", script.toogle1);
        script.toogle2 = EditorGUILayout.Toggle("toogle2", script.toogle2);
        EditorGUILayout.EndToggleGroup();


        EditorGUILayout.EndScrollView();
    }
}
#endif
```

在上述代码中，我们将脚本部分和 Editor 部分的代码合在一个文件中。如果需要拓展的面板比较复杂，建议分成两个文件存放，一个是脚本，另一个是 Editor 脚本。

3.7.2 EditorWindows 窗口

Unity 提供编辑器窗口,开发者可以自由拓展自己的窗口。Unity 编辑器系统自带的视图窗口其实也是用 EditorWindows 实现的。如图 3-26 所示,我们来制作一个简单的编辑窗口,它绘制元素时同样使用 EditorGUI 代码。由此可见,一个完美的编辑器 GUI 的知识是多么重要啊!

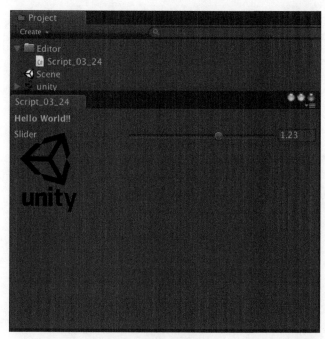

图 3-26 编辑窗口

如代码清单 3-24 所示,使用 `EditorWindow.GetWindow()` 方法即可打开自定义窗口,在 `OnGUI()` 方法中可以绘制窗口元素。请大家注意代码中 EditorWindows 窗口的生命周期。

代码清单 3-24 Script_03_24.cs 文件

```
using UnityEngine;
using UnityEditor;

public class Script_03_24 : EditorWindow
{
    [MenuItem("Window/Open My Window")]
    static void Init()
    {
        Script_03_24 window =(Script_03_24)EditorWindow.GetWindow(typeof(Script_03_24));
        window.Show();
    }

    private Texture m_MyTexture = null;
```

```csharp
    private float m_MyFloat = 0.5f;
    void Awake()
    {
        Debug.LogFormat("窗口初始化时调用");
        m_MyTexture = AssetDatabase.LoadAssetAtPath<Texture>("Assets/unity.png");
    }
    void OnGUI()
    {
        GUILayout.Label("Hello World!!", EditorStyles.boldLabel);
        m_MyFloat = EditorGUILayout.Slider("Slider", m_MyFloat, -5, 5);
        GUI.DrawTexture(new Rect(0,30,100,100),m_MyTexture);
    }
    void OnDestroy()
    {
        Debug.LogFormat("窗口销毁时调用");
    }
    void OnFocus(){
        Debug.LogFormat("窗口拥有焦点时调用");
    }
    void OnHierarchyChange()
    {
        Debug.LogFormat("Hierarchy 视图发生改变时调用");
    }
    void OnInspectorUpdate()
    {
        //Debug.LogFormat("Inspector 每帧更新");
    }
    void OnLostFocus()
    {
        Debug.LogFormat("失去焦点");
    }
    void OnProjectChange()
    {
        Debug.LogFormat("Project 视图发生改变时调用");
    }
    void OnSelectionChange()
    {
        Debug.LogFormat("在 Hierarchy 或者 Project 视图中选择一个对象时调用");
    }
    void Update()
    {
        //Debug.LogFormat("每帧更新");
    }
}
```

3.7.3 EditorWindows 下拉菜单

如图 3-27 所示，在 EditorWindows 编辑窗口的右上角，有个下拉菜单，我们也可以对该菜单中的选项进行拓展，不过这里需要实现 IHasCustomMenu 接口。相关代码如代码清单 3-25 所示。

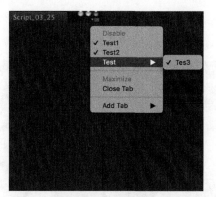

图 3-27 下拉菜单

代码清单 3-25　Script_03_25.cs 文件

```
using UnityEngine;
using UnityEditor;

public class Script_03_25 : EditorWindow, IHasCustomMenu
{
    void IHasCustomMenu.AddItemsToMenu(GenericMenu menu)
    {
        menu.AddDisabledItem(new GUIContent("Disable"));
        menu.AddItem(new GUIContent("Test1"), true,()=> {
            Debug.Log("Test1");
        });
        menu.AddItem(new GUIContent("Test2"), true,()=> {
            Debug.Log("Test2");
        });
        menu.AddSeparator("Test/");
        menu.AddItem(new GUIContent("Test/Tes3"),true,()=> {
            Debug.Log("Tes3");
        });

    }

    [MenuItem("Window/Open My Window")]
    static void Init()
    {
        Script_03_25 window =(Script_03_25)EditorWindow.GetWindow(typeof
            (Script_03_25));
        window.Show();
    }
}
```

上述代码中，我们通过 `AddItem()` 方法来添加列表元素，并且监听选择后的事件。

3.7.4　预览窗口

选择游戏对象或者游戏资源后，Inspector 面板下方将会出现它的预览窗口，但是有些资源是没有预览信息的，不过我们可以监听它的窗口方法来重新绘制它，如图 3-28 所示，相关代码如代码清单 3-26 所示。

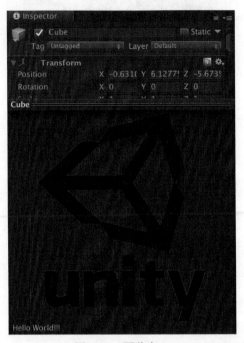

图 3-28　预览窗口

代码清单 3-26　Script_03_26.cs 文件

```
using UnityEngine;
using UnityEditor;

[CustomPreview(typeof(GameObject))]
public class Script_03_26 : ObjectPreview
{
    public override bool HasPreviewGUI()
    {
        return true;
    }
    public override void OnPreviewGUI(Rect r, GUIStyle background)
    {
        GUI.DrawTexture(r,AssetDatabase.LoadAssetAtPath<Texture>("Assets/unity.png"));
        GUILayout.Label("Hello World!!!");
    }
}
```

这段代码的原理就是继承 ObjectPreview 并且重写 OnPreviewGUI() 方法，接着就可以通过代码进行绘制了。[CustomPreview(typeof(GameObject))] 中的 GameObject 代表需要重新绘制的预览对象，也可以换成别的系统对象或自定义的脚本对象。

3.7.5 获取预览信息

有些资源是有预览信息的，比如模型资源。在预览窗口中，我们可以看到它的样式。如果需要在自定义窗口中显示它，就需要获取它的预览信息。如图 3-29 所示，选择一个游戏对象后，会在自定义窗口中显示它，相关代码如代码清单 3-27 所示。

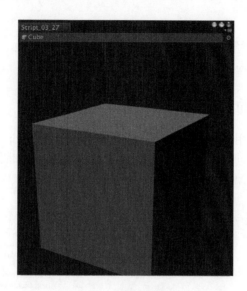

图 3-29 获取预览信息

代码清单 3-27　　Script_03_27.cs 文件

```
using UnityEngine;
using UnityEditor;

public class Script_03_27 : EditorWindow
{
    private GameObject m_MyGo;
    private Editor m_MyEditor;

    [MenuItem("Window/Open My Window")]
    static void Init()
    {
        Script_03_27 window =(Script_03_27)EditorWindow.GetWindow(typeof(Script_03_27));
        window.Show();
    }
    void OnGUI() {
        //设置一个游戏对象
        m_MyGo =(GameObject) EditorGUILayout.ObjectField(m_MyGo,
            typeof(GameObject), true);

        if(m_MyGo != null) {
            if(m_MyEditor == null) {
```

```
            //创建 Editor 实例
            m_MyEditor = Editor.CreateEditor(m_MyGo);
        }
        //预览它
        m_MyEditor.OnPreviewGUI(GUILayoutUtility.GetRect(500, 500),
            EditorStyles.whiteLabel);
    }
  }
}
```

在上述代码中，预览对象首先需要通过 `Editor.CreateEditor()` 拿到它的 `Editor` 实例对象，接着调用 `OnPreviewGUI()` 方法传入窗口的显示区域。

3.8 Unity 编辑器的源码

Unity 编辑器几乎都是用 C# 编写而成的，视图中也大量使用 EditorGUI 来编辑布局。例如，对于常见的 5 大布局视图，所有的代码都放在 UnityEditor.dll 中。如图 3-30 所示，打开 Unity 安装目录，在 Managed 子目录中存放着引擎所需要用到的所有 DLL 文件。

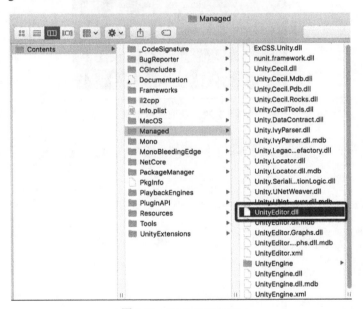

图 3-30　UnityEditor.dll

3.8.1 查看 DLL

拿到 UnityEditor.dll 以后，就可以通过第三方工具来分析和查看了，常用的工具包括 .NET Reflector 以及 ILSpy。此外，也可以使用 Unity 自带的 IDE（Visual Studio）来查看。

用 Visual Studio 打开任意 Unity 工程，接着在 Assembly-CSharp-Editor 中找到 UnityEditor，

双击打开它，然后在左上方的 Visibility 中选择 All members，在右上角的 Language 中选择 C#，此时源码都出来了，如图 3-31 所示。可以发现，面板最上方还有个搜索框，可以用来模糊搜索代码。

有兴趣的朋友还可以看看 UnityEngine.dll。其实 Unity 的 C# 版 API 接口都在这里，只是源码的核心功能都是在 C/C++ 中完成的，DLL 只负责中间调用接口而已。

图 3-31　查看 DLL 文件

阅读 UnityEditor.dll 源码是非常必要的，因为只有熟悉了 Unity 自己内部的编辑器开发，我们才能做出更完美的编辑器。但是 Unity 内部的有些方法设置了 sealed 属性，这样我们写的代码就无法访问它了，所以有时我们不得不使用反射来做点东西。反射可以访问私有成员变量、私有成员方法、私有静态变量和静态方法。

3.8.2　清空控制台日志

系统日志以及 `Debug.Log()` 产出的日志都输出在 Console 窗口中。在 Console 窗口的左上角，有个 Clean 按钮，它用于清空控制台日志。如果希望脚本可以灵活自动清空日志，就必须使用反射了。如图 3-32 所示，首先找到控制台的窗口类（`ConsoleWindows.cs`），接着在 `OnGUI()` 方法中可以看到：点击 Clean 按钮后，Unity 会执行 `LogEntries.Clear()` 方法。

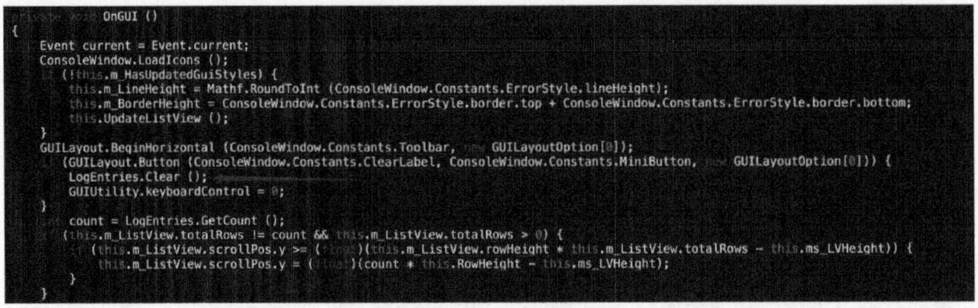

图 3-32 清空日志

接着，在反射中找到 LogEntries.cs 类，如图 3-33 所示，这个类标记了 sealed 属性，所以外部是无法直接访问到的，只能反射调用 Clear() 这个方法了。相关代码如代码清单 3-28 所示。

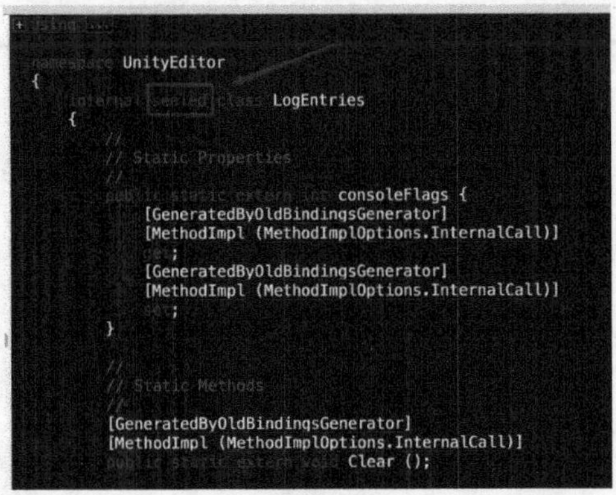

图 3-33 反射

代码清单 3-28　Script_03_28.cs 文件

```csharp
using UnityEngine;
using UnityEditor;
using System.Reflection;

public class Script_03_28
{
    [MenuItem("Tools/CreateConsole")]
    static void CreateConsole()
    {
        Debug.Log("CreateConsole");
    }

    [MenuItem("Tools/CleanConsole")]
    static void CleanConsole()
    {
        //获取assembly
        Assembly assembly = Assembly.GetAssembly(typeof(Editor));
```

```
        //反射获取 LogEntries 对象
        MethodInfo methodInfo = assembly.GetType("UnityEditor.LogEntries").
            GetMethod("Clear");
        //反射调用它的 Clear 方法
        methodInfo.Invoke(new object(), null);
    }
}
```

上述代码中，我们学习了如何通过反射调用 Unity 内部方法。此外，Unity 还有很多内部方法。

3.8.3 获取 EditorStyles 样式

EditorStyles 是编辑器用到的样式，但是 Unity 文档中并没有集中说明每个样式对应的效果。开发一个漂亮的编辑器，肯定会用到很多样式，我们可以借助反射的原理找出它们，这样以后使用起来就方便多了。如图 3-34 和图 3-35 所示，反射出 EditorSytles 的属性，找到 GUIStyle 并最终在 OnGUI 中预览出来。

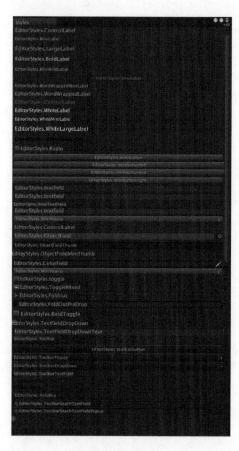

图 3-34　样式

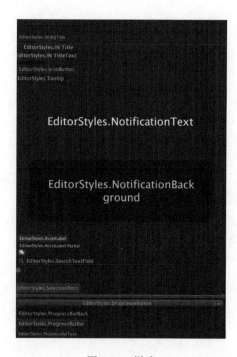

图 3-35　样式

如代码清单 3-29 所示，首先反射获取 EditorStyles 下的所有样式，然后在窗口中依次绘制出来。其中，BindingFlags 表示反射属性的类型。

代码清单 3-29　Script_03_29.cs 文件

```csharp
using UnityEngine;
using UnityEditor;
using System.Reflection;
using System.Collections.Generic;

public class Script_03_29 : EditorWindow
{
    static List<GUIStyle> styles = null;
    [MenuItem("Window/Open My Window")]
    public static void Test() {
        EditorWindow.GetWindow<Script_03_29>("styles");

        styles = new List<GUIStyle>();
        foreach(PropertyInfo fi in typeof(EditorStyles).GetProperties(BindingFlags.
            Static|BindingFlags.Public|BindingFlags.NonPublic))
        {
            object o = fi.GetValue(null, null);
            if(o.GetType() == typeof(GUIStyle)) {
                styles.Add(o as GUIStyle);
            }
        }
    }

    public Vector2 scrollPosition = Vector2.zero;
    void OnGUI()
    {
        scrollPosition = GUILayout.BeginScrollView(scrollPosition);
        for(int i = 0; i < styles.Count; i++) {
            GUILayout.Label("EditorStyles." +styles[i].name, styles[i]);
        }
        GUILayout.EndScrollView();
    }
}
```

通过上述代码，Unity 引擎内部的所有样式都反射出来了。如果以后做编辑器时需要用到这些样式，直接设置 style 名字即可。

3.8.4　获取内置图标样式

Unity 编辑器还内置了很多图标样式，这些样式也没有在文档中详细说明。如图 3-36 所示，系统一共内置了 2000 多个样式，有了这些图标，就可以随心所欲地拓展编辑器了。

3.8 Unity 编辑器的源码

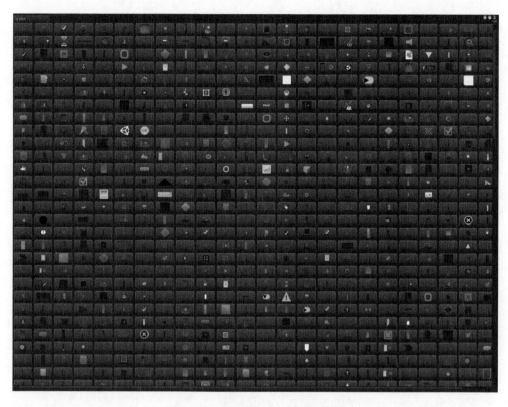

图 3-36 图标样式（另见彩插）

如代码清单 3-30 所示，首先通过 `Resources.FindObjectsOfTypeAll()` 方法查出所有贴图，接着使用 `EditorGUIUtility.IconContent()` 方法加载引擎所用到的图标。

代码清单 3-30　Script_03_30.cs 文件

```csharp
using UnityEngine;
using UnityEditor;
using System.Reflection;
using System.Collections.Generic;
using System;

public class Script_03_30 : EditorWindow
{
    [MenuItem("Window/Open My Window")]
    public static void OpenMyWindow() {
        EditorWindow.GetWindow<Script_03_30>("icons");
    }
    private Vector2 m_Scroll;
    private List<string> m_Icons = null;
    void Awake()
    {
        m_Icons = new List<string>();;
```

```
        Texture2D[] t = Resources.FindObjectsOfTypeAll<Texture2D>();
        foreach(Texture2D x in t) {
            Debug.unityLogger.logEnabled = false;
            GUIContent gc = EditorGUIUtility.IconContent(x.name);
            Debug.unityLogger.logEnabled = true;
            if(gc != null && gc.image != null) {
                m_Icons.Add(x.name);
            }
        }
        Debug.Log(m_Icons.Count);
    }
    void OnGUI()
    {
        m_Scroll = GUILayout.BeginScrollView(m_Scroll);
        float width = 50f;
        int count =(int)(position.width / width);
        for(int i =0; i< m_Icons.Count; i += count)
        {
            GUILayout.BeginHorizontal();
            for(int j =0; j < count; j++)
            {
                int index = i + j;
                if(index < m_Icons.Count)
                    GUILayout.Button(EditorGUIUtility.IconContent(m_Icons[index]),
                        GUILayout.Width(width), GUILayout.Height(30));
            }
            GUILayout.EndHorizontal();
        }
        EditorGUILayout.EndScrollView();
    }
}
```

通过上述代码,我们将引擎内所有的图标展示在窗口中。通过这些图标,可以让自定义编辑器的内容更加丰富。

3.8.5 拓展默认面板

有些资源系统没有提供面板,比如文件夹面板。如图 3-37 所示,选择任意文件夹,声明 [CustomEditor(typeof(UnityEditor.DefaultAsset))]后,即可开始拓展它。

图 3-37 默认面板

如代码清单 3-31 所示,在 OnInspectorGUI()方法中就可以拓展自定义文件夹面板了。

代码清单 3-31　Script_03_31.cs 文件

```
using UnityEngine;
using UnityEditor;

[CustomEditor(typeof(UnityEditor.DefaultAsset))]
public class Script_03_31 : Editor
{
    public override void OnInspectorGUI()
    {
        string path = AssetDatabase.GetAssetPath(target);
        GUI.enabled = true;
        if(path.EndsWith(string.Empty)){
            GUILayout.Label("拓展文件夹");
            GUILayout.Button("我是文件夹");
        }
    }
}
```

在上述代码中，UnityEditor.DefaultAsset 表示默认资源，也就是 Unity 引擎无法识别的资源类型。如果还有别的资源需要拓展面板，大家可以判断路径的后缀。

3.8.6　例子：查找 ManagedStaticReferences() 静态引用

在 Profiler 里，会经常看到某个资源在内存中，但无法被 GC 卸载掉，其原因可能是它被 static 静态引用了，也可能是资源的循环引用造成的。这样，我们就无法分辨出在哪里被引用了。在下面的代码中，我们使用 static 强引用一个游戏资源：

```
public class NewBehaviourScript : MonoBehaviour
{
    public static GameObject prefab;
    void Start() {
        prefab = Resources.Load<GameObject>("prefab");
    }
}
```

运行游戏后，通过反射来递归查询这个资源被代码哪里 static 强引用了。如图 3-38 所示，在导航栏菜单中选择"脚本 Static 引用"进行查询。

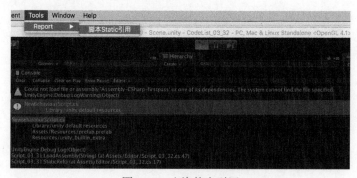

图 3-38　查询静态引用

如代码清单 3-32 所示，首先反射出 DLL 里的静态属性，找到内存中的对象，再利用 `EditorUtility.CollectDependencies()` 方法最终递归查找出资源被哪里 static 强引用了。

代码清单 3-32　Script_03_32.cs 文件

```csharp
using UnityEngine;
using UnityEditor;
using System.Reflection;
using System;
using System.Collections.Generic;
using System.Text;
using System.Collections;

public class Script_03_32 : Editor
{
    [MenuItem("Tools/Report/脚本 Static 引用")]
    static void StaticRef()
    {
        //静态引用
        LoadAssembly("Assembly-CSharp-firstpass");
        LoadAssembly("Assembly-CSharp");

    }

    static void LoadAssembly(string name)
    {
        Assembly assembly = null;
        try {
            assembly = Assembly.Load(name);
        }
        catch(Exception ex) {
            Debug.LogWarning(ex.Message);
        }
        finally{
            if(assembly != null) {
                foreach(Type type in assembly.GetTypes()) {
                    try {
                        HashSet<string> assetPaths = new HashSet<string>();
                        FieldInfo[] listFieldInfo = type.GetFields(BindingFlags.
                            Static | BindingFlags.NonPublic | BindingFlags.Public);
                        foreach(FieldInfo fieldInfo in listFieldInfo) {
                            if(!fieldInfo.FieldType.IsValueType) {
                                SearchProperties(fieldInfo.GetValue(null),
                                    assetPaths);
                            }
                        }
                        if(assetPaths.Count > 0) {
                            StringBuilder sb = new StringBuilder();
                            sb.AppendFormat("{0}.cs\n", type.ToString());
                            foreach(string path in assetPaths) {
                                sb.AppendFormat("\t{0}\n", path);
                            }
                            Debug.Log(sb.ToString());
                        }
```

```
            } catch(Exception ex){
                Debug.LogWarning(ex.Message);
            }
        }
    }
}

static HashSet<string> SearchProperties(object obj,HashSet<string> assetPaths)
{
    if(obj != null) {
        if(obj is UnityEngine.Object) {
            UnityEngine.Object[]depen = EditorUtility.CollectDependencies
                (new UnityEngine.Object[]{ obj as UnityEngine.Object });
            foreach(var item in depen) {
                string assetPath = AssetDatabase.GetAssetPath(item);
                if(!string.IsNullOrEmpty(assetPath)) {
                    if(!assetPaths.Contains(assetPath)) {
                        assetPaths.Add(assetPath);
                    }
                }
            }
        } else if(obj is IEnumerable) {
            foreach(object child in (obj as IEnumerable)) {
                SearchProperties(child,assetPaths);
            }
        }else if(obj is System.Object) {
            if(!obj.GetType().IsValueType) {
                FieldInfo[] fieldInfos = obj.GetType().GetFields();
                foreach(FieldInfo fieldInfo in fieldInfos) {
                    object o = fieldInfo.GetValue(obj);
                    if(o != obj) {
                        SearchProperties(fieldInfo.GetValue(obj),assetPaths);
                    }
                }
            }
        }
    }
    return assetPaths;
}
```

上述代码中，反射 Assembly-CSharp-firstpass 和 Assembly-CSharp 两个 DLL 文件，找到 `static` 静态对象后，递归查询被哪里 `static` 强引用到，最终输入被引用的对象位置。

需要注意的是，这个脚本只能找出代码中的静态引用。但是 `ManagedStaticReferences` 也可能是由资源的环形引用造成的。环形引用的意思就是 A 引用 B，B 引用 C，C 引用 A，这是一件非常恐怖的事，平时开发中一定要避免这种情况。

3.8.7 UIElements

以前的 Editor 开发编辑器只能使用 GUI 来开发，但是 GUI 有很多缺陷，比如显示区域必须写死在代码中，无法配置。而 Unity 2017 引入了新的概念，那就是 `UIElements`，它可以像 CSS

一样来布局界面。如图 3-39 所示，首先创建样式文件，将 style.uss 文件放入 Resources 目录下并且用记事本直接打开它，然后就可以配置 TextField 和 Button 的大小。UIElements 目前还处于实验阶段，未来可能还会进行调整。

图 3-39　设置样式

在代码中，通过 `AddStyleSheetPath(style)` 即可添加这个样式，后面添加的按钮以及文本输入框都会读取 style.uss 的样式配置，并且还可以设置循环自动换行。目前可以控制的元素包括 Button（按钮）、Toggle（勾选框）、Label（文本）、ScrollView（滚动区域）、TextField（输入框）和 EditorTextField（编辑文本输入框），如图 3-40 所示。相关代码如代码清单 3-33 所示。

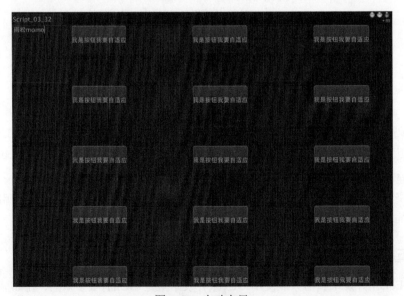

图 3-40　自动布局

代码清单3-33 Script_03_33.cs 文件

```
using UnityEngine;
using UnityEditor;
using UnityEditor.Experimental.UIElements;
using UnityEngine.Experimental.UIElements;
using UnityEngine.Experimental.UIElements.StyleEnums;

public class Script_03_33 : EditorWindow
{

    [MenuItem("UIElementsTest/Style")]
    public static void ShowExample()
    {
        Script_03_33 window = GetWindow<Script_03_33>();
        window.titleContent = new GUIContent("Script_03_33");
    }


    public void OnEnable()
    {
        var root = this.GetRootVisualContainer();
        //添加 style.uss 样式
        root.AddStyleSheetPath("style");

        var boxes = new VisualContainer();
        //设置自动换行
        boxes.style.flexDirection = FlexDirection.Row;
        boxes.style.flexWrap = Wrap.Wrap;
        for(int i = 0; i < 20; i++) {
            TextField m_TextField = new TextField();
            boxes.Add(m_TextField);
            Button button = new Button(delegate() {
                Debug.LogFormat("Click");
            });
            button.text = "我是按钮我要自适应";

            boxes.Add(button);
        }
        root.Add(boxes);
    }
}
```

上述代码中，我们添加 USS 样式，再创建的 `TextField` 和 `Button` 元素就会按样式中配置参数的大小显示。

3.8.8 查询系统窗口

Unity 系统的一些窗口做得很棒，有时需要仿照它来做自定义编辑器。例如，要查看 Profiler

窗口是如何绘制的，首先需要查询到它的窗口代码写在哪里，然后在编辑器中打开 Profiler 窗口，在代码中可以输出当前打开的所有窗口：

```
[MenuItem("Tool/GetEditorWindow")]
static void GetEditorWindow()
{
    foreach(var item in Resources.FindObjectsOfTypeAll<EditorWindow>())
    {
        Debug.Log(item.GetType().ToString());
    }
}
```

如图 3-41 所示，我们查出 Profiler 窗口的位置在 `UnityEditor.ProfilerWindow`。如图 3-42 所示，找到代码所在位置，就可以查看它是如何画出来的了，其中核心绘制代码都在 `OnGUI()` 方法中。

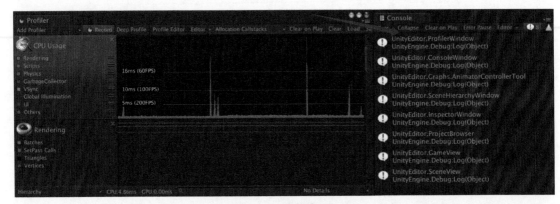

图 3-41　查询窗口

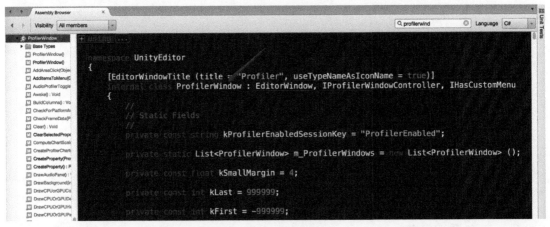

图 3-42　Profiler 窗口

3.8.9 自定义资源导入类型

将文本、FBX 模型、MP3 音乐等常用类型的资源拖入 Unity 引擎后即可直接识别，不过其他后缀名的文件 Unity 是无法识别的。Unity 2018 添加了自定义资源导入类型来识别自定义格式的资源。我们先创建一个自定义格式的文件 test.yusongmomo，默认情况下 Unity 是无法识别它的。

如代码清单 3-34 所示，首先声明 [ScriptedImporter(1, "yusongmomo")]（这表示该脚本用于监听后缀名是 yusongmomo 的自定义文件），接着在 OnImportAsset() 方法中就可以调用 Unity 自己的方法来对资源组合赋值了。

代码清单 3-34　Script_03_34.cs 文件

```
using UnityEngine;
using UnityEditor.Experimental.AssetImporters;
using System.IO;

//监听的后缀名
[ScriptedImporter(1, "yusongmomo")]
public class Script_03_33 : ScriptedImporter
{
    //监听自定义资源导入
    public override void OnImportAsset(AssetImportContext ctx)
    {
        //创建立方体对象
        var cube = GameObject.CreatePrimitive(PrimitiveType.Cube);
        //将参数提取出来
        var position = JsonUtility.FromJson<Vector3>(File.ReadAllText(ctx.assetPath));

        cube.transform.position = position;
        cube.transform.localScale = Vector3.one;
        //将立方体绑定到对象身上
        ctx.AddObjectToAsset("obj", cube);
        ctx.SetMainObject(cube);

        //添加材质
        var material = new Material(Shader.Find("Standard"));
        material.color = Color.red;
        ctx.AddObjectToAsset("material", material);

        var tempMesh = new Mesh();
        DestroyImmediate(tempMesh);
    }
}
```

在上述代码中，我们创建了立方体对象，并且将从自定义文件 test.yusongmomo 中获取的坐标信息赋值给它。如图 3-43 所示，该文件放入引擎中已经变成可识别资源了。

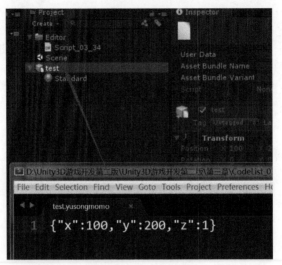

图 3-43　自定义资源

3.9　小结

本章中,我们了解了如何拓展编辑器。编辑器的 API 非常丰富,可以灵活地自由扩展。学习了 EditorGUI,可以用来编辑拓展界面,做出各式各样的编辑器窗口。我们还介绍了 Unity 的五大常用视图(Project 视图、Hierarchy 视图、Inspector 视图、Scene 视图和 Game 视图)的拓展,以及常用菜单栏的拓展。最后,我向大家介绍了如何查看 Unity 编辑器的源码。通过阅读源码,可以借鉴 Unity 内置编辑器的开发思路,这为我们日后开发优秀的编辑器打下良好的基础。

第 4 章

游戏脚本

游戏脚本是整个游戏的核心组件，使用它可以创建游戏对象，控制图形渲染，接受并处理用户输入事件，控制内存等。Unity 系统提供了很多脚本，它们拥有一套完整的生命周期。当然，开发者也可以创建自己的脚本。最新版本的 Unity 已经全面使用 C# 语言作为脚本的开发语言，对应也提供了强大的 API 接口。在开发模式下，它使用 Mono 来跨平台地编译和解析 C# 脚本。因为 Mono 是跨平台的，所以 Unity 的编辑器可以同时部署在 Windows、macOS 和 Linux 操作系统上。游戏发布后，Unity 还提供了自动将 DLL 转成 IL2CPP 的方式，这可以提升代码编译后的执行效率以及稳定性。对于这一切，开发者只需在设置界面简单操作一下即可。这实在是太方便了！

4.1 创建脚本

如图 4-1 所示，在 Project 视图中点击 Create→C# Script 菜单项，即可创建一个游戏脚本。脚本可以放在除 Editor 以外的任意目录或子目录下，因为 Editor 目录下的代码会被系统认为是编辑模式代码，打包后会被自动剥离。

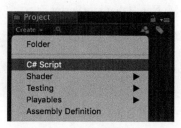

图 4-1　创建脚本

4.1.1 脚本模板

需要注意的是，菜单中还有两类脚本——Testing 和 Playables，前者是用来做单元测试的，后者是 Unity 新功能 TimeLine 中引入的全新概念，用于管理时间线上每一帧的动画、声音和视频等，第 7 章也会向大家详细介绍它。

脚本创建完毕后，会自动生成一套模板，如图 4-2 所示，其中 C# 脚本、Testing 脚本、Playables 脚本以及 Shader 都在 ScriptTemplates 目录下。我们可以修改脚本模板的格式，这样以后再创建脚本时，就会按照修改后的格式来。模板文件名前面的数字代表菜单栏的排序，如果想新增一套模板，可以按照这个格式加一套新的。

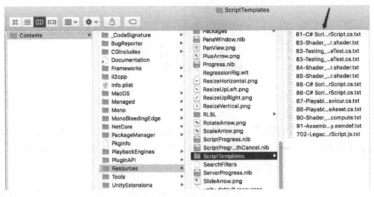

图 4-2　脚本模板

添加自定义模板其实很有意义。例如，程序使用一些框架来编写，它们的基础模板需要拓展，如果每次创建脚本后，都将其手动添加到代码中，那就太麻烦了，此时就可以使用自定义模板。

4.1.2　拓展脚本模板

前面介绍的添加模板的缺点就是无法进行版本化管理，项目组里的每个人都需要手动在本地安装的 Unity 目录下修改这个模板，未来如果要修改模板，也需要每个人单独改自己的，想想这确实有点麻烦。下面我们将介绍一种新的添加模板的方式，它可以很好地进行版本化管理。如图 4-3 所示，首先将代码模板 C# Script-MyNewBehaviourScript.cs.txt 放入 Editor/ScriptTemplates 目录下。

图 4-3　拓展脚本模板

该模板的代码如下：

```
using System.Collections;
using System.Collections.Generic;
using UnityEngine;

public class #NAME# : MonoBehaviour {

    void MyFunction()
    {

    }
}
```

如图 4-4 所示，在 Project 视图的 Create 菜单中添加 C# MyScript 菜单项。因为创建脚本时，需要监听用户输入的名字，所以代码需要继承 EndNameEditAction 来监听 Callback，最终根据用户输入的名称自动创建对应的模板类。相关代码如代码清单 4-1 所示。

图 4-4 拓展脚本模板

代码清单 4-1　Script_04_01.cs 文件

```
using UnityEngine;
using UnityEditor;
using System;
using System.IO;
using System.Text;
using UnityEditor.ProjectWindowCallback;
using System.Text.RegularExpressions;

public class Script_04_01
{
    //脚本模板所在的目录
    private const string MY_SCRIPT_DEFAULT = "Assets/Editor/ScriptTemplates/
        C# Script-MyNewBehaviourScript.cs.txt";

    [MenuItem("Assets/Create/C# MyScript", false, 80)]
    public static void CreatMyScript()
    {
        string locationPath = GetSelectedPathOrFallback();
        ProjectWindowUtil.StartNameEditingIfProjectWindowExists(0,
            ScriptableObject.CreateInstance<MyDoCreateScriptAsset>(),
            locationPath + "/MyNewBehaviourScript.cs",
            null,MY_SCRIPT_DEFAULT);
    }

    public static string GetSelectedPathOrFallback()
    {
        string path = "Assets";
        foreach(UnityEngine.Object obj in Selection.GetFiltered(typeof(UnityEngine.
            Object), SelectionMode.Assets))
        {
            path = AssetDatabase.GetAssetPath(obj);
            if (!string.IsNullOrEmpty(path) && File.Exists(path))
            {
                path = Path.GetDirectoryName(path);
                break;
            }
```

```csharp
        }
        return path;
    }
}

class MyDoCreateScriptAsset : EndNameEditAction
{
    public override void Action(int instanceId, string pathName, string resourceFile)
    {
        UnityEngine.Object o = CreateScriptAssetFromTemplate(pathName, resourceFile);
        ProjectWindowUtil.ShowCreatedAsset(o);
    }

    internal static UnityEngine.Object CreateScriptAssetFromTemplate(string pathName,
        string resourceFile)
    {
        string fullPath = Path.GetFullPath(pathName);
        StreamReader streamReader = new StreamReader(resourceFile);
        string text = streamReader.ReadToEnd();
        streamReader.Close();
        string fileNameWithoutExtension = Path.GetFileNameWithoutExtension(pathName);
        //替换文件名
        text = Regex.Replace(text, "#NAME#", fileNameWithoutExtension);
        bool encoderShouldEmitUTF8Identifier = true;
        bool throwOnInvalidBytes = false;
        UTF8Encoding encoding = new UTF8Encoding(encoderShouldEmitUTF8Identifier,
            throwOnInvalidBytes);
        bool append = false;
        StreamWriter streamWriter = new StreamWriter(fullPath, append, encoding);
        streamWriter.Write(text);
        streamWriter.Close();
        AssetDatabase.ImportAsset(pathName);
        return AssetDatabase.LoadAssetAtPath(pathName, typeof(UnityEngine.Object));
    }
}
```

这段代码的核心部分在 CreateScriptAssetFromTemplate() 回调方法中。这里可以拿到用户输入的名称以及文件将创建的目录，进行简单的字符串替换后，就会创建一个全新的模板脚本类。

4.2 脚本的生命周期

Unity 脚本有一套完整的生命周期，脚本需要挂在任意游戏对象上，并且同一个游戏对象可以挂不同的脚本，各脚本执行自己的生命周期，它们可以相互组合并且互不干预。学习脚本的生命周期之前，我们不得不引用文档中非常经典的一张图，如图 4-5 所示，这张图完整地描述了脚本的生命周期。

生命周期中的所有方法都是 Unity 系统自己回调的，不需要手动调用，主要有编辑脚本、初始化、物理碰撞事件、更新回调、渲染和销毁等。这些方法比较多，我们会在后面的例子中慢慢为大家讲解。

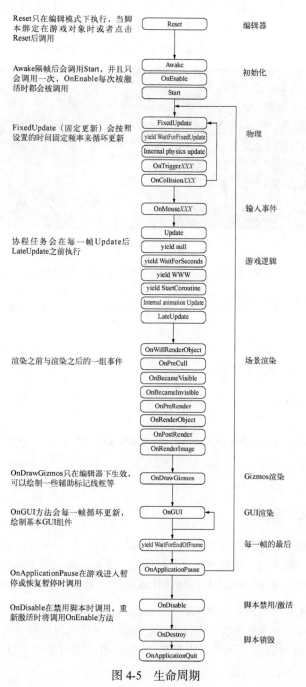

图 4-5 生命周期

4.2.1 脚本绑定事件

在编辑模式下，Unity 并没有提供脚本的绑定事件，但是我们可以通过生命周期中的 Reset() 方法来实现。Reset() 方法仅在非运行模式下才会生效，当把脚本挂在某个游戏对象上时，或者右击已经挂上脚本的对象，从弹出菜单中选择 Reset 菜单项（如图 4-6 所示）时，它就会执行。

图 4-6　脚本绑定事件

如代码清单 4-2 所示，在脚本中添加 Reset() 方法，就可以监听脚本绑定时的事件了。

代码清单 4-2　Script_04_02.cs 文件

```
using System.Collections;
using System.Collections.Generic;
using UnityEngine;

public class Script_04_02 : MonoBehaviour {

    #if UNITY_EDITOR
    void Reset()
    {
        Debug.LogFormat("GameObject:{0} 绑定 Script_04_02.cs 脚本", gameObject.name);
    }
    #endif
}
```

4.2.2 脚本初始化和销毁

脚本挂在游戏对象上，运行时就会立即执行初始化方法 Awake()，它是一个同步方法，而 Start() 方法会在下一帧执行。如果游戏对象被删除，或者挂在它身上的脚本被删除，就会执行 OnDestroy() 销毁方法。需要记住的是，初始化或销毁在脚本的生命周期中只会执行一次。

此外，游戏对象还有个状态，叫禁用状态。如图 4-7 所示，左上角的复选框控制整个游戏对象（包括绑定的所有脚本）的激活或禁用状态，下面脚本左边的复选框只控制某个脚本是否激活或禁用。在程序运行的过程中，可以多次设置激活/禁用，同时系统会分别回调生命周期中的 OnEnable() 和 OnDisable() 方法。

4.2 脚本的生命周期

图 4-7 隐藏状态

如代码清单 4-3 所示,在代码中添加脚本生命周期方法。该脚本绑定在任意对象后,运行游戏即可查看它们的执行顺序。

代码清单 4-3　Script_04_03.cs 文件

```
using System.Collections;
using System.Collections.Generic;
using UnityEngine;

public class Script_04_03 : MonoBehaviour
{
    void Awake()
    {
        Debug.Log("Awake 用于初始化并且永远只会执行一次");
    }

    void OnEnable()
    {
        Debug.Log("OnEnable 在脚本每次激活时执行一次");
    }

    void Start()
    {
        Debug.Log("Start 在初始化后的下一帧执行,并且永远只会执行一次");
    }

    void OnDisable()
    {
        Debug.Log("OnDisable 在脚本每次反激活后,执行一次");
    }

    void OnDestroy()
    {
        Debug.Log("OnDestroy 用于脚本反初始化并且永远只会执行一次");
    }

    void OnApplicationQuit()
```

```
        {
            Debug.Log("应用程序退出时执行一次");
        }
}
```

4.2.3 脚本更新与协程任务

在整个生命周期中，主要提供了如下 3 种更新方法。

- **Update()**：每一帧执行时，都会立即调用此方法。
- **LateUpdate()**：Update()方法执行后，都会调用此方法。
- **FixedUpdate()**：固定更新。默认情况下，系统每 0.02 秒调用一次，具体的间隔时间可以在 TimeManager 中配置。在导航菜单栏中选择 Editor→Project Settings→Time 菜单项，即可打开 Time Manager。

总体来说，Update()和 LateUpdate()属于立即更新，更新之间的频率是不固定的，比如某一帧有一个耗时操作时，就会影响到下一帧更新的时间，所以对更新频率要求比较稳定的物理系统就不太适合在这里处理更新。

FixedUpdate()虽然是固定更新，但是其实也是相对固定的，比如某一帧耗了好几秒，它依然会卡住。不过正常的程序会优化耗时操作，小范围的帧率波动是正常的，可以让它更新的时间间隔稍微长一点，这样它的更新是比较平滑的。在实际的开发中，例如以秒为单位的倒计时，并不需要每一帧去判断时间，所以用 FixedUpdate()就再合适不过了。

Unity 的脚本只支持单线程，不过它引入了 C#语言协程的概念，可以用来模拟多线程，而不是真正的多线程。举个实际点的例子，每等一秒就创建一个游戏对象，这在 Update()中写就比较麻烦，但是引入协程的概念后，就可以直接用 for 循环来写了。使用 StartCoroutine()方法，即可启动一个协程任务。在 for 循环中，我们使用 yield return 语句，告诉 Unity 需要等待多久再执行下一个循环。相关代码如代码清单 4-4 所示。

代码清单 4-4 Script_04_04.cs 文件

```
using System.Collections;
using System.Collections.Generic;
using UnityEngine;

public class Script_04_04 : MonoBehaviour
{
    void Start()
    {
        StartCoroutine(CreateCube());
    }
    IEnumerator CreateCube()
    {
        for(int i = 0; i < 100; i++) {
            GameObject.CreatePrimitive(PrimitiveType.Cube).transform.position =
                Vector3.one * i;
```

```
            yield return new WaitForSeconds(1f);
        }
    }
}
```

再回到脚本的生命周期中，`yield return new WaitForSeconds(1f)` 表示等一秒后再执行循环后面的方法。协程可以让代码写起来更简化，但是很容易出错。在平时开发中，我建议最好不要大量使用协程程序。

4.2.4 停止协程任务

在协程任务启动的过程中，如果需要重新启动它，必须停掉之前的协程。每次启动协程时，`StartCoroutine()` 将返回这个协程的对象，需要停止的时候使用 `StopCoroutine()` 传入对象即可。当然，也可以调用 `StopAllCoroutines()` 停止这个脚本所启动的所有协程任务。如代码清单4-5所示，启动 `CreateCube()` 协程方法，在循环中每隔一秒创建一个立方体对象。

代码清单4-5　Script_04_05.cs 文件

```csharp
using System.Collections;
using System.Collections.Generic;
using UnityEngine;

public class Script_04_05 : MonoBehaviour
{
    IEnumerator CreateCube()
    {
        for(int i = 0; i < 100; i++) {
            GameObject.CreatePrimitive(PrimitiveType.Cube).transform.position = 
                Vector3.one * i;
            yield return new WaitForSeconds(1f);
        }
    }

    private Coroutine m_Coroutine = null;

    void OnGUI()
    {
        if(GUILayout.Button("StartCoroutine")) {
            if(m_Coroutine != null) {
                StopCoroutine(m_Coroutine);
            }
            m_Coroutine = StartCoroutine(CreateCube());
        }
        if(GUILayout.Button("StopCoroutine")) {
            if(m_Coroutine != null) {
                StopCoroutine(m_Coroutine);
            }
        }
    }
}
```

4.2.5 使用 OnGUI 显示 FPS

GUI 是 Unity 4.6 版本之前的 UI 系统,因为其功能比较单一并且效率不高,已经被新版的 UGUI 所替代。如果想显示一些辅助信息或者调试按钮等,大多还会使用它。在这一节中,我们用示例说明如何使用 OnGUI 显示 FPS,如图 4-8 所示。

图 4-8 显示 FPS

FPS 值的含义就是 1 秒钟 Update 被执行了多少次。其计算原理就是先记一个初始时间,接着取当前时间减去初始时间,这期间 Update 执行的次数就是 FPS 了。

如代码清单 4-6 所示,在 Update()中获取每一秒所执行的次数,最终在 OnGUI()方法中将 FPS 打印在屏幕左上角。

代码清单 4-6 Script_04_06.cs 文件

```
using System.Collections;
using System.Collections.Generic;
using UnityEngine;

public class Script_04_06 : MonoBehaviour
{
    public float updateInterval = 0.5F;
    private float accum = 0;
    private int frames = 0;
    private float timeleft;
    private string stringFps;
    void Start()
    {
        timeleft = updateInterval;
    }
    void Update()
    {
        timeleft -= Time.deltaTime;
        accum += Time.timeScale / Time.deltaTime;
        ++frames;
        if(timeleft <= 0.0) {
            float fps = accum / frames;
            string format = System.String.Format("{0:F2} FPS", fps);
            stringFps = format;
            timeleft = updateInterval;
            accum = 0.0F;
            frames = 0;
        }
```

```
}
void OnGUI()
{
    GUIStyle guiStyle = GUIStyle.none;
    guiStyle.fontSize = 30;
    guiStyle.normal.textColor = Color.red;
    guiStyle.alignment = TextAnchor.UpperLeft;
    Rect rt = new Rect(40, 0, 100, 100);
    GUI.Label(rt, stringFps, guiStyle);
}
```

需要说明的是，FPS 值越高，游戏就越流畅。但是手机上如果 FPS 太高，可能会影响发热并且会费电，所以可以考虑降低 FPS。下面的代码强制设置 FPS 最高 30 帧：

```
Application.targetFrameRate = 30f;//强制设置FPS最高30帧
```

4.3 多脚本管理

Unity 脚本可以灵活地挂在多个游戏对象上，此时就衍生出一个问题：脚本多了，如何来管理，如何控制不同脚本执行的先后顺序。启动游戏后，Unity 会同时处理所有脚本。比如，执行脚本中的 `Awake()` 方法时，Unity 会先找到此时需要初始化的所有脚本，然后同时执行这些脚本的所有 `Awake()` 方法。学过编程的朋友都应该知道计算机处理是没有同时这个概念的，它们都是有先后顺序的，也就是说排在前面的脚本会优先执行。

4.3.1 脚本的执行顺序

脚本既可以在运行时动态添加在游戏对象上，也可以运行游戏前预制挂在游戏对象上。动态添加的脚本按添加的先后顺序决定执行顺序。但是静态脚本因为提前挂在了游戏对象上，所以初始化的顺序就不一样了。如图 4-9 所示，在 Script Execution Order 中可以设置脚本的执行顺序。

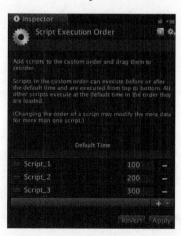

图 4-9　脚本排序

这就说明为什么在脚本生命周期中会提供 `Start()` 方法。例如，A 脚本先执行 B 脚本后执行，如果 A 脚本在自己的 `Awake()` 方法中获取 B 脚本的数据，那么可能就会出错。因为此时 B 脚本的初始化方法还没有执行，所以 `Awake()` 方法适合做初始化，而在 `Start()` 方法中才适合安全地访问其他脚本数据。

4.3.2 多脚本优化

脚本挂得越多，执行效率就越低。这些脚本都需要执行生命周期的方法，此时 Unity 需要遍历它们，然后再反射调用每个脚本的方法。如图 4-10 所示，一次全局的 `Update` 调用在 Unity 内部干了很多事情。

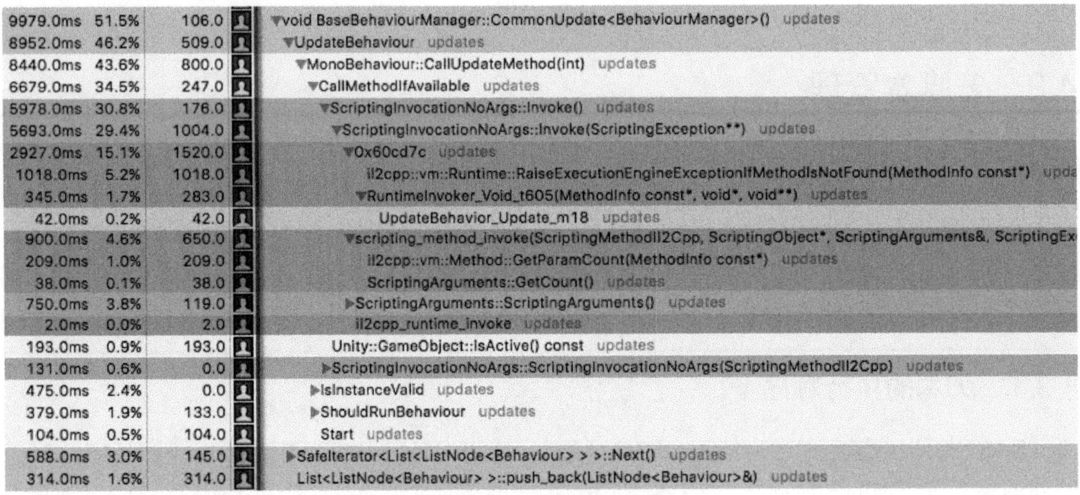

9979.0ms	51.5%	106.0	▼void BaseBehaviourManager::CommonUpdate<BehaviourManager>() updates
8952.0ms	46.2%	509.0	▼UpdateBehaviour updates
8440.0ms	43.6%	800.0	▼MonoBehaviour::CallUpdateMethod(int) updates
6679.0ms	34.5%	247.0	▼CallMethodIfAvailable updates
5978.0ms	30.8%	176.0	▼ScriptingInvocationNoArgs::Invoke() updates
5693.0ms	29.4%	1004.0	▼ScriptingInvocationNoArgs::Invoke(ScriptingException**) updates
2927.0ms	15.1%	1520.0	▼0x60cd7c updates
1018.0ms	5.2%	1018.0	il2cpp::vm::Runtime::RaiseExecutionEngineExceptionIfMethodIsNotFound(MethodInfo const*) upd
345.0ms	1.7%	283.0	▼RuntimeInvoker_Void_t605(MethodInfo const*, void*, void**) updates
42.0ms	0.2%	42.0	UpdateBehavior_Update_m18 updates
900.0ms	4.6%	650.0	▼scripting_method_invoke(ScriptingMethodIl2Cpp, ScriptingObject*, ScriptingArguments&, ScriptingEx
209.0ms	1.0%	209.0	il2cpp::vm::Method::GetParamCount(MethodInfo const*) updates
38.0ms	0.1%	38.0	ScriptingArguments::GetCount() updates
750.0ms	3.8%	119.0	▶ScriptingArguments::ScriptingArguments() updates
2.0ms	0.0%	2.0	il2cpp_runtime_invoke updates
193.0ms	0.9%	193.0	Unity::GameObject::IsActive() const updates
131.0ms	0.6%	0.0	▶ScriptingInvocationNoArgs::ScriptingInvocationNoArgs(ScriptingMethodIl2Cpp) updates
475.0ms	2.4%	0.0	▶IsInstanceValid updates
379.0ms	1.9%	133.0	▶ShouldRunBehaviour updates
104.0ms	0.5%	104.0	Start updates
588.0ms	3.0%	145.0	▶SafeIterator<List<ListNode<Behaviour> > >::Next() updates
314.0ms	1.6%	314.0	List<ListNode<Behaviour> >::push_back(ListNode<Behaviour>&) updates

图 4-10 脚本调用

所以我们能做的优化就是避免挂太多的脚本，避免在脚本中写入这种空方法，如果不需要用，就把它删除掉：

```
void Start()
{
}
```

4.4 脚本序列化

脚本可以通过序列化和反序列化来保存游戏数据，换句话说，就是脚本自身并没有保存数据，而是将数据保存在文件中，使用的时候不需要自己重新组织数据，而是通过语法直接访问即可。如图 4-11 所示，脚本挂在 `GameObject` 上，而 `GameObject` 属于 `Scene` 游戏场景，所以 Inspector 面板中写入的最终将保存在 Scene.unity 这个文件中。

4.4 脚本序列化

图 4-11 序列化

接着再复制一份对象，并将其起名为 Prefab。我们将它从 Hierarchy 视图拖入 Project 视图中，如图 4-12 所示，这份资源将变成预制体（Prefab.prefab）。如果在 Project 视图中赋值的话，所序列化的数据将会保存在 Prefab.prefab 文件中。但是如果在 Hierarchy 视图中赋值 Prefab 的话，数据将会保存在 Scene.unity 文件中。

图 4-12 Prefab

4.4.1 查看数据

Unity 可以设置资源编辑器中的保存格式，如图 4-13 所示，其中 Mixed 表示混合模型，而 Force Binary 表示二进制格式。这里推荐使用 Force Text，它表示所有资源可以使用文本打开，这个设置并不会影响最终游戏的发布效率，反而编辑模式下查看资源结构会更方便。

图 4-13 序列化结构

接着，把 Prefab.prefab 或者 Scene.unity 文件直接拖入文本编辑器中。如图 4-14 所示，序列化的数据和面板中设置的完全一样，其中 name 这一栏中文转成了 Unicode 编码，prefab 这栏保存的是对象的实例 id（这表示它的引用关系）。反序列化就更方便了，要想在代码中获取到脚本对象，可以直接通过 "." 来访问类对象上的数据。

图 4-14 序列化数据

如代码清单 4-7 所示，凡参与序列化的对象都需要在脚本中声明它的数据结构。此外，它还支持复杂的数据结构，比如自定义类对象和数组结构等。

代码清单 4-7　Script_04_07.cs 文件

```
using System.Collections;
using System.Collections.Generic;
using UnityEngine;

public class Script_04_07 : MonoBehaviour
{
    public int id;
    public string name;
    public GameObject prefab;
}
```

4.4.2　私有序列化数据

脚本中声明为 public 的对象都会支持序列化格式，但是声明为 public 的对象，外部可以直接访问它，而程序中有些数据仅希望在代码内部使用，这就需要支持私有化序列数据，此时可以使用 private 声明对象，例如：

```
[SerializeField]
private int id;
[SerializeField]
private string name;
[SerializeField]
private GameObject prefab;
```

这样又带来一个新问题，因为对象都设置为 private 数据，所以外部都无法访问，如果想使用拓展编辑器来编辑这些数据的话，可以使用 SerializedObject 和 SerializedProperty 来获取设置数据。如图 4-15 所示，在脚本中序列化一些数据。

图 4-15　私有序列化数据

如代码清单 4-8 所示，[SerializeField] 标记了所有私有属性对象，这样外部类就无法访问它们了。在代码中通过 serializedObject.FindProperty() 来访问私有属性，此时只需传入它的名称即可。

代码清单 4-8　Script_04_08.cs 文件

```csharp
using System.Collections;
using System.Collections.Generic;
using UnityEngine;
#if UNITY_EDITOR
using UnityEditor;
#endif

public class Script_04_08 : MonoBehaviour
{
    [SerializeField]
    private int id;
    [SerializeField]
    private string name;
    [SerializeField]
    private GameObject prefab;
}

#if UNITY_EDITOR
[CustomEditor(typeof(Script_04_08))]
public class ScriptInsector : Editor
{
    public override void OnInspectorGUI()
    {
        //更新最新数据
        serializedObject.Update();
        //获取数据信息
        SerializedProperty property = serializedObject.FindProperty("id");
        //赋值数据
        property.intValue = EditorGUILayout.IntField("主键", property.intValue);

        property = serializedObject.FindProperty("name");
        property.stringValue = EditorGUILayout.TextField("姓名", property.stringValue);

        property = serializedObject.FindProperty("prefab");
        property.objectReferenceValue = EditorGUILayout.ObjectField("游戏对象",
            property.objectReferenceValue,typeof(GameObject),true);

        //全部保存数据
```

```
            serializedObject.ApplyModifiedProperties();
        }
}
#endif
```

注意：如果 public 属性不想在面板中显示的话，可以设置[HideInInspector]。

4.4.3 serializedObject

serializedObject 只能在 Editor 中使用，它专门用于获取设置的序列化信息。通常，要开发复杂的编辑组件，都需要重写 OnInspectorGUI()方法，但是如果希望有些用源生的绘制结构，同时兼容有些自定义渲染的话，那该怎么办呢？如图 4-16 所示，"主键"这一栏才是自定义绘制的，而下面的 Targets 数组使用默认绘制方式，这样就实现组合兼容绘制了。

图 4-16 serializedObject

如代码清单 4-9 所示，我们使用 EditorGUILayout.PropertyField()方法即可使用源生方法绘制某个对象了，其余的则使用自定义绘制方法。

代码清单 4-9 Script_04_09.cs 文件

```
using System.Collections;
using System.Collections.Generic;
using UnityEngine;
#if UNITY_EDITOR
using UnityEditor;
#endif

public class Script_04_09 : MonoBehaviour
{
    [SerializeField]
    private int id;
    [SerializeField]
    private GameObject[] targets;
}

#if UNITY_EDITOR
[CustomEditor(typeof(Script_04_09))]
public class ScriptInsector : Editor
{
    public override void OnInspectorGUI()
    {
```

```
        //更新最新数据
        serializedObject.Update();
        //获取数据信息
        SerializedProperty property = serializedObject.FindProperty("id");
        //赋值数据
        property.intValue = EditorGUILayout.IntField("主键", property.intValue);
        //以默认样式绘制数组数据
        EditorGUILayout.PropertyField(serializedObject.FindProperty("targets"), true);
        //全部保存数据
        serializedObject.ApplyModifiedProperties();
    }
}
#endif
```

4.4.4 监听部分元素修改事件

在脚本面板中，参与绘制的元素都是在 `OnInspectorGUI()` 方法中绘制的。如图 4-17 所示，我们来监听 Targets 元素变化事件，将需要监听的 GUI 元素写在 `EditorGUI.BeginChangeCheck();` 后面如果中间有元素布局发生变化，就可以在 `if(EditorGUI.EndChangeCheck()){}` 方法中去处理。

如代码清单 4-10 所示，`GUI.changed` 可以判断 GUI 是否发生了变化，这段代码其实是在拓展编辑器中完成的。但是有时候，有些简单的脚本不一定需要写拓展编辑器代码，只用一个普通的 MonoBehaviour 脚本，可以写入 `OnValidate()` 方法即可。当编辑面板中的信息发生变化时，Unity 会回调这个方法。`OnValidate()` 方法的代码如下：

```
void OnValidate()
{
    Debug.Log("编辑面板信息发生变化");
}
```

图 4-17 监听面板变化

代码清单 4-10　Script_04_10.cs 文件

```csharp
using System.Collections;
using System.Collections.Generic;
using UnityEngine;
#if UNITY_EDITOR
using UnityEditor;
#endif

public class Script_04_10 : MonoBehaviour
{
    [SerializeField]
    private GameObject[] targets;
}

#if UNITY_EDITOR
[CustomEditor(typeof(Script_04_10))]
public class ScriptInsector : Editor
{
    public override void OnInspectorGUI()
    {
        //更新最新数据
        serializedObject.Update();
        //标记检查
        EditorGUI.BeginChangeCheck();
        EditorGUILayout.PropertyField(serializedObject.FindProperty("targets"), true);
        //标记检查发生变化
        if(EditorGUI.EndChangeCheck()) {
            Debug.Log("元素发生变化");
        }
        //判断面板元素是否任意发生改变
        if(GUI.changed) {

        }
        //全部保存数据
        serializedObject.ApplyModifiedProperties();

    }
}
#endif
```

4.4.5　序列化/反序列化监听

目前，序列化和反序列化都是在编辑面板中操作后 Unity 自动处理的，其实也可以对它进行监听，比如序列化数据之前做点什么或者序列化结束后做点什么。序列化目前是不支持字典的，如果想序列化字典，就可以利用序列化监听接口特性。如图 4-18 所示，使用 key 和 value 来保存名字和精灵对象。

4.4 脚本序列化

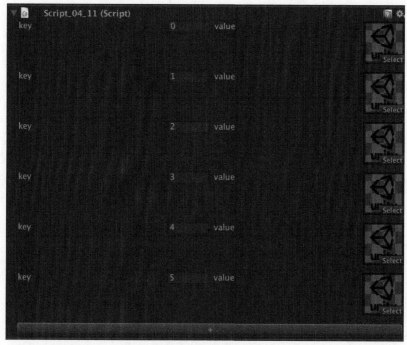

图 4-18 监听序列化

如代码清单 4-11 所示，代码中需要继承 ISerializationCallbackReceiver 接口，实现 OnBeforeSerialize() 和 OnAfterDeserialize() 方法来监听序列化和反序列化之前的事件。

代码清单 4-11　Script_04_11.cs 文件

```
using System.Collections;
using System.Collections.Generic;
using UnityEngine;
#if UNITY_EDITOR
using UnityEditor;
#endif

public class Script_04_11 : MonoBehaviour, ISerializationCallbackReceiver
{
    [SerializeField]
    private List<Sprite> m_Values = new List<Sprite>();
    [SerializeField]
    private List<string> m_Keys = new List<string>();

    public Dictionary<string,Sprite>spriteDic = new Dictionary<string, Sprite>();


    #region ISerializationCallbackReceiver implementation

    void ISerializationCallbackReceiver.OnBeforeSerialize()
    {
```

```csharp
            //序列化
            m_Keys.Clear();
            m_Values.Clear();
            foreach(KeyValuePair<string, Sprite> pair in spriteDic)
            {
                m_Keys.Add(pair.Key);
                m_Values.Add(pair.Value);
            }
        }

        void ISerializationCallbackReceiver.OnAfterDeserialize()
        {
            //反序列化
            spriteDic.Clear();
            if(m_Keys.Count != m_Values.Count) {
                Debug.LogError("m_Keys and m_Values 长度不匹配!!!");
            } else {
                for(int i = 0; i < m_Keys.Count; i++) {
                    spriteDic [m_Keys [i]] = m_Values [i];
                }
            }
        }
        #endregion
}

#if UNITY_EDITOR
[CustomEditor(typeof(Script_04_11))]
public class ScriptInsector : Editor
{
    public override void OnInspectorGUI()
    {
        //更新最新数据
        serializedObject.Update();
        SerializedProperty propertyKey = serializedObject.FindProperty ("m_Keys");
        SerializedProperty propertyValue = serializedObject.FindProperty("m_Values");

        int size = propertyKey.arraySize;

        GUILayout.BeginVertical();
        for(int i = 0; i < size; i++) {
            GUILayout.BeginHorizontal();
            SerializedProperty key = propertyKey.GetArrayElementAtIndex(i);
            SerializedProperty value = propertyValue.GetArrayElementAtIndex(i);
            key.stringValue = EditorGUILayout.TextField("key", key.stringValue);
            value.objectReferenceValue = EditorGUILayout.ObjectField("value",
                value.objectReferenceValue, typeof(Sprite), false);
            GUILayout.EndHorizontal();
        }
        GUILayout.EndVertical();

        GUILayout.BeginHorizontal();
        if(GUILayout.Button("+"))
        {
            (target as Script_04_11).spriteDic [size.ToString()] = null;
```

```
            }
            GUILayout.EndHorizontal();
            //全部保存数据
            serializedObject.ApplyModifiedProperties();
        }
    }
#endif
```

在上述代码中，我们将精灵的名称和精灵序列化在两个不同的列表中，使用时会将它们转成字典，这样通过名字就能取对应的精灵对象了。

上面介绍的序列化对象都比较简单。其实，开发中的数据可能比较复杂，此时就需要把对象放在不同的类中来管理，在 class 上面添加[System.Serializable]特性即可。如果不希望它序列化的话，可以添加[System.NonSerialized]表示它不参与序列化。相关代码如下：

```
[SerializeField]
public List<PlayerInfo> m_PlayerInfo;

[System.Serializable]
public class PlayerInfo
{
    public int id;
    public string name;
}
```

4.4.6 ScriptableObject

前面介绍了脚本的序列化，它只是脚本众多功能中很小的一部分，而且它是有限制的，必须绑定在游戏对象上。开发中有时候需要序列化一些编辑器数据，仅仅是数据，完全没必要依赖游戏对象使用游戏脚本，此时 ScriptableObject 就是最佳选择了。如图 4-19 所示，代码中使用特性[CreateAssetMenu]，就可以在 Create 菜单中出现它的创建栏了。

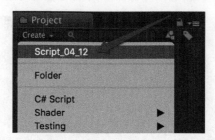

图 4-19 菜单项

创建 ScriptableObject 后，其实就是只创建带序列化的资源，因为它只保存数据，所以更轻量级一些。如图 4-20 所示，同样也可以在 Inspector 面板中对它进行编辑，如果需要拓展，就和之前的拓展方法一样。

图 4-20　编辑资源

如代码清单 4-12 所示，继承 `ScriptableObject` 后，需要在代码中添加需要序列化的数据结构。如果需要使用类对象，那和前面介绍的序列化脚本的方式类似。

代码清单 4-12　Script_04_12.cs 文件

```csharp
using System.Collections;
using System.Collections.Generic;
using UnityEngine;
[CreateAssetMenu]
public class Script_04_12 : ScriptableObject
{
    [SerializeField]
    public List<PlayerInfo> m_PlayerInfo;

    [System.Serializable]
    public class PlayerInfo
    {
        public int id;
        public string name;
    }
}
```

New Script_04_12.asset 就是创建的资源，可以把它放在 Resources 目录下，这样运行期间就可以读取它了，如图 4-21 所示。接着，将 Main 脚本挂在任意对象上运行游戏即可。

图 4-21　读取资源

如代码清单 4-13 所示，首先使用 `Resources.Load<T>` 传入上面定义的泛型类对象，然后通过对象访问对应的属性即可。

代码清单 4-13　Main.cs 文件

```
using System.Collections;
using System.Collections.Generic;
using UnityEngine;

public class Main : MonoBehaviour {

    void Start() {
        Script_04_12 script = Resources.Load<Script_04_12>("New Script_04_12");
        Debug.LogFormat("name : {0} id : {1}", script.m_PlayerInfo [0].name,
            script.m_PlayerInfo [0].id);
    }
}
```

如代码清单 4-14 所示，[CreateAssetMenu] 菜单需要在编辑器中点击 Create 按钮才能创建，如果想批量生成它，就不方便了，此时也可以通过代码来自行创建并对其赋值，最终调用 AssetDatabase.SaveAssets() 即可保存了。

代码清单 4-14　Create.cs 文件

```
using System.Collections;
using System.Collections.Generic;
using UnityEngine;
using UnityEditor;

public class Create
{
    [MenuItem("Assets/Create ScriptableObject")]
    static void CreateScriptableObject()
    {
        //创建 ScriptableObject
        Script_04_12 script = ScriptableObject.CreateInstance<Script_04_12>();
        //赋值
        script.m_PlayerInfo = new List<Script_04_12.PlayerInfo>();
        script.m_PlayerInfo.Add(new Script_04_12.PlayerInfo(){ id = 100, name =
            "Test" });

        //将资源保存到本地
        AssetDatabase.CreateAsset(script, "Assets/Resources/Create Script_04_12.asset");
        AssetDatabase.SaveAssets();
        AssetDatabase.Refresh();
    }
}
```

在游戏开发中，编辑配置相关的数据是海量的，强烈推荐使用 ScriptableObject，这样设置和读取数据会非常方便。通过代码可以直接访问属性变量，所以代码写起来就很舒服。

4.4.7　脚本的 Attributes 特性

C# 语言提供了 Attributes 特性，Unity 自己的 API 中很多也使用了它。运行时，它可以让类、方法、对象和枚举执行一些别的方法。我们需要使用 [] 来声明标签。Unity 预制了很多自己的标签，大家可以在 API 中查询到，而 API 都在 UnityEngine.Attributes 和 UnityEditor.Attributes 命名空间下。其实，Attributes 也可以自定义拓展。相关代码如代码清单 4-15 所示。

代码清单 4-15　Script_04_13.cs 文件
```
using UnityEngine;
public class Script_04_13 : MonoBehaviour
{
    [RangeInt(1,100)]
    public int rangeInt;
}
```

如图 4-22 所示，自定义 Attributes 特性即可拓展它的面板。我们在代码中设置了整型取值的范围，并且需要分别继承 PropertyAttribute 和 PropertyDrawer。我们可以在 OnGUI 中做整型数的限制。

图 4-22　拓展

如代码清单 4-16 所示，当数据在面板中发生修改时，就会回调到 OnGUI() 方法中。由于在 RangeIntAttribute 中约束了数据的取值范围，所以就可以限制 Range Int 的数据在 1 和 100 之间了。

代码清单 4-16　RangeIntAttribute.cs 文件
```
using System.Collections;
using System.Collections.Generic;
using UnityEngine;

#if UNITY_EDITOR
using UnityEditor;
#endif
public sealed class RangeIntAttribute : PropertyAttribute
{
    public readonly int min;

    public readonly int max;

    public RangeIntAttribute(int min, int max)
    {
        this.min = min;
        this.max = max;
    }
}

#if UNITY_EDITOR
```

```csharp
[CustomPropertyDrawer(typeof(RangeIntAttribute))]
public sealed class RangeIntDrawer : PropertyDrawer
{
    public override float GetPropertyHeight(SerializedProperty property, GUIContent label)
    {
        return 100; //设置面板的高度
    }
    public override void OnGUI(Rect position, SerializedProperty property, GUIContent label)
    {
        RangeIntAttribute attribute = this.attribute as  RangeIntAttribute;
        property.intValue = Mathf.Clamp(property.intValue, attribute.min, attribute.max);
        EditorGUI.HelpBox(new Rect(position.x, position.y, position.width, 30),
            string.Format("范围{0}~{1}",attribute.min, attribute.max),MessageType.Info);

        EditorGUI.PropertyField(new Rect(position.x, position.y +35, position.width,
            20), property, label);
    }
}
#endif
```

4.4.8 单例脚本

Unity 的脚本也属于一种特殊的 C# 类，它不能 new 出来，而需要绑定在游戏对象，而且脚本会随着游戏对象的删除而自动被释放掉。如果代码逻辑都写在了脚本中，那么出现切换场景一类的情形时，脚本的逻辑就走不了了。虽然 Unity 提供了 DontDestroyOnLoad() 方法，它可以在切换场景时不卸载游戏对象，但是总不能给所有对象都添加这个属性吧。总之，游戏中大量的逻辑代码是不需要写在脚本中的，除了依赖脚本生命周期中的回调方法以外。

有些功能比较单一且需要用到脚本生命周期方法的类，就比较适合使用单例脚本了。单例脚本的特点是它必须依赖游戏对象，并且必须保证这个游戏对象不能被卸载掉。

如代码清单 4-17 所示，Global 脚本在它自己的 static 构造方法中创建对象并且设置 DontDestroyOnLoad()，这样就能保证它自己不被主动卸载掉，并且构造方法只会执行一次。

代码清单 4-17　Global.cs 文件

```csharp
using System.Collections;
using System.Collections.Generic;
using UnityEngine;

public class Global : MonoBehaviour
{
    public static Global instance;

    static Global()
    {
        GameObject go = new GameObject("#Globa#");
        DontDestroyOnLoad(go);
        instance = go.AddComponent<Global>();
```

```
    }
    public void DoSomeThings()
    {
        Debug.Log("DoSomeThings");
    }

    void Start()
    {
        Debug.Log("Start");
    }
}
```

Global 脚本不需要在编辑模式下绑定在某个对象上,运行时直接获取它的实例就能操作它了。具体使用方法如下所示:

```
//调用单例脚本的方法
Global.instance.DoSomeThings();
```

4.4.9 定时器

协程任务是可以做定时器的,但是有个最大的问题,那就是必须用在脚本中,但是我们游戏的逻辑大部分都在 C# 代码中,所以需要封装一个不依赖于脚本实现的定时器。

如代码清单 4-18 所示,利用协程程序来做定时器。给 WaitTime() 传入定时时间以及时间结束后的回调方法,外部代码就可以处理定时解决的事件了。

代码清单 4-18　WaitTimeManager.cs 文件

```
using System.Collections;
using System.Collections.Generic;
using UnityEngine;
using UnityEngine.Events;

public class WaitTimeManager
{
    private static TaskBehaviour m_Task;
    static WaitTimeManager()
    {
        GameObject go = new GameObject("#WaitTimeManager#");
        GameObject.DontDestroyOnLoad(go);
        m_Task = go.AddComponent<TaskBehaviour>();
    }

    //等待
    static public Coroutine WaitTime(float time,UnityAction callback)
    {
        return m_Task.StartCoroutine(Coroutine(time,callback));
    }
    //取消等待
    static public void CancelWait(ref Coroutine coroutine)
    {
```

```
            if(coroutine != null) {
                m_Task.StopCoroutine(coroutine);
                coroutine = null;
            }
        }

        static IEnumerator Coroutine(float time,UnityAction callback){
            yield return new WaitForSeconds(time);
            if(callback != null) {
                callback();
            }
        }
        //内部类
        class TaskBehaviour : MonoBehaviour{}
}
```

无论是脚本还是类，在需要定时器的地方调用它即可，相关代码如下：

```
//开启定时器
Coroutine coroutine = WaitTimeManager.WaitTime(5f,delegate {
    Debug.Log("等待 5 秒后回调");
});

//等待过程中取消它
WaitTimeManager.CancelWait(ref coroutine);
```

4.4.10 CustomYieldInstruction

上一节的定时器满足了经过多少秒回调的需求，但是游戏可能还需要一种定时回调。例如，一共 10 秒结束，但是每过 1 秒回调一下，图 4-23 显示了输出结果。

图 4-23　自定义定时器

如代码清单 4-19 所示，继承 CustomYieldInstruction 后，就可以处理自定义协程程序了。其中，keepWaiting 表示是否继续让协程等待。通过判断时间，即可算出是否抛出间隔事件来实现这种间隔定时器。

代码清单 4-19　Script_04_16.cs 文件

```csharp
using System.Collections;
using UnityEngine;
using UnityEngine.Events;

public class Script_04_16 : MonoBehaviour {

    IEnumerator Start()
    {
        //10 秒后结束
        yield return new CustomWait(10f,1f,delegate() {
            Debug.LogFormat("每过一秒回调一次 : {0}",Time.time);

        });
        Debug.Log("十秒结束");
    }




    public class CustomWait : CustomYieldInstruction
    {
        public override bool keepWaiting
        {
            get
            {
                //此方法返回 false 表示协程结束
                if(Time.time - m_StartTime >= m_Time){
                    return false;
                }else if(Time.time - m_LastTime >= m_Interval){
                    //更新上一次间隔时间
                    m_LastTime = Time.time;
                    m_IntervalCallback();
                }
                return true;
            }
        }

        private UnityAction m_IntervalCallback;
        private float m_StartTime;
        private float m_LastTime;
        private float m_Interval;
        private float m_Time;

        public CustomWait(float time, float interval, UnityAction callback)
        {
            //记录开始时间
            m_StartTime = Time.time;
            //记录上一次间隔时间
            m_LastTime = Time.time;
            //记录间隔调用时间
            m_Interval = interval;
```

```
            //记录总时间
            m_Time = time;
            //间隔回调
            m_IntervalCallback = callback;
        }
    }
}
```

4.4.11 工作线程

Unity 之前是完全不支持多线程的，从 Unity 2018 起开放了工作线程，我们可以使用它的 Job System 做一些耗时的事情。例如，场景中有很多对象需要同步坐标，此时就可以考虑将同步坐标放在工作线程中。如图 4-24 所示，在 Profiler 中可以看到工作线程的状态。

图 4-24 工作线程

如代码清单 4-20 所示，只需要继承 `IJobParallelForTransform` 接口，并且实现 `Execute()` 方法，就可以在工作线程中处理耗时操作了，是不是很酷呢？

代码清单 4-20　Script_04_17.cs 文件

```
using Unity.Collections;
using Unity.Jobs;
using UnityEngine;
using UnityEngine.Jobs;

public class Script_04_17 : MonoBehaviour {

    public Transform[] cubes;

    void Update()
    {
        //按下鼠标左键后
        if(Input.GetMouseButtonDown(0))
        {
            //随机设置坐标
            NativeArray<Vector3> position = new NativeArray<Vector3>(cubes.Length,
                Allocator.Persistent);
```

```
        for(int i = 0; i < position.Length; i++){
            position[i] = Vector3.one * i;
        }
        //设置Transform
        TransformAccessArray transformArray = new TransformAccessArray(cubes);
        //启动工作线程
        MyJob job = new MyJob(){position = position};
        JobHandle jobHandle = job.Schedule(transformArray);
        //等待工作线程结束
        jobHandle.Complete();
        Debug.Log("工作线程结束");
        //结束
        transformArray.Dispose();
        position.Dispose();
    }
}

struct MyJob : IJobParallelForTransform
{
    //只读属性
    [ReadOnly]public NativeArray<Vector3> position;

    public void Execute(int index, TransformAccess transform)
    {
        //工作线程中设置坐标
        transform.position = position[index];
    }
}
```

注意，主线程和工作线程的数据需要使用 Native 来传递。工作线程只需要读数据，并不需要设置数据，所以需要标记 [ReadOnly]。

4.5 脚本编译

在编辑器下，每次修改完代码后，就会自动开始编译，最终所有代码将编译成 DLL 文件。如图 4-25 所示，DLL 最终将编译在 Project/Library/ScriptAssemblies 目录下，一共是 4 个 DLL 文件。

图 4-25 编译 DLL 文件

4.5.1 编译规则

脚本分运行时脚本和编辑时脚本两大类,运行时脚本最终会编译进游戏包中,而编辑时脚本仅用于编辑器模式下,不会被打进游戏包。脚本存放的目录决定了它将编译在哪个 DLL 文件中。如图 4-26 所示,Plugins 目录下的脚本将优先编译,然后才编译其他目录,最后会编译 Editor 目录下的脚本,所以最终就会编译出 4 份 DLL 文件。

- PS1.CS:编译在 Assembly-CSharp-firstpass.dll。
- PE1.CS:编译在 Assembly-CSharp-Editor-firstpass.dll。
- A1.CS:编译在 Assembly-CSharp.dll。
- E1.CS:编译在 Assembly-CSharp-Editor.dll。

图 4-26 脚本目录

DLL 的编译顺序是 Assembly-CSharp-firstpass.dll>Assembly-CSharp-Editor-firstpass.dll Assembly-CSharp.dll>Assembly-CSharp-Editor.dll,先编译的脚本无法访问后编译的脚本,所以 PS1.CS 脚本无法访问其他的所有脚本,而 E1.CS 可以访问其他任意脚本。

4.5.2 优化编译

游戏可能有很多 C#代码,这就会造成随便改点代码,就要等 Unity 编译 DLL 好久,这样开发效率将大打折扣。游戏代码我们大致可以分成两类,框架类代码和逻辑性代码。框架类代码一旦成熟后,并不需要经常修改并且它不需要访问逻辑性代码,而逻辑性代码的改动频率是极其高的。所以,我们可以把框架类代码放在 Plugins 目录下,这样当改动非 Plugins 目录下的逻辑性代码时,就不会额外编译 Plugins 目录下的代码了,编译就快了。可是即便把逻辑性代码都放在 Plugins 目录外,如果代码量非常大,还是会造成编译慢的问题,此时我们还可以把部分 CS 代码预先编译成 DLL,这样编译速度就更快了。

4.5.3 编译 DLL

.NET 可以把 C/C++语言编译进 DLL,但是游戏发布后,有的平台是识别不了的,例如移动平台,此时如果编译 DLL 时只能编译 C# 代码,则需要在 macOS 系统中打开终端,输入编译指令,DLL 编译完后直接拖到项目中就可以了:

```
mcs -r:/Applications/Unity/Unity.app/Contents/Managed/UnityEngine.dll
-target:library -out:test.dll *.cs
```

下面简要介绍上述命令中部分项的含义。

- UnityEngine.dll：编译所依赖的 DLL 文件。
- -target:library：生成 Libraryl 类型。
- -out:test.dll：最终生成 DLL 的保存目录。
- *.cs：表示当前目录下的所有 C# 代码。如果有多个目录多个文件，可以以空格分开。

4.5.4 脚本跨平台

Unity 自己提供了两个核心的 DLL 库，Unity 编辑器只支持 Windows、macOS 和 Linux 这 3 个平台，所以代码大部分都是由 C# 编写的并且编译在 UnityEditor.dll 中，然后通过 Mono 实现了跨平台。运行时由于它兼容的平台非常多（目前已经有 20 多个平台），并不是所有平台都能运行 DLL 的。再说，底层渲染方法也不能使用 C# 来调用，所以 Unity 只把 C# 接口封装到了 UnityEngine.dll 中。至于更底层的内部实现，则是由这个 DLL 再去调用 C++ 来完成的。拿移动平台来说，编译 C++ 的方式是不同的，Android 需要编译成 .so，iOS 则编译在 .a 中。所以，Unity 会针对每个平台编译出这份核心库，从而就实现了跨平台。现在 Unity 还支持 IL2CPP，它可以把 DLL 代码转成 C++ 来执行，这从效率上又能提高一个台阶。而开发者还是使用 C# 语言开发的，打包时才自动转换成 IL2CPP，整个过程是毫无感知的，太棒了！虽然 Unity 源码我们拿不到，但是 UnityEngine.dll 和 UnityEditor.dll 这两个 DLL 文件是可以反编译出来了，有兴趣的读者可以好好看看里面的代码，很有意思！

4.5.5 程序集定义

Unity 2018 提供了程序集定义，它可以更进一步地优化编译速度。用户可以指定某个文件夹（包括子文件夹）单独生成 DLL，如此将进一步简化编译代码的数量。如图 4-27 所示，在不同的文件夹下分别创建 Assembly。这表示 A_DLL 文件夹下的代码将编译在 A_Assembly 中，而 B_DLL 文件夹下的代码将编译在 B_Assembly 中，并且在 Library/ScriptAssemblies 文件夹下可以直接查看到这两个 DLL 文件。

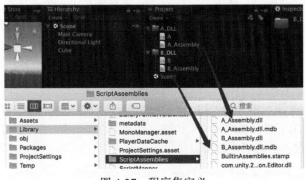

图 4-27　程序集定义

外部代码是可以直接访问这两个 DLL 的，但是它们之间是无法直接访问的，除非设置依赖关系。如图 4-28 所示，在 A_Assembly 中引用 B_Assembly，这样 A_DLL 目录下的代码都可以访问 B_DLL 目录下的代码了。

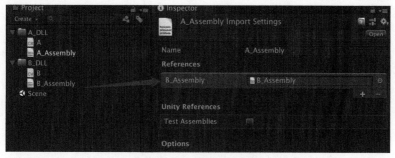

图 4-28 程序集定义依赖

通过程序集定义这个功能，我们可以很好地将代码模块化分类管理，将每个独立的功能划分到一个程序集中，相互编译互不影响。

4.5.6 日志

Unity 提供了 `Debug` 类来打印日志，常用的就是如下几种：

```
Debug.Log("Log");
Debug.LogError("LogError");
Debug.LogWarning("LogWarning");
```

在开发阶段，多打印日志可以方便地查看程序的行为。但是一旦发布以后，一定要把日志关闭掉，因为它会有一些额外的开销。如下代码所示，可以在初始化的位置设置条件编译，在非编辑模式下运行时，则关闭掉所有日志的输出：

```
#if !UNITY_EDITOR
    Debug.unityLogger.logEnabled = false;
#endif
```

另外，错误日志并不是我们主动打印出来的。错误日志的现场往往是非常珍贵的，我们需要尽可能地将错误日志保存下来。如果是移动平台，那么保存和提取日志其实挺不方便的，所以可以监听错误以及异常，并且及时将其打印在屏幕上。如代码清单 4-21 所示，监听 `Application.logMessageReceived` 事件即可捕获错误日志，最终在 `OnGUI()` 方法中将它们打印在屏幕上。

代码清单 4-21 Script_04_18.cs 文件

```
using System;
using System.Collections.Generic;
using UnityEngine;

public class Script_04_18 : MonoBehaviour {
```

```csharp
//错误详情
private List<String> m_logEntries = new List<String>();
//是否显示错误窗口
private bool m_IsVisible = false;
//窗口显示区域
private Rect m_WindowRect = new Rect(0, 0, Screen.width, Screen.height);
//窗口滚动区域
private Vector2 m_scrollPositionText = Vector2.zero;

private void Start()
{
    //监听错误
    Application.logMessageReceived += (condition, stackTrace, type) =>
    {
        if (type == LogType.Exception || type == LogType.Error)
        {
            if (!m_IsVisible)
            {
                m_IsVisible = true;
            }
            m_logEntries.Add(string.Format("{0}\n{1}", condition, stackTrace));
        }
    };

    //创建异常以及错误
    for (int i = 0; i < 10; i++)
    {
        Debug.LogError("momo");
    }
    int[] a = null;
    a[1] = 100;
}

void OnGUI()
{
    if(m_IsVisible){
        m_WindowRect = GUILayout.Window(0, m_WindowRect, ConsoleWindow, "Console");
    }
}

//日志窗口
void ConsoleWindow(int windowID)
{
    GUILayout.BeginHorizontal();
    if (GUILayout.Button("Clear", GUILayout.MaxWidth(200)))
    {
        m_logEntries.Clear();
    }
    if (GUILayout.Button("Close", GUILayout.MaxWidth(200)))
    {
        m_IsVisible = false;
    }
    GUILayout.EndHorizontal();
```

```
m_scrollPositionText = GUILayout.BeginScrollView(m_scrollPositionText);
foreach (var entry in m_logEntries)
{
    Color currentColor = GUI.contentColor;
    GUI.contentColor = Color.red;
    GUILayout.TextArea(entry);
    GUI.contentColor = currentColor;
}
GUILayout.EndScrollView();
}
```

如图 4-29 所示，错误的调用栈已经打印在屏幕上了。如果错误比较多，那么右侧需要一个滚动条。此外，还提供了 Clear 和 Close 操作。

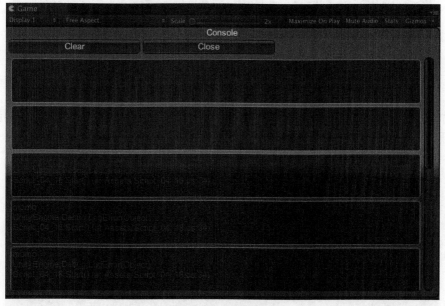

图 4-29　错误日志

4.6　脚本调试

Unity 2018 将彻底删除之前的 MonoDevelop，全面替换成 Visual Studio。Visual Studio 的功能比 MonoDevelop 更加全面，这一节我们就来学习 Mac 下它是如何调试的。打开 Visual Studio 后，在导航菜单栏中选择 Run→Attach to Process 菜单项，点击 Attach 按钮来关联 Unity Editor，如图 4-30 所示。

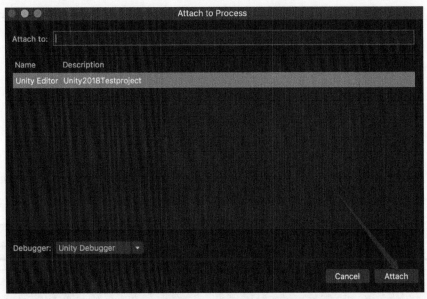

图 4-30　Attach to Process 窗口

接着，在代码中设置好断点，运行 Unity 即可。如图 4-31 所示，在程序断点后，将鼠标放在上面就可以看值了。Visual Studio 工具还有很多功能，读者可以自行来摸索。

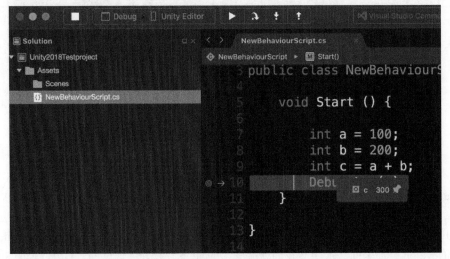

图 4-31　调试

另外，添加断点时还有个小技巧：当给程序设置断点时，需要查询一些类中的成员变量。如图 4-32 所示，这里可以直接输入 gameObject.name 或者 gameObject.transform.position 等能访问到的变量即可。

图 4-32　查询

4.7　小结

　　本章重点学习了游戏脚本，首先介绍了脚本的生命周期、多脚本管理以及多脚本执行的生命周期，然后讨论了脚本可以用于序列化和反序列化。在 Inspector 面板中，可以很方便地自定义拓展布局。如果只序列化数据，可以使用轻量级的 `ScriptableObject` 来处理。此外，Unity 提供了强大的 API 接口，Unity 2018 中已全面使用 Visual Studio 来代替 MonoDevelop，它更加强大、好用。

第 5 章

UGUI 游戏界面

Unity 4.6 推出了全新的 UI 系统，称为 UGUI。通过这几年的发展，UGUI 的功能已经十分健壮了，大量的游戏开始使用它开发游戏界面了。UGUI 提供的基础元素包括文本、图片、按钮、滑动条和滚动组件等，配合 UI 还提供了强大的 EventSystem 事件系统来管理 UI 元素。从 Unity 2017 开始还引入了图集的概念，让 UGUI 的功能更加全面，用起来更灵活。作为跨平台游戏引擎，其发布设备的分辨率是极其不固定的，不过 UGUI 在自适应处理上显然下了功夫，提供了锚点、布局、对齐方式和 Canvas，专门用来解决分辨率不同所带来的自适应屏幕问题。通过本章的学习，我们来一点点解开 UGUI 的神秘面纱。

5.1 基础元素

UI 是游戏开发中非常重要的一部分，所有交互都需要在其中完成。界面中的元素是多元化的，例如文本、图片和按钮互相组合，搭配排列出完美的界面。UI 提供了 Rect Transform 组件，用来设置锚点以及对齐方式。布局方式对于 UI 元素都是通用的，它的参数也比之前的 Transform 多了很多。本节中，我们开始学习 UGUI 的基础元素。

5.1.1 Text

在导航菜单栏中选择 Create→UI→Text 菜单项，即可创建一个文本组件。如图 5-1 所示，文本需要使用 TTF 字体。电脑上可以找到很多 TTF 字体库，大多都会有无用的字符，所以我建议最好使用 FontCreator 软件来精简一下字体库，而且有些文本数字和中文需要两种不同的字体，这样做起来就很麻烦，此时可以将两种字体合并，这样使用一个 Text 组件就可以同时显示正确的中文加数字了。

Text 组件提供了横向、纵向自动换行的功能，分为横向与纵向两种。请注意 Raycast Target 复选框，如果 UI 元素不需要点击事件，一定不要勾选它。因为 UGUI 的事件系统会遍历出所有带 Raycast Target 的组件，这无疑会带来一些不必要的开销。还有 UGUI 默认的材质我们是无法修改的，但是可以重写它，只需要将新的材质拖入 Material 处即可。

图 5-1 文本组件

5.1.2 描边和阴影

在 Text 游戏对象上添加 Outline 和 Shadow，即表示支持文本的描边和阴影。如图 5-2 所示，可以设置它们的颜色以及描边的距离。

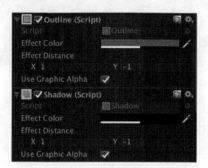

图 5-2 描边和阴影

注意：描边的原理就是在原有 Text 组件的基础上在上、下、左、右各多画了一遍，所以它的效率是很低的。阴影会比描边好很多，因为它只需要多画一遍，所以能用阴影就不要用描边。

5.1.3 动态字体

UGUI 动态字体的原理是根据传入的字体以及字体的大小生成到一张纹理上，最终将纹理上的字体显示出来。这就带来一个问题：如图 5-3 所示，Text 中设置了 3 个完全相同的字体，只是字体的大小不一样，然而在纹理中却生成了 3 份。所以，游戏中是不太建议使用动态字体的。

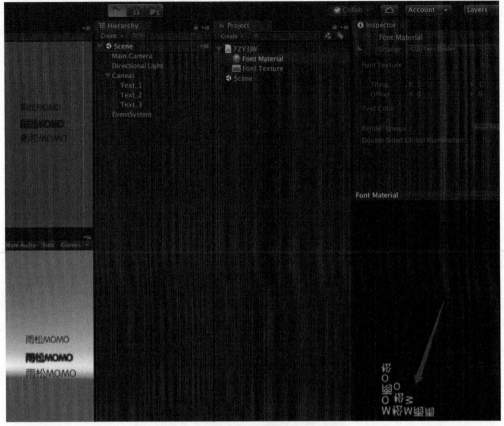

图 5-3 动态字体

除了聊天和起名等必须由用户自己主动输入的文字外，游戏中大量的文字实际上并不需要使用动态字体，此时可以使用 SDF 字体。它的原理就是用位图来保存矢量信息，记录到边的最短距离，最后再用 Shader 还原回来。TextMesh Pro（目前已免费）这个插件就使用了这个原理，我强烈建议在游戏中使用此插件。

5.1.4 字体花屏

前面提到过，UGUI 的动态字体会动态生成材质，它开始是 256×256（像素），然后根据使用字体的情况慢慢扩大，直到 4096×4096（像素）。当文字太多不够放的时候，会触发 UGUI 内部重建字体贴图命令，接着就可能造成文字花屏了。若要解决这个问题，就要监听它内部重建的事件，然后整理刷新一下当前场景中的所有字体即可。

如代码清单 5-1 所示，监听 `Font.textureRebuilt` 字体重建事件，记录当前待重建的字体，在下一帧更新的时候，调用 `FontTextureChanged()` 方法刷新所有 `Text` 文本。

代码清单 5-1　Script_05_01.cs 文件

```csharp
using UnityEngine;
using System.Text;
using UnityEngine.UI;

public class Script_05_01 : MonoBehaviour
{
    //标记某个字体发生了重建
    private Font m_NeedRebuildFont = null;

    void Start()
    {
        //监听字体贴图重建事件
        Font.textureRebuilt += delegate(Font font)
        {
            m_NeedRebuildFont = font;
        };
    }

    void Update()
    {
        if(m_NeedRebuildFont)
        {
            //找到当前场景中的所有 Text,重新刷新一下
            Text [] texts = GameObject.FindObjectsOfType<Text>();
            if(texts!=null){
                foreach(Text text in texts)
                {
                    if(text.font == m_NeedRebuildFont){
                        text.FontTextureChanged();
                    }
                }
            }
            m_NeedRebuildFont = null;
        }
    }
}
```

这里使用 GameObject.FindObjectsOfType<Text>()方法取到当前 Hierarchy 视图中的所有字体,依次遍历后,重新刷新它即可。

5.1.5　Image 组件

Image 组件用来显示图片。如图 5-4 所示,图片一共分为如下 4 种格式。

- Simple:直接显示图片。
- Sliced:通过九宫格方式显示图片,可用 SpriteEditor 来编辑九宫格的区域。
- Tiled:平铺图片。
- Filled:像技能 CD 一样,可以旋转图片。

Preserve Aspect 复选框表示是否强制等比例显示图片。单击下方的 Set Native Size 按钮后，可重新格式化图片大小。

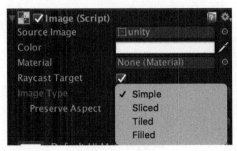

图 5-4　Image 组件

5.1.6　Raw Image 组件

Image 组件只能显示 Sprite，Raw Image 组件既可以显示任意 Texture，也可以使用 Sprite，不过它还是以 Texture 方式显示的。Image 组件使用 Sprite 时，可以使用 Atlas 来合并批次，但是 Raw Image 组件却不能，每个 Raw Image 就占一次 DrawCall。所以，一般不太建议使用 Raw Image 组件。

但是有时候又不能不使用 Raw Image 组件，比如 Render Texture，需要将摄像机渲染到纹理中，就必须使用它。如图 5-5 所示，Raw Image 组件的参数也比较简单，不像 Image 那样有很多类型，它挂上图片就可以使用了。Raw Image 属于原始显示图片的组件，初始化速度也是最快的。但是由于它无法合并 DrawCall，所以大量 UI 系统不建议使用它。

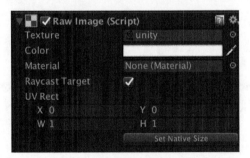

图 5-5　Raw Image 组件

5.1.7　Button 组件

Button 组件必须依赖 Image 组件。按钮有普通、点击、抬起和悬浮这几种状态，切换各状态可以改变它的颜色，也可以更换 Sprite 图片样式，再或者使用 Animation 动画系统来控制各状态。Button 组件直接提供了点击方法来监听点击事件。图 5-6 显示了 Button 组件的参数。

图 5-6 Button 组件

如代码清单 5-2 所示，调用 onClick.AddListener()方法就可以给按钮添加监听点击的事件了。

代码清单 5-2　Script_05_02.cs 文件

```
using UnityEngine;
using UnityEngine.UI;

public class Script_05_02 : MonoBehaviour
{
    public Button button;

    void Start()
    {
        button.onClick.AddListener(delegate() {
            Debug.Log("click");
        });
    }
}
```

需要说明的是，按钮不仅可以通过代码来添加点击事件，也可以通过在按钮组件面板中来添加。选择 OnClick 事件，在下拉框中选择对应的方法名，前提是该方法需要为 public 属性的公有方法。开发中，在代码中对按钮进行监听更加灵活，不建议在面板中添加点击方法。

5.1.8　Toggle 组件

Toggle 组件就像单项选择一样，选择其中一个选项，剩下的会自动取消选择，如图 5-7 所示。记得要把所有 Toggle 对象都关联进同一个 ToggleGroup 里。下面我们来监听一下它的选择/取消选择事件。

图 5-7 Toggle 组件

如代码清单 5-3 所示，Toggle 组件可以使用 onValueChanged() 方法来监听切换事件。

代码清单 5-3　Script_05_03.cs 文件

```
using UnityEngine;
using UnityEngine.UI;

public class Script_05_03 : MonoBehaviour
{
    public Toggle []toogles;

    void Start()
    {
        foreach(var toogle in toogles) {
            toogle.onValueChanged.AddListener(delegate(bool selected) {
                Debug.LogFormat("toogle = {0} selected = {1}",toogle.name,selected);
            });
        }
    }
}
```

如果运行时需要用脚本单独改变其中的某一个，可以使用 toogle.isOn 来设置它的选中状态。

5.1.9　Slider 组件

Slider 组件就是一个滑块在进度条上左右拖动，游戏中经常会使用它来做人物的血条。如图 5-8 所示，滑动 Slider 组件，可以更改血条的显示状态。

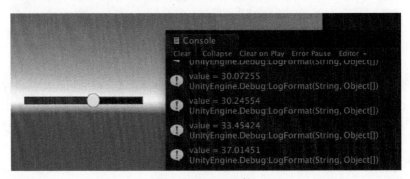

图 5-8　Slider 组件

如代码清单 5-4 所示，使用 onValueChanged() 方法监听 Slider 组件的滚动事件，并且可以取到滚动的进度。

代码清单 5-4　Script_05_04.cs 文件

```
using UnityEngine;
using UnityEngine.UI;

public class Script_05_04 : MonoBehaviour
```

```
{
    public Slider slider;

    void Start()
    {
        //设置取值范围的最小值/最大值
        slider.minValue = 0;
        slider.maxValue = 100;

        slider.onValueChanged.AddListener(delegate(float value) {
            Debug.LogFormat("value = {0}",value);
        });
    }
}
```

如果运行时需要用脚本更改它，则可以使用 `slider.value` 来设置新的值。另外，Slider 和 Toggle 都是使用 `onValueChanged()` 来监听事件的，但是它们传递的 `UnityEvent` 是不同的，并且回调的参数也是不同的。

5.1.10　Scrollbar & ScrollView 组件

Scroll Rect 组件由 Scrollbar 组件（拖动条）和 ScrollView 组件（滑动区域）组成。它的原理就是使用 Scroll Rect 组件设置一个滑动区域，然后挂上 Scrollbar 组件监听滑动的事件，这和 Silder 组件类似，如图 5-9 所示。

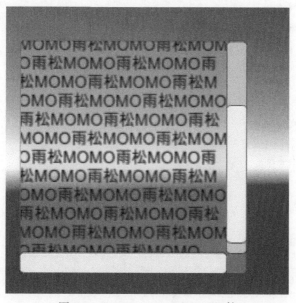

图 5-9　Scrollbar & ScrollView 组件

5.1.11 使用 ScrollRect 组件制作游戏摇杆

ScrollRect 是 UGUI 提供的基础拖动组件,给它的 Content 绑定一个滑块就能工作了。利用这个特性来做游戏摇杆再合适不过了,如图 5-10 所示。

图 5-10 游戏摇杆

如代码清单 5-5 所示,继承 ScrollRect 后,重写 OnDrag()方法来监听滑动摇杆事件。其中,contentPostion.magnitude 计算滑动摇杆的长度,我们需要让摇杆保持在圆形区域内。

代码清单 5-5 Script_05_05.cs 文件

```
using UnityEngine;
using UnityEngine.UI;

public class Script_05_05 : ScrollRect
{
    protected float mRadius=0f;

    protected override void Start()
    {
        base.Start();
        //计算摇杆半径
        mRadius = (transform as RectTransform).sizeDelta.x * 0.5f;
    }

    public override void OnDrag(UnityEngine.EventSystems.PointerEventData eventData)
    {
        base.OnDrag(eventData);
        var contentPostion = this.content.anchoredPosition;
        if(contentPostion.magnitude > mRadius){
            contentPostion = contentPostion.normalized * mRadius ;
            SetContentAnchoredPosition(contentPostion);
        }
    }
}
```

在上述代码中，mRadius 表示摇杆滑动圆形区域的半径，contentPostion.normalized 则表示摇杆的单位向量，两者相乘，即可得出摇杆最终所在的位置。

5.2 事件系统

UGUI 所有的事件系统都是依赖 EventSystem 组件完成的，Unity 新版的事件系统已经全面代替之前的 SendMessage 系统了。操作的事件是非常庞大的。鼠标、键盘和手势能产生太多事件了，而且不同的 UI 操作事件还不太一样。就拿按钮来说，点击事件、按下事件和抬起事件都各不相同。可想而知，所有的 UI 系统加在一起能产生的事件将有多少。如图 5-11 所示，需要在全局添加一个 EventSystem 对象，运行起来后，在面板中还可以查看操作的一些详细信息，便于日后调试。

事件系统不仅会抛出点击一类的事件，还可以取到最基本的操作信息，例如鼠标在屏幕中的坐标、滑动开始的坐标以及滑动结束的坐标等。新版的 EventSystem 不仅供 UI 使用，3D 游戏对象也可以使用它，并且使用方法都比较接近，后面介绍 3D 游戏开发时会向大家详细介绍。

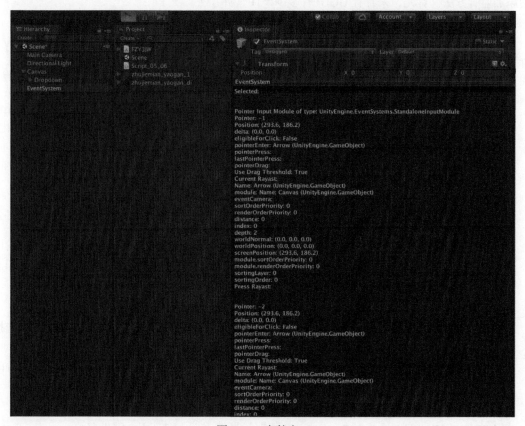

图 5-11 事件窗口

5.2.1 UI 事件

UI 事件依赖于 Graphic Raycaster 组件，如图 5-12 所示，它必须绑定在 Canvas 组件上，表示这个 Canvas 下所有 UI 元素支持的事件。比如游戏中同时有很多 Canvas，如果想让游戏中某些 UI 不可接收点击事件，那么可以考虑把部分 Canvas 上的 Graphic Raycaster 组件设置成 enable=false，或者直接删掉 Graphic Raycaster 组件即可。

图 5-12 Graphic Raycaster 组件

其实 UGUI 已经帮我们封装好了一些 UI 元素的事件，比如前面提到的 Button、Toggle、Slider 等，但像 Image、Text 这种特别基础的 UI 元素是没有事件封装的，如果非要监听的话，只能手动添加监听方法。首先，来看看 UGUI 有多少事件监听方法。

- `IPointerEnterHandler` - `OnPointerEnter`：进入该区域时调用。
- `IPointerExitHandler` - `OnPointerExit`：离开该区域时调用。
- `IPointerDownHandler` - `OnPointerDown`：按下时调用。
- `IPointerUpHandler` - `OnPointerUp`：抬起时调用。
- `IPointerClickHandler` - `OnPointerClick`：按下并且抬起时调用，好比按钮的点击。
- `InitializePotentialDragHandler` - `OnInitializePotentialDrag`：拖动初始化。
- `IBeginDragHandler` - `OnBeginDrag`：拖动开始时调用，并且可以取到拖动的方向，而 `OnInitializePotentialDrag` 只表示滑动初始化，无法取到方向。
- `IDragHandler` - `OnDrag`：滑动持续时调用。
- `IEndDragHandler` - `OnEndDrag`：滑动结束时调用。
- `IDropHandler` - `OnDrop`：落下时调用。
- `IScrollHandler` - `OnScroll`：鼠标滚轮持续时调用。
- `IUpdateSelectedHandler` - `OnUpdateSelected`：选择时续调用，只针对 Selectable 起作用。
- `ISelectHandler` - `OnSelect`：选择后调用，只针对 Selectable 起作用。
- `IDeselectHandler` - `OnDeselect`：取消选择，由于只能选择一个 Selectable，当选择新的后，之前选择的就会回调取消选择事件。
- `IMoveHandler` - `OnMove`：选择后，可监听上下左右 WSAD 方向键。如果访问 eventData.moveDir，可以取到具体移动的方向。
- `ISubmitHandler` - `OnSubmit`：按钮按下事件。
- `ICancelHandler` - `OnCancel`：按钮取消事件，按下时按 Esc 键可取消。

监听的方法是非常全面的，可根据自己的需求依次实现上面的接口。现在我们根据这个来举个例子，监听 Image 或者 Text 的点击事件。如图 5-13 所示，代码中继承需要监听的 `IPointerClickHandler` 接口，接着就可以重写 `OnPointerClick()` 点击方法了。当点击这个 UI 元素时，就会回调一次此方法。

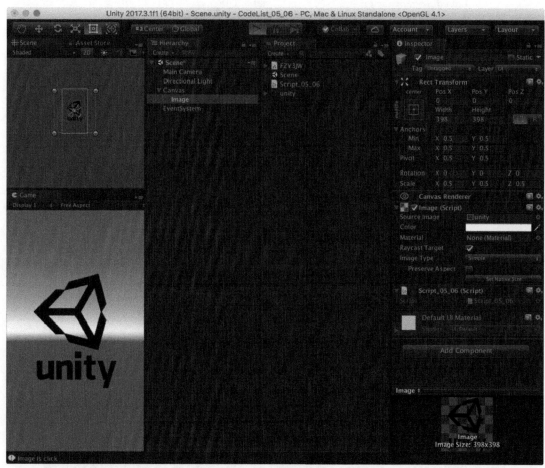

图 5-13 点击事件

如代码清单 5-6 所示，继承 `IPointerClickHandler` 接口并且实现 `OnPointerClick()` 方法，此时的 `gameObject` 就表示点击的按钮对象。如果需要更多详细的点击信息，可以访问回调中的 `eventData` 对象。

代码清单 5-6　Script_05_06.cs 文件

```
using UnityEngine;
using UnityEngine.UI;
using UnityEngine.EventSystems;
```

```
public class Script_05_06 : MonoBehaviour,IPointerClickHandler
{
    #region IPointerClickHandler implementation
    public void OnPointerClick(PointerEventData eventData)
    {
        Debug.LogFormat("{0} is click", gameObject.name);
    }
    #endregion
}
```

注意：这里只举例了按钮点击，前面介绍的很多别的事件都可以按照这种方式来添加监听。

5.2.2　UI 事件管理

上一节我们讲了如何给一个普通的 UI 元素添加点击事件，可是 UI 中需要响应的事件太多了，总不能给每个元素都添加一个脚本来处理点击按钮后的逻辑吧，所以每个 UI 界面都应该有一个类来统一处理事件。如图 5-14 所示，我们用 `OnClick()` 方法统一处理按钮、文本和图片元素的点击事件。

图 5-14　事件管理

如代码清单 5-7 所示，在统一位置给所有 UI 组件添加监听，调用 `UGUIEventListener.Get()` 方法传入需要监听的对象，文本或者图片。

代码清单 5-7　Script_05_07.cs 文件

```
using UnityEngine;
```

```csharp
using UnityEngine.UI;
using UnityEngine.EventSystems;

public class Script_05_07 : MonoBehaviour
{
    public Button button1;
    public Button button2;

    public Text text;
    public Image image;

    void Awake()
    {
        button1.onClick.AddListener(delegate() {
            OnClick(button1.gameObject);
        });

        button2.onClick.AddListener(delegate() {
            OnClick(button2.gameObject);
        });

        UGUIEventListener.Get(text.gameObject).onClick = OnClick;
        UGUIEventListener.Get(image.gameObject).onClick = OnClick;
    }

    void OnClick(GameObject go)
    {
        if(go == button1.gameObject) {
            Debug.Log("点击按钮1");
        } else if(go == button2.gameObject) {
            Debug.Log("点击按钮2");
        }else if(go == text.gameObject) {
            Debug.Log("点击文本框");
        }else if(go == image.gameObject) {
            Debug.Log("点击图片");
        }
    }
}
```

按钮点击后，能干的事太多了，比如打开一个新界面，或者刷新当前界面中的某个元素，再或者发送网络请求等，我们总不能把逻辑都写在 OnClick() 这个监听方法中吧，所以需要在控制层接收事件，再将事件传递给模块层，等模块层处理完毕后再通知 UI 刷新显示层，这就是经典的 MVC 设计思路了。UGUI 并没有对文本和图片元素提供点击事件，不过我们可以继承 EventSystems.EventTrigger 来实现 OnPointerClick() 点击方法。当然，还可以实现拖动、按下、抬起等一些事件。相关代码如代码清单 5-8 所示。

代码清单 5-8　UGUIEventListener.cs 文件

```csharp
using System.Collections;
using System.Collections.Generic;
using UnityEngine;
```

```csharp
using UnityEngine.Events;
public class UGUIEventListener : UnityEngine.EventSystems.EventTrigger
{
    public UnityAction<GameObject> onClick;

    public override void OnPointerClick(UnityEngine.EventSystems.PointerEventData
        eventData)
    {
        base.OnPointerClick(eventData);
        if(onClick != null) {
            onClick(gameObject);
        }
    }

    ///<summary>
    ///获取或者添加 UGUIEventListener 脚本来实现对游戏对象的监听
    ///</summary>
    static public UGUIEventListener Get(GameObject go)
    {
        UGUIEventListener listener = go.GetComponent<UGUIEventListener>();
        if(listener == null)
            listener = go.AddComponent<UGUIEventListener>();
        return listener;
    }
}
```

5.2.3 UnityAction 和 UnityEvent

UnityAction 是 Unity 自己实现的事件传递系统，就像 C# 里的委托和事件一样，它属于函数指针，可以把方法传递到另外一个类中去执行。而 UnityEvent 负责管理 UnityAction，它提供了 AddListener、RemoveListener 和 RemoveAllListeners。UnityAction 只能调用自己，但是 UnityEvent 可以同时调用多个 UnityAction，而且 UnityEvent 还提供面板上的赋值操作。如图 5-15 所示，可以设置 Event 的游戏对象，以及脚本中的方法名。

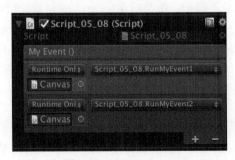

图 5-15　UnityEvent

使用事件是非常好的做法，因为它可以大量减少代码中的耦合，避免类与类之间相互直接调用，也避免后期修改带来的维护成本。例如，游戏中的网络请求、发送请求和处理请求的结果希望在同一个类中处理，那么就可以把处理结果用函数指针传入网络模块，等网络消息处理结束后，再执行当初传入的函数指针。因为发送消息和处理消息完全不依赖网络模块，所以未来就算将发送和处理消息的代码删掉，网络模块的代码也不会受到影响。如代码清单5-9所示，如果事件需要带参数，就得继承UnityEvent并且声明参数的数量以及数据类型。

代码清单5-9　Script_05_08.cs文件

```csharp
using UnityEngine;
using UnityEngine.UI;
using UnityEngine.Events;

public class MyEvent : UnityEvent<int,string>{}

public class Script_05_08 : MonoBehaviour
{
    public UnityAction<int,string> action1;
    public UnityAction<int,string> action2;
    public MyEvent myEvent = new MyEvent();

    public void RunMyEvent1(int a,string b)
    {
        Debug.Log(string.Format("RunMyEvent1 ,{0} , {1}", a, b));
    }
    public void RunMyEvent2(int a,string b)
    {
        Debug.Log(string.Format("RunMyEvent2 ,{0} , {1}", a, b));
    }

    void Start()
    {
        //也可以使用+=，但是+=操作执行多次后，如果没有对应的-=，就会有隐患
        action1 = RunMyEvent1;
        action2 = RunMyEvent2;

        myEvent.AddListener(action1);
        myEvent.AddListener(action2);

        //如果需要删除的话，就执行Remove
        //myEvent.RemoveListener(action1);
        //myEvent.RemoveListener(action2);
        //myEvent.RemoveAllListeners();
    }

    void Update(){
        if(Input.GetKeyDown(KeyCode.A)) {
            Debug.Log("按下键盘A");
            action1.Invoke(0,"a");
            action2.Invoke(1,"b");
        }
```

```
    if(Input.GetKeyDown(KeyCode.B)) {
        Debug.Log("按下键盘 B");
        myEvent.Invoke(100,"a & b");
    }
  }
}
```

注意：UnityAction 可添加泛型带参数，但是 UnityEvent 如果要带参数，需要再写个继承类。

5.2.4　RaycastTarget 优化

UGUI 的点击事件也基于射线，这在前面章节中也提过。如果不需要响应事件，千万不要在 Image 和 Text 组件上勾选 RaycastTarget。UI 事件会在 EventSystem 的 Update() 方法中调用 Process 时触发。UGUI 会遍历屏幕中所有 RaycastTarget 是 true 的 UI，接着就会发射线，并且排序找到玩家最先触发的那个 UI，再抛出事件给逻辑层去响应，这样无形中就会带来很多开销。

如图 5-16 所示，可以拓展一下 Scene 视图，将所有勾选过 RaycastTarget 的 UI 用线框的方式显示出来，这样就可以及时把不需要响应点击事件的 UI 全部取消勾选。

如代码清单 5-10 所示，其工作原理就是重写了 OnDrawGizmos() 方法，同时把场景中的所有 UI 组件找出来，如果勾选了 RaycastTarget，计算出元素的 4 个顶点，最终用 Gizmos.DrawLine() 绘制出来即可。

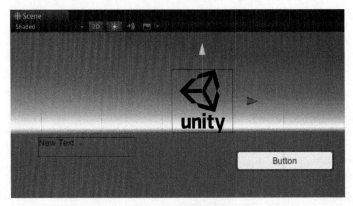

图 5-16　辅助线框

代码清单 5-10　Script_05_09.cs 文件

```
using UnityEngine;
using UnityEngine.UI;
using UnityEngine.Events;
public class Script_05_09 : MonoBehaviour
{
```

```
#if UNITY_EDITOR
static Vector3[] fourCorners = new Vector3[4];
void OnDrawGizmos()
{
    foreach(MaskableGraphic g in GameObject.FindObjectsOfType<MaskableGraphic>())
    {
        if(g.raycastTarget)
        {
            RectTransform rectTransform = g.transform as RectTransform;
            rectTransform.GetWorldCorners(fourCorners);
            Gizmos.color = Color.blue;
            for(int i = 0; i < 4; i++)
                Gizmos.DrawLine(fourCorners[i], fourCorners[(i + 1) % 4]);
        }
    }
}
#endif
}
```

如果想在 Scene 视图中暂时取消显示辅助线，单击视图左上角的"Gizmos"下拉菜单，取消勾选脚本即可，如图 5-17 所示。

图 5-17　取消显示辅助线

5.2.5　渗透 UI 事件

如图 5-18 所示，在按钮的上面放了一个 Image 组件，按钮的事件监听在 Image 组件上，如果需要 Button 也响应事件，就要把事件从 Image 中传递下去，此时请将脚本挂在 Image 对象上。

图 5-18　渗透事件

如代码清单 5-11 所示,继承需要监听的类,并且实现对应的方法。在方法中,可以取到 eventData 对象,使用 EventSystem.current.RaycastAll()方法找到所有能传递点击事件的对象,最终使用 ExecuteEvents.Execute()方法将事件传递给需要的对象。

代码清单 5-11　Script_05_10.cs 文件

```csharp
using UnityEngine;
using UnityEngine.UI;
using UnityEngine.Events;
using UnityEngine.EventSystems;
using System.Collections.Generic;

public class Script_05_10 : MonoBehaviour, IPointerClickHandler ,IPointerDownHandler,
    IPointerUpHandler
{
    //监听按下
    public void OnPointerDown(PointerEventData eventData)
    {
        PassEvent(eventData,ExecuteEvents.pointerDownHandler);
    }

    //监听抬起
    public void OnPointerUp(PointerEventData eventData)
    {
        PassEvent(eventData,ExecuteEvents.pointerUpHandler);
    }

    //监听点击
    public void OnPointerClick(PointerEventData eventData)
    {
        PassEvent(eventData,ExecuteEvents.submitHandler);
        PassEvent(eventData,ExecuteEvents.pointerClickHandler);
    }


    //把事件传递下去
    public void PassEvent<T>(PointerEventData data,ExecuteEvents.EventFunction<T>
        function)
        where T : IEventSystemHandler
    {
        List<RaycastResult> results = new List<RaycastResult>();
        EventSystem.current.RaycastAll(data, results);
        GameObject current = data.pointerCurrentRaycast.gameObject ;
        for(int i =0; i< results.Count;i++)
        {
            if(current!= results[i].gameObject)
            {
                ExecuteEvents.Execute(results[i].gameObject, data,function);
                //如果只想响应渗透下去的第一个游戏对象,使用 break 语句跳出这个循环即可
                //break;
            }
        }
    }
}
```

5.2.6 例子——新手引导聚合动画

新手引导的事件最好与业务逻辑分开，不然就需要在每个界面里再写一遍新手引导的事件，最好的方法就是引导的 UI 在最前面再做一个全屏透明层以挡住所有的事件。当事件点在这层 UI 上并且将事件渗透下来的同时，处理点击后的逻辑。

这一节中，我们做引导点击的聚合动画。如图 5-19 所示，四周的灰色面板会向目标图标慢慢聚合，将脚本挂在背景的面板对象上，此时需要添加一个 Shader，使用 distance() 方法来判断两个点的距离，然后再设置这个像素点 color 是否显示。

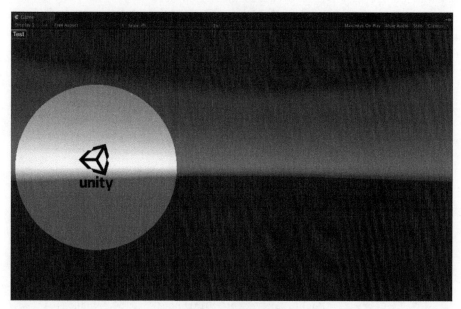

图 5-19 聚合动画

如代码清单 5-12 所示，首先计算出点击区域中心的位置，并且传入 Shader 中，然后声明 Silder 属性来控制圆形遮罩的区域，最终在 Update() 方法中动态修改它，形成一个动画效果。

代码清单 5-12　Script_05_11.cs 文件

```
using UnityEngine;
using UnityEngine.UI;
using UnityEngine.Events;
using UnityEngine.EventSystems;
using System.Collections.Generic;

public class Script_05_11 : MonoBehaviour
{
    //需要聚合的对象（例子中的Unity图标）
    public Image target;
    //Canvas对象
    public Canvas canvas;
```

```csharp
private Vector4 m_Center;
private Material m_Material;
private float m_Diameter; //直径
private float m_Current =0f;

Vector3[] corners = new Vector3[4];

void Awake()
{

    target.rectTransform.GetWorldCorners(corners);
    m_Diameter = Vector2.Distance(WordToCanvasPos(canvas,corners [0]),
        WordToCanvasPos(canvas,corners [2])) / 2f;

    float x =corners [0].x + ((corners [3].x - corners [0].x) / 2f);
    float y =corners [0].y + ((corners [1].y - corners [0].y) / 2f);

    Vector3 center = new Vector3(x, y, 0f);
    Vector2 position = Vector2.zero;
    RectTransformUtility.ScreenPointToLocalPointInRectangle(canvas.transform
        as RectTransform, center, canvas.GetComponent<Camera>(), out position);



    center = new Vector4(position.x,position.y,0f,0f);
    m_Material = GetComponent<Image>().material;
    m_Material.SetVector("_Center", center);



    (canvas.transform as RectTransform).GetWorldCorners(corners);
    for(int i = 0; i < corners.Length; i++) {
        m_Current = Mathf.Max(Vector3.Distance(WordToCanvasPos(canvas,corners
            [i]), center),m_Current);
    }

    m_Material.SetFloat("_Silder", m_Current);
}


float yVelocity = 0f;
void Update() {
    float value = Mathf.SmoothDamp(m_Current, m_Diameter, ref yVelocity, 0.3f);
    if(!Mathf.Approximately(value, m_Current)) {
        m_Current = value;
        m_Material.SetFloat("_Silder", m_Current);
    }
}

void OnGUI(){
    if(GUILayout.Button("Test")){
        Awake();
```

```
        }
    }

    Vector2 WorldToCanvasPos(Canvas canvas,Vector3 world){
        Vector2 position = Vector2.zero;
        RectTransformUtility.ScreenPointToLocalPointInRectangle(canvas.transform
            as RectTransform, world, canvas.GetComponent<Camera>(), out position);
        return position;
    }
}
```

为了让动画更加自然,可以使用 Mathf.SmoothDamp() 方法来控制它。如代码清单 5-13 所示,这是修改了默认的 Shader 文件来处理动画的效果。

代码清单 5-13 Default_Mask.shader 文件

```
Shader "UI/Default_Mask"
{
    Properties
    {
        [PerRendererData] _MainTex("Sprite Texture", 2D) = "white" {}
        _Color("Tint", Color) = (1,1,1,1)

        _StencilComp("Stencil Comparison", Float) = 8
        _Stencil("Stencil ID", Float) = 0
        _StencilOp("Stencil Operation", Float) = 0
        _StencilWriteMask("Stencil Write Mask", Float) = 255
        _StencilReadMask("Stencil Read Mask", Float) = 255

        _ColorMask("Color Mask", Float) = 15

        [Toggle(UNITY_UI_ALPHACLIP)] _UseUIAlphaClip("Use Alpha Clip", Float) = 0

        //------------------add---------------------
          _Center("Center", vector) = (0, 0, 0, 0)
          _Silder("_Silder", Range (0,1000)) = 1000 //sliders
        //------------------add---------------------
    }

    SubShader
    {
        Tags
        {
            "Queue"="Transparent"
            "IgnoreProjector"="True"
            "RenderType"="Transparent"
            "PreviewType"="Plane"
            "CanUseSpriteAtlas"="True"
        }

        Stencil
        {
```

```
            Ref [_Stencil]
            Comp [_StencilComp]
            Pass [_StencilOp]
            ReadMask [_StencilReadMask]
            WriteMask [_StencilWriteMask]
        }

        Cull Off
        Lighting Off
        ZWrite Off
        ZTest [unity_GUIZTestMode]
        Blend SrcAlpha OneMinusSrcAlpha
        ColorMask [_ColorMask]

        Pass
        {
            Name "Default"
        CGPROGRAM
            #pragma vertex vert
            #pragma fragment frag
            #pragma target 2.0

            #include "UnityCG.cginc"
            #include "UnityUI.cginc"

            #pragma multi_compile __ UNITY_UI_ALPHACLIP

            struct appdata_t
            {
                float4 vertex : POSITION;
                float4 color : COLOR;
                float2 texcoord : TEXCOORD0;
                UNITY_VERTEX_INPUT_INSTANCE_ID
            };

            struct v2f
            {
                float4 vertex : SV_POSITION;
                fixed4 color : COLOR;
                float2 texcoord : TEXCOORD0;
                float4 worldPosition : TEXCOORD1;
                UNITY_VERTEX_OUTPUT_STEREO

            };

            fixed4 _Color;
            fixed4 _TextureSampleAdd;
            float4 _ClipRect;
            //-------------------add---------------------
            float _Silder;
            float2 _Center;
            //-------------------add---------------------
            v2f vert(appdata_t IN)
            {
```

```
            v2f OUT;
            UNITY_SETUP_INSTANCE_ID(IN);
            UNITY_INITIALIZE_VERTEX_OUTPUT_STEREO(OUT);
            OUT.worldPosition = IN.vertex;
            OUT.vertex = UnityObjectToClipPos(OUT.worldPosition);

            OUT.texcoord = IN.texcoord;

            OUT.color = IN.color * _Color;
            return OUT;
        }

        sampler2D _MainTex;

        fixed4 frag(v2f IN) : SV_Target
        {
            half4 color = (tex2D(_MainTex, IN.texcoord) + _TextureSampleAdd) *
                IN.color;

            color.a *= UnityGet2DClipping(IN.worldPosition.xy, _ClipRect);

            #ifdef UNITY_UI_ALPHACLIP
            clip(color.a - 0.001);
            #endif
            //-------------------add---------------------
            color.a*=(distance(IN.worldPosition.xy,_Center.xy) > _Silder);
            color.rgb*= color.a;
            //-------------------add---------------------
            return color;
        }
    ENDCG
        }
    }
}
```

在上述代码中，我们在 `frag()` 方法中通过 `distance` 判断当前像素点与圆形中心点的距离。如果当前像素点在圆形区间内，则保持透明。

5.3　Canvas 组件

Canvas 组件是 UI 的基础画布，所有 UI 元素都必须放在 Canvas 对象下面，并且它支持嵌套。Canvas 支持 3 种绘制方式——Overlay（最上层）、Camera 和 World Space（3D 布局），其中用得最多的是 Camera，它可以把正交摄像机投影出来的 UI 元素绘制在 Canvas 面板上。

5.3.1　自适应屏幕

自适应屏幕需要缩放画布，如图 5-20 所示，Canvas 的同级需要绑定 Canvas Scaler 组件。接着，需要确认开发的分辨率，也就是说，UI 美术人员必须按这个屏幕尺寸来出对应的图片和布局。以移动平台为例,现在主流的手机大多数是 16∶9 的分辨率比例,不能把分辨率设置得太高,得考虑兼容低配的手机。这里我设置的分辨率是 1136×640,目标分辨率等比例缩放即可,最后将屏幕相对模式设置成 Expand。

图 5-20　自适应屏幕

这里 Expand 表示 Canvas 下的 UI 始终保持在屏幕内,当屏幕宽度变窄后,它会整体缩放高度来保持自适应。在 Screen Match Mode 下拉框中,还可以选择 Match Width Or Height 和 Shrink,其中前者表示始终保持宽度或高度来自适应高度或宽度,后者表示当分辨率变化时,始终保持原始比例,超出屏幕部分会被裁切掉(将显示不全)。所以,在游戏开发中,界面自适应首先应该选择 Expand 模式。

游戏中至少需要两个摄像机,一个是 3D 摄像机(主摄像机),另一个就是 UI 摄像机了。我们需先渲染 3D 摄像机,然后再渲染 UI 摄像机,这样 UI 就会盖在 3D 场景的前面。我们可以设置 UICamera 里的 Depth 值。如图 5-21 所示,Depth 在 3D 摄像机上设置的是 -1,在 UI 摄像机设置的就是 0(或者大于 -1)了。另外,UICamera 需要设置正交摄像机,此时只需在 Projection 中选择 Orthographic 即可。Clipping Planes 的最小值建议设置成 0,不然有时候某些 UI 会被剔除掉。

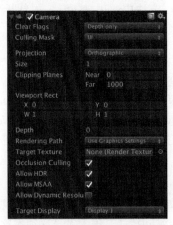

图 5-21　UI 摄像机的设置

5.3.2 锚点对齐方式

目前，UI 的自适应是整体缩放。有时候，我们可能需要将某些 UI 挂靠在屏幕的 4 个边角上。如图 5-22 所示，锚点的对齐方式一共有 9 种。由于 UI 对象是可以多层嵌套的，请记住这里面设置的是相对它的父对象的对齐方式。如图 5-23 所示，设置好锚点挂靠后，无论如何修改屏幕分辨率，边角挂靠的按钮永远都会自适应地挂在 4 个角上。

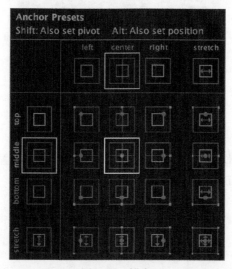

图 5-22 锚点

图 5-23 锚点挂靠

5.3.3 背景图全屏

背景图拉伸分两种，一种是允许图片变形，另一种就是不允许图片拉伸变形，可以自动裁切。如图 5-24 所示，设置图片的 Rect Transform 锚点的横向、纵向都支持自动拉伸，这样图片永远都会保持和屏幕大小一致，但是图片会被拉伸变形。

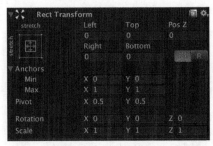

图 5-24　全屏拉伸

如果允许图片适当地裁切，保证它不会变形，那么可以给 Image 添加 Aspect Ratio Fitter 组件，并且在 Aspect Mode 中选择 Envelope Parent，如图 5-25 所示。

图 5-25　填充屏幕

如图 5-26 和图 5-27 所示，将分辨率修改到极端情况，Rect Transform 锚点设置的图片虽然显示全了，但是整体都被拉伸了。而 Aspect Ratio Fitter 设置的图片被裁切掉了部分内容，图片整体并没有发生变形。

图 5-26　填充屏幕

图 5-27　填充屏幕

5.3.4 布局组件

GUI 提供了一组 Layout 组件，其中 Horizontal Layout Group（横向布局）、Vertical Layout Group（纵向布局）、Grid Layout Group（表格布局）的用法非常简单。这里举个例子，如图 5-28 所示，给 Layout 对象绑定 Horizontal Layout Group（横向布局）组件和 Content Size Fitter 组件，接着给 Text 对象也绑定上 Content Size Fitter 组件，此时动态变化的文字背景布局将自适应屏幕。

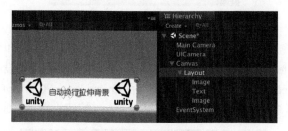

图 5-28　自动拉伸

Content Size Fitter 组件可以重新计算子对象的 RectTransform 区域，如果子对象很多的话，它的效率并不是很高。可能是出于效率的考虑，Content Size Fitter 需要等一帧才能算出正确的区域。如果想立即取到正确的区域，可以使用如下代码：

```
LayoutRebuilder.ForceRebuildLayoutImmediate(rectTransform);
```

5.3.5 Canvas 优化

UGUI 会自动合并批次，原理是它会把一个 Canvas 下的所有元素合并在一个 Mesh 里。如果 Canvas 下的元素很多，任意一个元素发生位置、大小的改变，就需要重新合并所有元素的 Mesh。如果元素非常多的话，可能就会造成卡顿。

一个比较好的做法就是每个 UI 界面都设置成一个 Canvas。如果这个界面下的元素比较多，可以考虑多套几个 Canvas。尤其是会频繁改变位置大小的元素，这样就可以降低它合并 Mesh 的开销。但是 Canvas 套得太多也不好，Mesh 合并是降低了，但是 DrawCall 又上去了，因为每个 Canvas 都会单独占用一个 DrawCall。

5.4 Atlas

前面，我们讲了每个 Canvas 会自动合并下面所有元素到一个 Mesh 里。Mesh 虽然可以合并在一起，但是如果贴图是分开的，那么每个贴图依然会多占用一个 DrawCall。为了减少 DrawCall，我们可以把多张图片合并在一个图片中，这称为 Atlas（图集）。

5.4.1 创建 Atlas

创建 Atlas 之前，请先确保 Sprite Packer 已启用。如图 5-29 所示，在 Editor Settings 页面中，

设置 Sprite Packer 中的 Mode 为 Always Enabled。

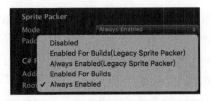

图 5-29　启动 Sprite Packer

接着，在 Project 视图中选择 Create→Sprite Atlas 命令创建图集。如图 5-30 所示，可以将单张 Sprite 或者整个 Sprite 文件夹下的所有图片生成图集，其中图片类型设置成 Sprite(2D and UI)。然后，将它们拖入 Objects for Packing 处。接着，在 Override for PC, Mac & Linux Standalone 处设置图集的大小和贴图压缩格式。最后，单击 Pack Preview 按钮，即可生成图集。

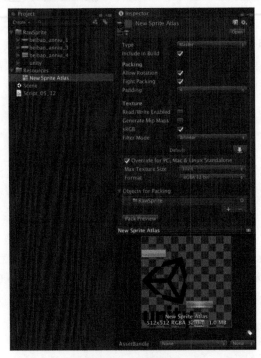

图 5-30　创建图集

5.4.2　读取 Atlas

Atlas 可以把很多 Sprite 合并在一起。假如有一个 Image 元素，运行期间需要更换它的图片。如图 5-31 所示，代码中首先需要加载这个图集，然后通过名字去读取 Sprite，这里的名字就是 Sprite 在 Project 视图中的文件名。

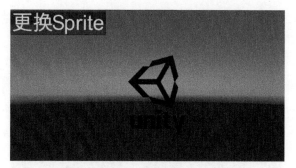

图 5-31 读取 Sprite

有了图集的概念,当需要更换 Sprite 时,就需要先读取图集了。如代码清单 5-14 所示,我们首先使用 Resources.Load()方法读取 Atlas,接着可以使用 m_SpriteAtlas.GetSprite()方法读取该图集上的某张 Sprite,最终将其赋值给 Image 组件即可。

代码清单 5-14　Script_05_12.cs 文件

```
using UnityEngine;
using UnityEngine.UI;
using UnityEngine.Events;
using UnityEngine.EventSystems;
using System.Collections.Generic;
using UnityEngine.U2D;
public class Script_05_12 : MonoBehaviour
{
    public Image image;

    private SpriteAtlas m_SpriteAtlas = null;

    void Start()
    {
        m_SpriteAtlas = Resources.Load<SpriteAtlas> ("New Sprite Atlas");
    }

    void OnGUI()
    {
        if(GUILayout.Button ("<size=80>更换 sprite</size>")) {
            image.sprite = m_SpriteAtlas.GetSprite ("unity");
        }
    }
}
```

在上述代码中,我们通过点击屏幕中的 GUI 按钮来动态更换 Image 图片。

5.4.3　Variant

此外,Atlas 还可以设置 Variant,也就是可以引用另外一个 Atlas 的信息。如图 5-32 所示,新建一个 Sprite Atlas,其中类型选择 Variant,Master Atlas 选择之前的 Sprite Atlas 即可。另外,还可以重新设置 Atlas 的缩放，如果将 Scale 设置为 0.5，表示是原图集的一半。在 Override for PC,Mac & Linux Standalone 处，还可以设置图集的压缩格式。

图 5-32　Variant

5.4.4　多图集管理

　　使用图集就是为了优化效率，避免过多的 DrawCall，但是图集也有大小限制。以移动平台为例，图集尽量不要超过 1024。如果图片太多，就要考虑把它们放在多个图集上。如果不同图集下的图片发生叠层的现象，那么 DrawCall 必然又会上去了。所以说，图集管理是很重要的。

　　尽可能地把复用性很强的图片都放在一个公共图集下，每个 UI 系统可以有一个自己的私有图集。由于战斗部分是最容易发生卡顿的地方，所以战斗下的 UI 尽可能都合并在一个图集上。一定要将 UI 动静分离，频繁发生改变的 UI 元素要套上 Canvas。

　　这里需要说明的是，并不是所有图片都适合使用图集。比如，游戏中的图标资源，几千个图标都放在一个图集的话，如果游戏中只需要同时显示少部分的图标，那么整个图集都会被载入内存里。图标的出现率又那么高，所以这么大的图集内存几乎就没机会释放了，所以像图标一类的图片就不太合适使用图集。

5.5　UGUI 实例

　　目前 UGUI 的内容已经讲完了，但是实际开发中这些是远远不够的。本节开始向大家介绍实际开发中用到的一些典型 UI 技术实例。

5.5.1　置灰

　　UI 置灰这个功能在游戏中很常见，比较方便的就是同时置灰某个节点下的所有 UI 元素。置灰图片的效果还可以，但是文字就不太好了，因为文字类型比较多，例如描边、渐变和阴影等。直接置灰的效果未必好，可以整体修改文字的颜色。如果需要置灰文字，可以和美术人员商量一种颜色。例如，置灰将所有文字改成"白色"，不过还得考虑文字描边、阴影的情况。如图 5-33 所示，在 UI 节点的顶层挂上 UIGray 脚本，可以在编辑模式下勾选 isGray 复选框来预览效果。

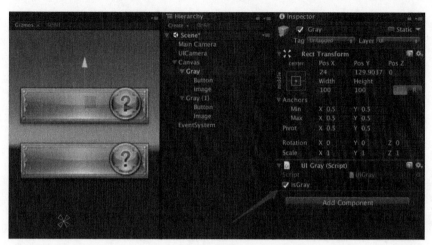

图 5-33　置灰

置灰的话，首先需要给节点下的所有 UI 元素换成一种带置灰的 Shader，还原置灰就是把这些 Shader 取消即可。

如代码清单 5-15 所示，预先在工程中设置好一个材质并且绑定好置灰的 Shader，运行时批量遍历某个节点下的 Image 组件，并且动态替换它们的材质即可实现置灰效果。

代码清单 5-15　UIGray.cs 文件

```
using UnityEngine;
using System.Collections;
using System.Collections.Generic;
using UnityEngine.UI;

#if UNITY_EDITOR
using UnityEditor;
#endif

[DisallowMultipleComponent]
public class UIGray : MonoBehaviour
{
    private bool _isGray = false;
    public bool isGray
    {
        get{ return _isGray;}
        set
        {
            if(_isGray != value)
```

```csharp
            {
                _isGray = value;
                SetGray(isGray);
            }
        }
    }

    static private Material _defaultGrayMaterial;
    static private Material grayMaterial
    {
        get
        {
            if(_defaultGrayMaterial == null)
            {
                _defaultGrayMaterial = new Material(Shader.Find("UI/Gray"));
            }
            return _defaultGrayMaterial;
        }
    }

    void SetGray(bool isGray)
    {
        int i =0, count = 0;
        Image [] images = transform.GetComponentsInChildren<Image>();
        count = images.Length;
        for(i =0; i< count; i++)
        {
            Image g = images[i];
            if(isGray)
            {
                g.material = grayMaterial;
            }else
            {
                g.material  = null;
            }
        }
    }
}
#if UNITY_EDITOR
[CustomEditor(typeof(UIGray))]
public class UIGrayInspector : Editor
{
    public override void OnInspectorGUI()
    {
        base.OnInspectorGUI();
        UIGray gray = target as UIGray;
        gray.isGray = GUILayout.Toggle(gray.isGray ," isGray");
        if(GUI.changed)
        {
            EditorUtility.SetDirty(target);
        }
    }
}
#endif
```

首先在 Shader 中找到 frag() 方法, 接着计算颜色时乘上置灰的颜色即可:

```
fixed4 frag(v2f IN) : SV_Target
{
    half4 color = (tex2D(_MainTex, IN.texcoord) + _TextureSampleAdd) * IN.color;

    color.a *= UnityGet2DClipping(IN.worldPosition.xy, _ClipRect);

    #ifdef UNITY_UI_ALPHACLIP
    clip(color.a - 0.001);
    #endif

    float gray = dot(color.xyz, float3(0.299, 0.587, 0.114));
    color.xyz = float3(gray, gray, gray);
    return color;
}
```

5.5.2 粒子特效与 UI 的排序

在 Canvas 下 UI 元素是按照 Hierarchy 视图下的层级顺序排序的, 如果有多个 Canvas, 可以根据每个 Canvas 的 Order in Layer 来排序。如果只有 UI 元素, 一般可以不用在意这个规则。但是如果两个 UI 之间需要叠加特效, 那就需要重新进行排序了。

如图 5-34 所示, 在 Canvas 节点下创建 3 个 UIOrder, 分别设置它们的深度为 0、1、2, 这样特效就被夹在两个 UI 元素之间了。由于特效可能由多个粒子组成, 所以需要遍历所有粒子, 然后一起设置它们的 Order in Layer 了。相关代码如代码清单 5-16 所示。

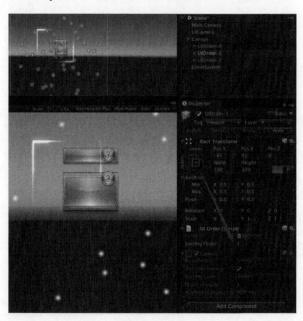

图 5-34 排序

代码清单 5-16　UIOrder.cs 文件

```csharp
using UnityEngine;
using System.Collections;
using UnityEngine.UI;
using System.Collections.Generic;
using UnityEngine.EventSystems;

[AddComponentMenu("UI/UIOrder")]
public class UIOrder : MonoBehaviour
{

    [SerializeField]
    private int _sortingOrder=0;
    public int sortingOrder
    {
        get{
            return _sortingOrder;
        }
        set{
            if(_sortingOrder !=value){
                _sortingOrder = value;
                Refresh();
            }
        }
    }

    private Canvas _canvas = null;
    public Canvas canvas
    {
        get
        {
            if(_canvas == null)
            {
                _canvas = gameObject.GetComponent<Canvas>();
                if(_canvas==null)
                    _canvas = gameObject.AddComponent<Canvas>();
                _canvas.hideFlags = HideFlags.NotEditable;
            }
            return _canvas;
        }
    }

    public void Refresh()
    {
        canvas.overrideSorting = true;
        canvas.sortingOrder = _sortingOrder;
        foreach(ParticleSystemRenderer pariicle in transform.GetComponentsInChildren
            <ParticleSystemRenderer>(true))
        {
            pariicle.sortingOrder = _sortingOrder;
        }
    }
```

```
#if UNITY_EDITOR
void OnValidate()
{
    Refresh();
}

void Reset()
{
    Refresh();
}
#endif
```
}

粒子特效也提供了 sortingOrder，直接设置它即可。另外，Unity 还提供了 sortingLayer 来区分不同排序的层，所以正确的排序应该是，优先看 sortingLayer，然后再看 sortingOrder。

5.5.3　Mask & RectMask2D 裁切

UI 裁切推荐使用 RectMask2D 组件，将其挂在父对象上，它会自动把所有超出父对象区域的子对象都裁切掉。它是按照矩形范围来裁切的，所以效果更好一些。另外，UGUI 还提供了一种按图片样式来裁切的方式。如果要裁切一个圆形，就需要提供一个圆形的效果图，此时就可以将它以裁切成与圆形一样的效果，如图 5-35 所示。

图 5-35　裁切

5.5.4 粒子的裁切

此外，在 UI 上经常需要添加粒子特效，例如滑动列表中就需要同时考虑 UI 和特效的裁切。前面提到的 RectMask2D 和 Mask 只能裁切 UI，但是不能裁切特效，如果要裁切特效的话，就要通过 Shader 剔除裁切部分的特效。如图 5-36 所示，将特效和 Image 都放在 SuperMask 下统一进行裁切。

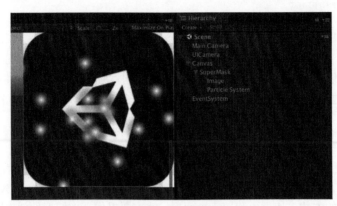

图 5-36 裁切特效

如代码清单 5-17 所示，首先计算出裁切区域 4 个点的坐标，接着将它们传递进 Shader 中进行判断，超出区域的直接显示成透明的即可。

代码清单 5-17 UISuperMask.cs 文件

```
using UnityEngine;
using System.Collections;
using UnityEngine.UI;

[DisallowMultipleComponent]
[RequireComponent(typeof(Image))]
[AddComponentMenu("UI/UISuperMask")]
public class UISuperMask : Mask
{
    private float m_LastminX=-1f,m_LastminY=-1f,m_LastmaxX=-1f,m_LastmaxY=-1f;
    private float m_MinX=0f,m_MinY=0f,m_MaxX=0f,m_MaxY=0f;
    private Vector3[] m_Corners = new Vector3[4];
    private Image m_Image;

    void GetWorldCorners()
    {
        //避免每次都计算
        if(!Mathf.Approximately(m_LastminX,m_MinX) ||
           !Mathf.Approximately(m_LastminY,m_MinY)||
           !Mathf.Approximately(m_LastmaxX,m_MaxX)||
           !Mathf.Approximately(m_LastmaxY,m_MaxY))
        {
            RectTransform rectTransform = transform as RectTransform;
```

```csharp
            rectTransform.GetWorldCorners(m_Corners);

            m_LastminX = m_MinX;
            m_LastminY = m_MinY;
            m_LastmaxX = m_MaxX;
            m_LastmaxY = m_MaxY;


            m_MinX = m_Corners[0].x;
            m_MinY = m_Corners[0].y;
            m_MaxX = m_Corners[2].x;
            m_MaxY = m_Corners[2].y;
        }
    }

    protected override void OnRectTransformDimensionsChange()
    {
        base.OnRectTransformDimensionsChange();
        Refresh();
    }

    public void Refresh()
    {
        GetWorldCorners();
        if(Application.isPlaying)
        {
            foreach(ParticleSystemRenderer system in transform.GetComponentsInChildren
                <ParticleSystemRenderer>(true))
            {
                SetRenderer(system);
            }
        }
    }


    void SetRenderer(Renderer renderer)
    {
        if(renderer.sharedMaterial) {
            Shader shader = Resources.Load<Shader>("SuperMask/Alpha Blended
                Premultiply");
            renderer.material.shader = shader;
            Material m = renderer.material;
            m.SetFloat("_MinX",m_MinX);
            m.SetFloat("_MinY",m_MinY);
            m.SetFloat("_MaxX",m_MaxX);
            m.SetFloat("_MaxY",m_MaxY);
        }
    }

}
```

裁切的原理就是将裁切区域告诉 Shader。可是粒子特效用到的 Shader 很多，所以需要动态地更换适合当前粒子裁切的 Shader。Unity 内置的 Shader 都可以在官网上下载，在原 Shader 的基础上来拓展，在代码清单 5-18 所示的 Shader 文件中，请注意 add 注释所添加的部分。

代码清单 5-18　Alpha Blended Premultiply.shader 文件

```
//Upgrade NOTE: replaced 'mul(UNITY_MATRIX_MVP,*)' with 'UnityObjectToClipPos(*)'

Shader "UI/Mask/Particles/Alpha Blended Premultiply" {
    Properties {
        _MainTex ("Particle Texture", 2D) = "white" {}
        _InvFade ("Soft Particles Factor", Range(0.01,3.0)) = 1.0
        //------------------add---------------------
         _MinX("Min X", Float) = -10
         _MaxX("Max X", Float) = 10
         _MinY("Min Y", Float) = -10
         _MaxY("Max Y", Float) = 10
        //------------------add---------------------
    }

    Category {
        Tags { "Queue"="Transparent" "IgnoreProjector"="True" "RenderType"="Transparent" }
        Blend One OneMinusSrcAlpha
        ColorMask RGB
        Cull Off Lighting Off ZWrite Off

        SubShader {
            Pass {

                CGPROGRAM
                #pragma vertex vert
                #pragma fragment frag
                #pragma multi_compile_particles

                #include "UnityCG.cginc"

                sampler2D _MainTex;
                fixed4 _TintColor;

                struct appdata_t {
                    float4 vertex : POSITION;
                    fixed4 color : COLOR;
                    float2 texcoord : TEXCOORD0;
                };

                struct v2f {
                    float4 vertex : SV_POSITION;
                    fixed4 color : COLOR;
                    float2 texcoord : TEXCOORD0;
                    #ifdef SOFTPARTICLES_ON
                    float4 projPos : TEXCOORD1;
                    #endif
                    //------------------add---------------------
                    float3 vpos : TEXCOORD2;
                    //------------------add---------------------
                };
```

```
            float4 _MainTex_ST;
            //-------------------add---------------------
            float _MinX;
            float _MaxX;
            float _MinY;
            float _MaxY;
            //-------------------add---------------------
            v2f vert(appdata_t v)
            {
                v2f o;
                o.vertex = UnityObjectToClipPos(v.vertex);
                #ifdef SOFTPARTICLES_ON
                o.projPos = ComputeScreenPos(o.vertex);
                COMPUTE_EYEDEPTH(o.projPos.z);
                #endif
                //-------------------add---------------------
                o.vpos = v.vertex.xyz;
                //-------------------add---------------------
                o.color = v.color;
                o.texcoord = TRANSFORM_TEX(v.texcoord,_MainTex);
                return o;
            }

            sampler2D_float _CameraDepthTexture;
            float _InvFade;

            fixed4 frag(v2f i) : SV_Target
            {
                #ifdef SOFTPARTICLES_ON
                float sceneZ = LinearEyeDepth(SAMPLE_DEPTH_TEXTURE_PROJ
                    (_CameraDepthTexture, UNITY_PROJ_COORD(i.projPos)));
                float partZ = i.projPos.z;
                float fade = saturate(_InvFade * (sceneZ-partZ));
                i.color.a *= fade;
                #endif

                fixed4 col = i.color * tex2D(_MainTex, i.texcoord) * i.color.a;
                //-------------------add---------------------
                col.a *= (i.vpos.x >= _MinX);
                col.a *= (i.vpos.x <= _MaxX);
                col.a *= (i.vpos.y >= _MinY);
                col.a *= (i.vpos.y <= _MaxY);
                col.rgb *= col.a;
                //-------------------add---------------------
                return col;
            }
            ENDCG
        }
    }
}
```

上述代码中，MinX、MaxX、MinY 和 MaxY 就是裁切的区域。由于 Shader 中不太建议使用 if...else 来做判断（效率非常低），所以尽可能直接计算出颜色来。

5.5.5 粒子自适应

当分辨率发生变化后,UI 上的粒子特效就会有问题。例如,一个贴合按钮周围发光的特效,分辨率变化后,就无法贴合按钮了。如图 5-37 所示,特效师按照 1136×640 的比例将特效做好后放在 UI 界面中。也就是说,我们可以修改分辨率让特效自适应贴合按钮。

图 5-37 特效自适应

在代码清单 5-19 中,我们首先需要计算当前分辨率与开发分辨率的比例。自适应屏幕使用的是 Expand 模式,这表示只有屏幕变窄的时候自适应需要重新计算。

代码清单 5-19 UIParticleScale.cs 文件

```
using UnityEngine;
using System.Collections.Generic;

public class UIParticleScale : MonoBehaviour
{
    struct ScaleData
    {
        public Transform transform;
        public Vector3 beginScale;
    }

    const float DESIGN_WIDTH = 1136f;//开发时分辨率的宽度
    const float DESIGN_HEIGHT = 640f;//开发时分辨率的高度

    private Dictionary<Transform,ScaleData> m_ScaleData = new Dictionary<Transform,
        ScaleData>();

    void Start()
    {
        Refresh();
    }

    void Refresh()
    {
        float designScale = DESIGN_WIDTH / DESIGN_HEIGHT;
        float scaleRate = (float)Screen.width/(float)Screen.height;

        foreach(ParticleSystem p in transform.GetComponentsInChildren
```

```
            <ParticleSystem>(true)) {
        if(!m_ScaleData.ContainsKey(p.transform)) {
            m_ScaleData [p.transform] = new ScaleData(){ transform = p.transform,
                beginScale = p.transform.localScale };
        }
    }
    foreach(var item in m_ScaleData)
    {
        if(scaleRate<designScale)
        {
            float scaleFactor = scaleRate / designScale;
            item.Value.transform.localScale = item.Value.beginScale * scaleFactor;
        }else{
            item.Value.transform.localScale   = item.Value.beginScale;
        }
    }
}

///<summary>
///子节点发生变化时重新刷新深度
///</summary>
void OnTransformChildrenChanged()
{
    Refresh();
}
#if UNITY_EDITOR
    //编辑模式下修改分辨率后在Update()中刷新
    private void Update()
    {
        Refresh();
    }
#endif
}
```

上述代码中，我们在Awake()方法中计算特效原始的缩放系数。当屏幕变窄后，重新计算新的缩放系数，并将其赋值给每个特效。

特效虽然挂在了Canvas下，但是它还是由UI摄像机渲染出来的，所以UI摄像机需要设置成开发时的比例，也就是3.2。计算方法是屏幕高度的一半，也就是640/2=320，但是由于UI的单位参考像素是100，所以320还需要除以100，也就是320/100=3.2了。如图5-38所示。

图 5-38 摄像机自适应

5.5.6 滑动列表嵌套

通常，我们会根据 Scroll Rect 来拓展滑动列表组件。但是 Unity 的事件比较霸道，如果两个 Scroll Rect 嵌套在一起，后面的就会挡住前面的，如图 5-39 所示。如果在横向滑动的区域，操作上下滑动是无法响应的。为了解决这个问题，我们可以判断滑动的方向是否和滑动区域一致，如果在横向区域操作竖向滑动，那么直接将事件抛给父列表执行。

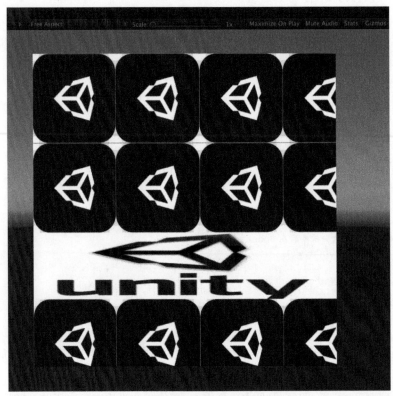

图 5-39 滑动嵌套

如代码清单 5-20 所示，在 OnBeginDrag() 方法中判断滑动的方向，从而决定是否将事件传递或者过滤掉。

代码清单 5-20 CustomScrollRect.cs 文件

```
using System.Collections;
using System.Collections.Generic;
using UnityEngine;
using UnityEngine.EventSystems;
using UnityEngine.UI;

public class CustomScrollRect : ScrollRect {
```

```csharp
//父 CustomScrollRect 对象
private CustomScrollRect m_Parent;

public enum Direction
{
    Horizontal,
    Vertical
}
//滑动方向
private Direction m_Direction = Direction.Horizontal;
//当前操作方向
private Direction m_BeginDragDirection = Direction.Horizontal;

protected override void Awake()
{
    base.Awake();
    //找到父对象
    Transform parent = transform.parent;
    if(parent){
        m_Parent = parent.GetComponentInParent<CustomScrollRect>();
    }
    m_Direction = this.horizontal ? Direction.Horizontal : Direction.Vertical;
}


public override void OnBeginDrag(PointerEventData eventData)
{
    if(m_Parent){
        m_BeginDragDirection = Mathf.Abs(eventData.delta.x) > Mathf.Abs
            (eventData.delta.y) ? Direction.Horizontal : Direction.Vertical;
        if(m_BeginDragDirection != m_Direction){
            //当前操作方向不等于滑动方向，将事件传给父对象
            ExecuteEvents.Execute(m_Parent.gameObject, eventData, ExecuteEvents.
                beginDragHandler);
            return;
        }
    }

    base.OnBeginDrag(eventData);
}
public override void OnDrag(PointerEventData eventData)
{
    if(m_Parent) {
        if(m_BeginDragDirection != m_Direction){
            //当前操作方向不等于滑动方向，将事件传给父对象
            ExecuteEvents.Execute(m_Parent.gameObject, eventData, ExecuteEvents.
                dragHandler);
            return;
        }
    }
    base.OnDrag(eventData);
}

public override void OnEndDrag(PointerEventData eventData)
```

```
    {
        if(m_Parent){
            if(m_BeginDragDirection != m_Direction){
                //当前操作方向不等于滑动方向,将事件传给父对象
                ExecuteEvents.Execute(m_Parent.gameObject, eventData, ExecuteEvents.
                    endDragHandler);
                return;
            }
        }
        base.OnEndDrag(eventData);
    }
    public override void OnScroll(PointerEventData data)
    {
        if(m_Parent){
            if(m_BeginDragDirection != m_Direction){
                //当前操作方向不等于滑动方向,将事件传给父对象
                ExecuteEvents.Execute(m_Parent.gameObject, data, ExecuteEvents.
                    scrollHandler);
                return;
            }
        }
        base.OnScroll(data);
    }
}
```

5.5.7　UI 模板嵌套

　　Unity 并没有提供 Prefab 的嵌套，有些 UI 的底框面板都是一样的，但是每个界面都得单独做一遍，后面改起来非常麻烦。其实可以给某个对象设置不被保存状态，这样当 Prefab 被应用的时候，并不会被保存在自身的 Prefab 中，使用中采取运行时读取的方式，这样 Prefab 的状态就不会被嵌套破坏。如图 5-40 所示，我们给按钮设置了不可被保存状态，当界面被保存的时候，界面的 Prefab 并不会引用按钮。

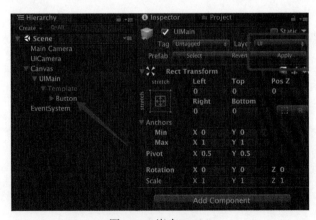

图 5-40　嵌套 Prefab

如代码清单 5-21 所示，我们在脚本中添加 `Reset()` 方法，它表示绑定此脚本时，设置下面所有对象不被保存。

代码清单 5-21　Template.cs 文件

```
using System.Collections;
using System.Collections.Generic;
using UnityEngine;

public class Template : MonoBehaviour {

    //绑定在某个脚本上，设置它不可被保存
    void Reset()
    {
        foreach(var item in GetComponentsInChildren<RectTransform>(true))
        {
            //过滤掉自身
            if(item.gameObject != gameObject) {
                item.gameObject.hideFlags = HideFlags.DontSave;
            }
        }
    }
}
```

另外，应用 Prefab 的时候是可以监听事件的。凡是设置成 `HideFlags.DontSave` 的 `GameObject`，我们建议保存 Prefab 时再清空一下。下面的代码用来监听保存事件：

```
[InitializeOnLoadMethod]
static void OnInitializeOnLoadMethod() {

    //监听 Prefab 被保存的事件
    PrefabUtility.prefabInstanceUpdated = delegate (GameObject go){
        Debug.LogFormat("path:{0}",AssetDatabase.GetAssetPath(PrefabUtility.
            GetPrefabParent(go)));
    };
}
```

在静态方法的上方标记 `[InitializeOnLoadMethod]`，表示此方法会在初始化中执行。接着，调用 `PrefabUtility.prefabInstanceUpdated` 就可以监听 Prefab 被保存的事件了。

5.5.8　UI 特效与界面分离

Unity 的特效最好不要直接挂在 Prefab 上，因为特效文件依赖的关系非常复杂，一旦挂在 Prefab 上，以后在构建 AssetBundle 的时候可能产生很多冗余。代码中直接加载是最好的，但是为了预览方便，可以使用上述模板嵌套的方式。如图 5-41 所示，将 Resources 目录下的资源拖入 Res Preview 脚本中来序列化加载路径，运行时自动加载出来。为了预览方便，拖入后需要在 `OnEnable()` 方法中自动加载出来。

158 第 5 章 UGUI 游戏界面

图 5-41 特效 Prefab

如代码清单 5-22 所示,我们封装了 Load()方法以在 Awake()中调用,它用于运行时创建特效。

代码清单 5-22 ResPreview.cs 文件

```
using UnityEngine;
public class ResPreview : MonoBehaviour
{

    ///--------序列化信息面板--------
    [SerializeField]
    private string m_LoadPath = string.Empty;
    [SerializeField]
    private bool m_IsInitLoad = true;


    ///--------序列化信息面板--------
    ///<summary>
    ///资源是否已经完成加载
    ///</summary>
    public bool IsLoad { get; private set;}

    void Awake()
    {
        IsLoad = false;
        if(m_IsInitLoad) {
            Load();
        }
    }

    ///<summary>
    ///加载资源
    ///</summary>
    public void Load()
    {
        GameObject prefab = Resources.Load<GameObject>(m_LoadPath);
```

```
            if(prefab) {
                GameObject go = Instantiate<GameObject>(prefab);
                go.transform.SetParent(transform, false);
                go.name = prefab.name;
#if UNITY_EDITOR
                foreach(Transform t in go.GetComponentsInChildren<Transform>()) {
                    t.gameObject.hideFlags = HideFlags.NotEditable | HideFlags.DontSave;
                }
#endif
            }

            IsLoad = true;
        }
    }
```

如代码清单 5-23 所示,将特效资源拖入 Inspector 面板中时,只序列化保存它的路径,并且实例化特效到 Hierarchy 视图中来预览即可。

代码清单 5-23　ResPreviewInspector.cs 文件

```
using UnityEngine;
using UnityEditor;
using System.IO;


[CanEditMultipleObjects]
[CustomEditor(typeof(ResPreview))]
public class ResPreviewInspector : Editor
{

    [MenuItem("GameObject/ResPreview",false,12)]
    static void LoadResPreview()
    {
        ResPreview respreview = new GameObject("ResPreview").AddComponent<ResPreview>();
        respreview.transform.SetParent(Selection.activeTransform, false);
        Selection.activeTransform = respreview.transform;
    }


    void OnEnable()
    {
        if(target != null) {
            if(!Application.isPlaying) {
                ClearHierarchy();
                (target as ResPreview).Load();
            }
        }
    }


    public override void OnInspectorGUI()
    {
        serializedObject.Update();
```

```csharp
            GUILayout.Label("请把Resources目录下的Prefab拖入");
            string loadPath = serializedObject.FindProperty("m_LoadPath").stringValue;
            EditorGUI.BeginChangeCheck();
            GameObject prefab = Resources.Load<GameObject>(loadPath);
            GameObject newPrefab =  EditorGUILayout.ObjectField("Prefab", prefab,
                typeof(GameObject)) as GameObject;
            if(EditorGUI.EndChangeCheck()) {
                string resPath;
                bool isResFolder = IsResourcesFolder(newPrefab, out resPath);
                if(!isResFolder){
                    EditorUtility.DisplayDialog("提示", "必须拖曳Resources目录下的Prefab",
                        "知道了");
                    ClearHierarchy();
                }
                serializedObject.FindProperty("m_LoadPath").stringValue = resPath;

                if(isResFolder) {
                    serializedObject.ApplyModifiedProperties();
                    if(!Application.isPlaying) {
                        (target as ResPreview).Load();
                    }
                }
            }

            serializedObject.FindProperty("m_IsInitLoad") .boolValue = EditorGUILayout.
                Toggle("是否Awake加载",serializedObject.FindProperty("m_IsInitLoad").
                boolValue);
            GUILayout.Space(18f);
            GUILayout.BeginHorizontal();
            if(GUILayout.Button("Refresh",GUILayout.Width(80))){
                (target as ResPreview).Load();
            }
            if(GUILayout.Button("Clear",GUILayout.Width(80))){
                ClearHierarchy();
            }
            GUILayout.EndHorizontal();

            serializedObject.ApplyModifiedProperties();

        }




        protected bool IsResourcesFolder(Object o,out string resPath)
        {
            if(o) {
                string path = AssetDatabase.GetAssetPath(o);
                bool beFirst = true;
                string tmp=string.Empty;
                DirectoryInfo dir = new DirectoryInfo(path);
                while(dir != null) {
                    if(dir.Name == "Resources") {
                        resPath = tmp;
```

```
                return true;
            }
            tmp= tmp.Insert(0,beFirst? Path.GetFileNameWithoutExtension
                (dir.Name):dir.Name+"/");
            if(beFirst)beFirst = false;
            dir = dir.Parent;
        }
    }
    resPath = string.Empty;
    return false;
}


private void ClearHierarchy()
{
    Transform transform = (target as ResPreview).transform;
    if(transform != null)
    {
        while(transform.childCount > 0)
        {
            DestroyImmediate(transform.GetChild(0).gameObject);
        }
    }
}
```

5.5.9 输入事件

UGUI 提供了 `InputField` 类来管理输入事件。如图 5-42 所示，监听输入文字的变化，当出现 a 时，将其替换成*号，并且更新输入字符的数量，相关代码如代码清单 5-24 所示。其中，`onValueChanged` 用于监听输入后的事件，`onValidateInput` 用于精准监听每次输入的字符。

图 5-42　输入

代码清单 5-24　Script_05_20.cs 文件

```
using UnityEngine;
using UnityEngine.UI;

public class Script_05_20 : MonoBehaviour {
    public InputField inputField;
    public Text tips;
    void Start() {
        //监听输入事件
        inputField.onValueChanged.AddListener((string content) => {
            tips.text = string.Format("已经输入{0}个字符", content.Length);
        });

        //监听输入文字的变化,当出现a时,将其替换成*号
        inputField.onValidateInput += delegate(string input, int charIndex,
            char addedChar)
        {
            if(addedChar == 'a'){
                addedChar = '*';
            }
            return addedChar;
        };
    }
}
```

5.5.10　按钮不规则点击区域

按钮通常是矩形的,但是有时候需要它响应不规则点击区域,比如地图块的交界处。如图 5-43 所示,首先将按钮 Image 和 Text 组件的 Raycast Target 取消勾选,接着在最下面添加一个 UIPolygon 组件,然后点击 Edit Collider 即可编辑它的点击区域了。

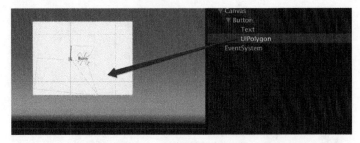

图 5-43　不规则点击区域

如代码清单 5-25 所示,首先利用 `PolygonCollider2D` 的编辑功能确定不规则点击区域,接着调用 `OverlapPoint()` 方法判断是否点击在不规则区域中。

代码清单 5-25　UIPolygon.cs 文件

```
using UnityEngine;
using UnityEngine.UI;
#if UNITY_EDITOR
```

```csharp
using UnityEditor;
#endif
[RequireComponent(typeof(PolygonCollider2D))]
public class UIPolygon : Image
{
    private PolygonCollider2D _polygon = null;
    private PolygonCollider2D polygon
    {
        get{
            if(_polygon == null)
                _polygon = GetComponent<PolygonCollider2D>();
            return _polygon;
        }
    }
    //设置只响应点击,不进行渲染
    protected UIPolygon()
    {
        useLegacyMeshGeneration = true;
    }
    protected override void OnPopulateMesh(VertexHelper vh)
    {
        vh.Clear();
    }
    public override bool IsRaycastLocationValid(Vector2 screenPoint, Camera eventCamera)
    {
        return polygon.OverlapPoint(eventCamera.ScreenToWorldPoint(screenPoint));
    }

#if UNITY_EDITOR
    protected override void Reset()
    {
        //重置不规则区域
        base.Reset();
        transform.position = Vector3.zero;
        float w = (rectTransform.sizeDelta.x *0.5f) + 0.1f;
        float h = (rectTransform.sizeDelta.y *0.5f) + 0.1f;
        polygon.points = new Vector2[]
        {
            new Vector2(-w,-h),
            new Vector2(w,-h),
            new Vector2(w,h),
            new Vector2(-w,h)
        };
    }
#endif
}
#if UNITY_EDITOR
[CustomEditor(typeof(UIPolygon), true)]
public class UIPolygonInspector : Editor
{
    public override void OnInspectorGUI()
    {
        //什么都不写,用于隐藏面板的显示
    }
}
#endif
```

上述代码利用 PolygonCollider2D 组件先将不规则区域设置好，接着在 IsRaycastLocation-Valid() 方法中判断是否点击在不规则区域中。

5.5.11 更换默认 Shader

UI 元素不需要指定材质但是却能显示正确，其原因是 UGUI 提供了默认材质以及 Shader。默认是可以兼容任何复杂情况，但是未必是最优方案。例如，大部分 UI 是不需要渲染背面的，并且不需要设置颜色。在 Shader 中删除 Cull Off 表示剔除 UI 背景。而在 Shader 中不再乘以颜色：

```
half4 color = (tex2D(_MainTex, IN.texcoord) + _TextureSampleAdd);//* IN.color;
```

如果修改的 Shader 还需要绑定到新材质上并赋予每个 UI 元素，那就太麻烦了，我们希望改掉默认的 Shader。如图 5-44 所示，在 Unity 官网上可以下载到所有默认的 Shader，然后将其拖入项目中修改，最终绑定到 Always Included Shaders 处即可。以后所有用到 UI 的地方都将默认开启背景剔除，不再乘以颜色。另外，如果有些 UI 确实需要显示背景或者调节颜色，就单独写一个 Shader 赋予它。毕竟这样的需求是少数。

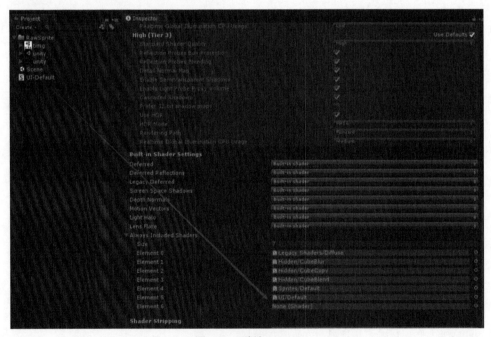

图 5-44 默认 Shader

5.5.12 小地图优化

小地图一般由美术人员提供一张比较大的图片，程序使用 Mask 裁切并将其显示在屏幕的左上角或右上角。虽然显示上裁切掉了，但是还会造成额外的渲染开销。其实，修改图片的 UV Rect

同样可以达到裁切的效果。如图 5-45 所示，当角色移动时，动态修改 UV 的区域，这样就可以保证屏幕上只渲染一块很小的地图空间了。

图 5-45　UV Rect

5.5.13　查看 UGUI 源码

除了 `Canvas` 和 `RectTransform` 这两个非常核心的类以外，UGUI 的源码（源码地址：https://bitbucket.org/Unity-Technologies/ui）都是开源的，编辑模式和运行模式下的代码都由 C# 编写完成，源码都托管在 Bitbucket 上。多阅读它的代码，对我们了解其内部工作原理有非常大的帮助。

5.6　小结

本章中，我们重点学习了 UGUI 游戏界面，其中包括所有 UI 基础元素的创建以及用法。UGUI 提供了强大的事件系统，可以很方便地监听点击、滑动和抬起等事件。多事件的管理和分发可以使用 Unity 提供的 UnityAction 和 UnityEvent。在优化方面，RaycastTarget 组件要尽可能地从 Text 和 Image 中去掉本章还学习了分辨率自适应的方法、强大的 Canvas 组件和 UI 的布局，以及 UI 的排序和多 Canvas 的嵌套，还有图集优化和管理等。最后，我们学习了 UI 的置灰、粒子的排序以及裁切。

第 6 章

2D 游戏开发

在第 5 章中，我们讲了 Unity 的 UI 系统，这里可能会有朋友问能否直接用 UGUI 来开发 2D 游戏。原则上是可以的，不过最好不要那么做。UGUI 最大的特点是利用 Canvas 画布动态合并 Mesh，来保证 UI 的渲染效率。对于 2D 游戏，假设控制角色移动，如果角色本身是 UGUI 元素，那么它一旦发生了移动，每一帧都会触发 Canvas 合并 Mesh，这势必会带来没必要的性能开销。在 2D 游戏中，Unity 提供了 Sprite Renderer 组件，它可以配合 Animator 组件来控制播放 2D 精灵动画。此外，还有强大的物理引擎和碰撞事件等。

6.1 Sprite

Sprite Renderer 是 2D 游戏开发中最基础的组件，它使用的资源是 Sprite 类型资源。前面我们讲过，Sprite 资源同样可以用在 UI 上，不过这里需要将它绑定在 Sprite Renderer 组件上。这里的 Sprite Renderer 和 UI 一样，都可以使用 Atlas 进行图集合并。关于 Atlas 合并图集的功能，第 5 章已经介绍过，这里不再赘述。如图 6-1 所示，Sprite 一共有 3 个模式可以选择。

- Single：表示它是一个单张图。
- Multiple：可以把一张图分拆成多个 Sprite。我们既可以自动分拆，也可以手动在 Sprite Editor 中编辑出每个 Sprite 的区域。
- Polygon：自定义 Sprite 的多边形形状。例如，在不需要美术人员修改的情况下，可以让它变成圆形。

Sprite 中可能有部分是透明的，如果按矩形渲染的话，会造成透明区域填充率的浪费，这时我们可以设置 Mesh Type 为 Tight，这样 Unity 会自动把这张图生成网格后再渲染。Extrude Edges 可以调节网格的数量。如图 6-2 所示，在 Scene 中选择 Wireframe 渲染方式，即可看到网格的信息。

图 6-1　Sprite 资源

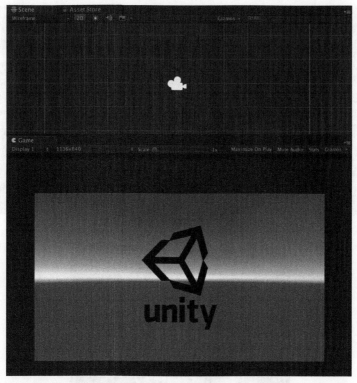

图 6-2　Sprite 网格

6.1.1　2D 摄像机与分辨率自适应

首先，我们需要确认开发分辨率。前面讲过，以移动平台为例，主流的分辨率比例是 16∶9，我们暂定开发分辨率是 1136×640。如图 6-3 所示，先设置 Orthographic 正交摄像机，这里面 Size 的含义是屏幕的一半，也就是 640/2 = 320。由于 Sprite 默认的 Pixels Per Unit 设置的是 100，所以 320/100 = 3.2 了。

图 6-3　摄像机参数

如果 Unity 当前分辨率大于开发分辨率，它会自动缩放。但是如果当前分辨率小于开发分辨率，就需要我们手动处理了。如图 6-4 和图 6-5 所示，动态调节 Free Aspect 来自适应分辨率，保证显示区域在屏幕区域中。

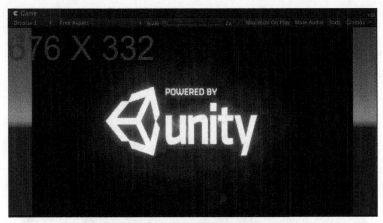

图 6-4 分辨率自适应

图 6-5 分辨率自适应

如代码清单 6-1 所示，首先获取当前分辨率，并且计算出它与开发分辨率的比例系数，最终设置 orthographicSize 即可。

代码清单 6-1　Script_06_01.cs 文件

```
using System.Collections;
using System.Collections.Generic;
using UnityEngine;

public class Script_06_01 : MonoBehaviour {

    public Camera camera;

    void Update()
    {
```

```
        float designWidth = 1136f;//开发时分辨率中的宽度
        float designHeight = 640f;//开发时分辨率中的高度
        float designOrthographicSize=3.2f;//开发时正交摄像机的大小
        float designScale  =  designWidth/designHeight;
        float scaleRate  =  (float)Screen.width/(float)Screen.height;
        if(scaleRate<designScale)
        {
            float scale = scaleRate / designScale;
            camera.orthographicSize = 3.2f / scale;
        }else{
            camera.orthographicSize = 3.2f;
        }
    }

    void OnGUI()
    {
        GUILayout.Label(string.Format("<size=60><color=red>{0} X {1} </color></size>",
            Screen.width, Screen.height));
    }
}
```

这里整体自适应摄像机的区域，开发中我们也可以选择不自适应摄像机，而选择自适应相机中的元素。

6.1.2 Sprite Renderer 排序

如图 6-6 所示，单张图可以使用 Order in Layer 来排序，但是游戏角色可能不一定由一张图组成，比如脑袋、胳膊、腿特效由好几部分组成，此时就需要使用 Sorting Group 组件了。它可以同时生效游戏对象节点下的所有图片，并且保持它们是同一个 Sorting Order。另外，排序始终是先看 Sorting Layer，然后才看 Sorting Order 的。在 Mask Interaction 下拉框中可以选择裁切区域，裁切区域在它的外面或者在它的里面。

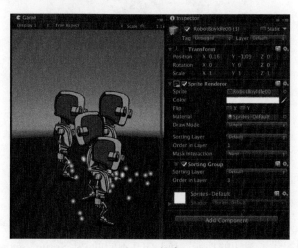

图 6-6　排序

2D 横版卷轴的游戏人物是可以上下移动的，比如屏幕中心位置有一棵树，人物在树上面，树应该挡住人，如果人物在树下面，就应该人挡住树，所以需要动态调节人或树的 Order in Layer。Unity 提供了一套自动计算深度的方案，如图 6-7 所示，在 GraphicsSettings 面板中，设置 Custom Axis 后，将 Transparency Sort Axis 的 Y 轴设置成 1，表示 Sprite Renderer 的 Y 轴坐标相差大于 1 时，自动重新设置与目标对象的排序。

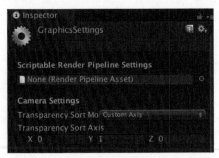

图 6-7　排序轴向

注意，如果需要自动排序，那么参与排序的所有 Sprite Renderer 的深度必须是相同的。如图 6-8 所示，在 Hierarchy 视图中改变人物的 Y 轴坐标，它已经自动调整深度了。

图 6-8　自动排序

本例的整个工程可以参考 CodeList_06_02（详见本书源码）。

6.1.3　裁切

Sprite Renderer 也提供了裁切的功能。和 UGUI 不同的是，Sprite Renderer 中不需要被裁切的元素必须挂在裁切组件下面。如图 6-9 所示，Sprite Mask 可以设置裁切的样板图，Custom Range 用于设置裁切的区域，只有 Front 和 Back 之前的 Sprite Renderer 组件才可以被裁切。

图 6-9　裁切

注意：Sprite Mask 是不能裁切带特效的 Sprite Renderer 的。如果要裁切的话，可以参考 5.5.4 节，将裁切区域传入 Shader，然后处理它。

本例的整个工程可以参考 CodeList_06_03。

6.2　Sprite 动画

Unity 提供的动画系统非常强大。2D 游戏和 3D 游戏中的动画控制原理是相同的，而且都可以在 Unity 中编辑动画，不过只能编辑一些简易的动画。由于动画编辑器没有骨骼信息，它只能编辑每一帧节点的 Transform 信息，只能整体旋转、缩放和平移动画，却不能改变自身的属性。例如，飘带一类的自身可以随风摇摆，这种效果 Unity 编辑器就做不了了。如果对 2D 动画有更高的要求，可以尝试用 Spine 或者 Unity 收购的插件 Anima2D（目前已免费）。

6.2.1　创建 2D 动画

在 Project 视图中全选需要创建的 Sprite，如果是首次创建动画，可直接拖入 Hierarchy 视图中，它会自动提示创建动画以及动画系统需要用到的 AnimationController 资源。后面如果只添加新动画，可以单击鼠标右键，在弹出的快捷菜单中选择 Create→Animation 命令，即可将选中的 Sprite 全部创建成动画，如图 6-10 所示。这是按 Sprite 顺序来的一组标准帧动画。如果为了节省图片的内存，或者考虑换装功能，也可以把人物的脑袋、身体、胳膊、腿和武器等都拆成散图，此时就需要使用编辑器将每一帧的散图组合拼接在一起，从而实现最终的动画效果。

172　第 6 章　2D 游戏开发

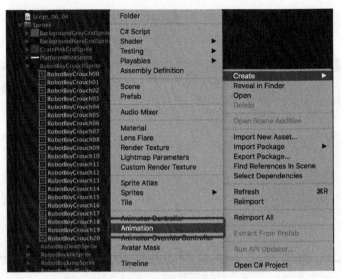

图 6-10　创建动画

在 Hierarchy 视图中选择刚刚创建的 Sprite Renderer 对象，它自动绑定上了 Animator 组件，双击打开它的 AnimationController，默认的动画已经添加进去了。此时选中它的同时，在导航菜单栏中选择 Windows→Animation 命令即可弹出编辑窗口。

如图 6-11 所示，打开动画文件后即可看到帧序列信息，拖动时间线可以编辑每一帧的信息。如果每一帧是由多张 Sprite 拼合而成的，可以再添加 Transform，编辑每个 Sprite 的旋转、缩放和平移信息，编辑完毕后保存即可。

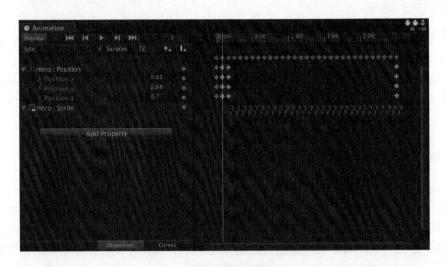

图 6-11　编辑动画

6.2.2 2D 动画控制器

动画已经做出来了，但还缺少一个东西去管理动画。Unity 提供了一套非常强大的动画控制工具，那就是 AnimationController，它可以编辑动画与动画之间的行为信息。作为一个强大的状态机，它还提供了状态层、子状态机以及混合树等功能，后面讲 3D 部分的时候会重点向大家介绍。

首先，我们将刚刚做好的 3 个动画拖入状态机中，如图 6-12 所示，其中 Entry 表示状态机的入口（无法删除），将它连线到 Idle 表示首先播放 Idle 动画。Any State 表示任意状态，例如角色死亡了，无论当前处于什么状态，都可立即进入死亡状态。任何状态都可以直接切换到 Any State，重新开始新的状态。Exit 表示退出当前状态机，如果有子状态机，则表示返回上一层状态机。

下面我们通过一个例子来介绍。首先，让角色进入 Idle 状态，按下 WSAD 键控制它移动，按下空格键让角色进入跳跃状态。在没有按键响应的时候，Idle 应该是需要循环播放的，我们在 Idle 处连线并将其指向自己。在左边的 Paramerters 面板中，可以添加变量条件，类型包括 int、bool、floot 和 trigger，Idle 到 Run 的连线就可以配置这个变量，达成条件后即可切换状态，播放 Run 动画。运行游戏后，可以直接在表面中设置变量，方便编辑者查看效果。

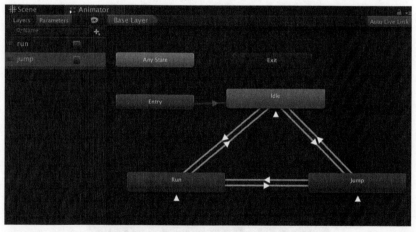

图 6-12 状态机

用鼠标单击一下 Idle 到 Run 的连线，即可出现状态机条件面板，如图 6-13 所示，其中 Has Exit Time 表示切换状态的时候是否需要等上一个动画播放完过度动画，过度动画的时候在 Settings 下拉菜单中配置。在最下面的 Conditions 处，可以添加条件参数，这里面的 run 就是上面我们配置的 bool 变量，表示这个值等于 true 的时候切换动画。

另外，还要注意 Transitions 菜单下的 Solo 和 Mute 复选框。由于状态机编辑到后面会越来越复杂，不排除需要一些特殊的跳转条件。如果选中 Solo 复选框，表示即使当前别的条件达成了，也只能进 Solo 的状态；如果选中 Mute 复选框，表示即使当前条件达成了，也不能进入 Mute 的状态。

图 6-13　状态机条件

如图 6-14 所示，状态机编辑完毕后，我们即可通过代码来动态设置状态的满足条件，从而实现人物移动以及跳跃。

图 6-14　控制角色

如代码清单 6-2 所示，在 Update() 中判断按键方向，按下空格键时让角色跳起来。控制状态切换就是运行时对 jump 和 run 这两个 bool 参数赋值。

代码清单 6-2　Script_06_04.cs 文件

```
using UnityEngine;

public class Script_06_04 : MonoBehaviour
{
```

```csharp
public Animator animator;
public SpriteRenderer heroRenderer;
private State m_State = State.Idle;

void Awake()
{
    Idle();
}

void Update()
{

    float y = Input.GetAxis("Vertical");
    float x = Input.GetAxis("Horizontal");

    //纵向移动
    if(y > 0)
    {
        Run(Direction.Up, Vector3.up);
    }else if(y < 0)
    {
        Run(Direction.Down, Vector3.down);
    }
    //横向移动
    if(x > 0)
    {
        Run(Direction.Right, Vector3.right);
    }else if(x < 0)
    {
        Run(Direction.Left, Vector3.left);
    }

    //奔跑状态中，松手后回归待机状态
    if(m_State == State.Run)
    {
        if(Mathf.Approximately(x, 0f) && Mathf.Approximately(y, 0f))
        {
            Idle();
        }
    }

    //处理跳跃
    if(Input.GetKey(KeyCode.Space)) {
        Jump();
    }else if(Input.GetKeyUp(KeyCode.Space)) {
        Idle();
    }

}

void Idle()
{
    SetState(State.Idle);
}
```

```
void Run(Direction dir,Vector3 postion)
{
    SetState(State.Run);

    heroRenderer.flipX = (dir == Direction.Left);
    heroRenderer.transform.position += (postion * 0.1f);
}

void Jump()
{
    SetState(State.Jump);
}

void SetState(State newState)
{
    m_State = newState;
    animator.SetBool("run", m_State == State.Run);
    animator.SetBool("jump", m_State == State.Jump);
}

enum State
{
    Idle=1,
    Run,
    Jump,
}

enum Direction
{
    None=1,
    Up,
    Down,
    Left,
    Right
}
}
```

6.3 Tile 地图

Unity 提供全套的 Tile 工具来编辑 2D 地图。使用 Tile 编辑地图，可以极大地减少内存占用。美术人员只需要提供地图小格式元素，并且保证可以相互拼接接口，策划人员最终将这些小格子元素编辑拼在一起。灵活的拼接可以生成很多不同的 2D 地图。

6.3.1 创建 Tile

创建 Tile 之前，需要和美术人员确认每个 Tile 的大小，一般是 32 或 64。如图 6-15 所示，因为资源的大小是 32，所以这里的 Pixels Per Unit 需要设置成 32。每个 Tile 还是建议美术人员出单张的散图，这样方便以后单独修改，拿到 Unity 后可以将它们重新合并成一张图集。

6.3 Tile 地图

图 6-15　Tile 资源

接着，在 Hierarchy 视图中创建 TileMap 游戏对象。选择 Create→2D Object→ TileMap 命令，此时场景中将出现一个 Grid 对象并且它下面套着 TileMap 对象。如图 6-16 所示，Grid 就是 Tile 的画布了，它可以设置画布每个单元的大小。由于前面 Pixels Per Unit 我们已经设置成 32，所以这里默认填 1 就可以了。Cell Gap 可以设置每个单元之间的间距，一般都设置成 0 就好。最后的 Cell Swizzle 用于设置 Grid 布局的朝向，而 XYZ 表示斜角为 45° 的 2D 游戏。

图 6-16　Grid 画布

Grid 对象下面就是 Tilemap 了。如图 6-17 所示，它可以设置动画帧率、锚点和朝向等信息。下面的 Tilemap Renderer 组件和 Sprite Renderer 很像，可以设置排序和遮罩等。Tile 的排序也很重要，后面我们会详细说明。

图 6-17　Tilemap

6.3.2 Tile Palette

接着，需要将可编辑的 Sprite 汇总在 Tile Palette 面板中，此时在导航菜单栏中选择 Window→Tile Palette 命令即可。如图 6-18 所示，Create New Palette 可以创建多个调色板。可能游戏中每个场景风格不一样，多 Palette 面板就可以把相同风格的 Tile 汇总在一起，最后将需要编辑的 Sprite 资源直接拖入这个面板就可以自动生成 Tile 了。同一场景中可能会有多个 Tilemap，例如背景层和前景层，在 Active Tilemap 下拉框中选择当前场景创建的所有 Tilemap，方便相互切换编辑。调色板的元素也是可以编辑的，单击右上角的 Edit 按钮即可。如果想删掉一些不想要的 Tile，可以点击上方的橡皮擦来删除它，最后按 Command+S 键保存即可。

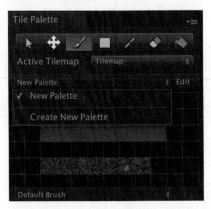

图 6-18　创建 Palette

6.3.3 编辑 Tile

准备工作已经就绪，下面开始编辑 Tile。善用 Tile Palette 工具栏，可更快捷地编辑 Tile。如图 6-19 所示，Tile Palette 工具栏中，工具从左到右依次如下。

- 点选工具：可以选择某一个 Tile。
- 移动工具：当使用点击工具选择一个 Tile 时，可使用移动工具移动它的位置。
- 画笔工具：选择一个 Tile，即可在 Scene 视图中刷它的区域了。
- 区域工具：按下鼠标左键可同时刷多个区域。
- 吸图工具：吸取 Scene 中某个 Tile 的图，方便下次使用新吸取的图来编辑。
- 橡皮工具：Scene 中删除掉不需要的 Tile。
- 批量填充工具：可以大规模填充 Tile。

图 6-19　Tile Palette 工具栏

如图 6-20 所示，在 Tile Palette 中选择一个 Tile 元素，即可在 Scene 视图中拖动鼠标来刷它的区域了。

图 6-20　编辑 Tile

本例的整个工程可以参考 CodeList_06_05。

6.3.4　多 Tile 编辑与排序

在普通的 2D 游戏场景中，至少是需要 2 个 Tile 层，我们称它为前景层和后景层。比如一棵树，上半部分是树叶（前景层），下半部分是树干（后景层），这样人物在这棵树附近上下移动的时候，走到树下面应该自己要挡住树干，走到树上面应该被树叶挡住自己。

如图 6-21 所示，在 Hierarchy 视图中可以创建多个 Tilemap，在 Tile Palette 面板中可以切换当前需要编辑的 Tilemap，在 Scene 视图下面可以选择 Focus On Tilemap，这样就可以凸出显示正在编辑的 Tilemap 了。

图 6-21 多 Tile 层

要排序的话，可以设置 Order in Layer。将 Tilemap 背景层设置成 0，人物 SpriteRenderer 设置成 1，前景层设置成 2，如图 6-22 所示，这样控制人物在屏幕中移动时，它就会渲染在这两个层之间了。

图 6-22 排序

本例中的整个工程可以参考 CodeList_06_06。

6.3.5 拓展 Tile Palette

Unity 提供了 Tile Palette 的拓展编辑功能，可以重写画笔的任意行为。如图 6-23 所示，可以添加画笔类型，新的画笔要继承 GridBrush，如果画笔需要复用，还支持可添加配置参数的画笔。

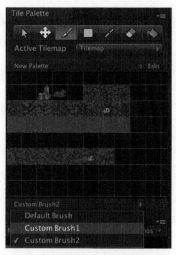

图 6-23 拓展画笔

如图 6-24 所示，Editor 目录下的 CustomBrush 就是重写的画笔类。如果需要配置多参数的话，可以创建 Custom Brush1 和 Custom Brush2，它们就是自定义的画笔类，最后将不同的信息保存进去。

图 6-24 配置画笔

如图 6-25 所示，我们来重写画笔的绘制方法，在 Scene 视图中用鼠标拖动 Tile 的同时，将 Tile 所在格子的坐标信息绘制出来。

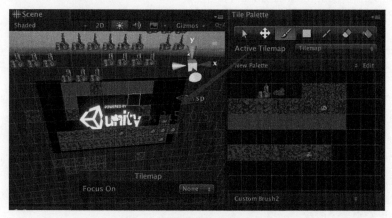

图 6-25 重写画笔的绘制方法

如代码清单 6-3 所示，在自定义的画笔类中重写 Paint() 方法，就可以监听笔刷的事件了。在 OnPaintSceneGUI() 方法中可以重写绘制方法，这里将坐标信息显示在屏幕中。

代码清单 6-3　CustomBrush.cs 文件

```csharp
using UnityEngine;
using UnityEditor;
using System.Collections.Generic;
using System;
using UnityEngine.Tilemaps;

[CustomGridBrush(false,true,false, "Custom Brush")]
public class CustomBrush : GridBrush {
    //序列化数据
    public string name;
    public override void Paint(GridLayout grid, GameObject brushTarget, Vector3Int
        position)
    {
        if(EditorUtility.DisplayDialog("重要提示", string .Format("确认笔刷: {0} {1}",
            grid.name,position), "ok")) {
            base.Paint(grid, brushTarget, position);
        }
    }

    [MenuItem("Assets/Create/CustomBrush")]
    public static void CreateCustomBrush()
    {
        string path = EditorUtility.SaveFilePanelInProject("Save CustomBrush",
            "New CustomBrush", "Asset", "Save CustomBrush", "Assets");
        if(path == "")
            return;
        AssetDatabase.CreateAsset(ScriptableObject.CreateInstance<CustomBrush>(),
            path);
    }
}
[CustomEditor(typeof(CustomBrush))]
public class CustomBrushEditor : GridBrushEditor
{
    protected override void OnEnable()
    {
        base.OnEnable();
        //获取序列化的信息
        Debug.Log((target as CustomBrush).name);
    }
    public override void OnPaintSceneGUI(GridLayout gridLayout, GameObject brushTarget,
        BoundsInt position, GridBrushBase.Tool tool, bool executing)
    {
        base.OnPaintSceneGUI(gridLayout, brushTarget, position, tool, executing);
        Handles.color = Color.red;
        GUIStyle style = new GUIStyle();
        style.normal.textColor = Color.red;
        style.fontSize = 20;
        Handles.Label(gridLayout.CellToWorld(new Vector3Int(position.x, position.y,
```

```
0)),position.center.ToString(),style);
    }
}
```

Handles 类是 Unity 提供的渲染辅助类，Handles.Label() 表示在场景中渲染出一个文本信息。

6.3.6 拓展 Tile

Tile 实际上就是一个序列化的 assets 资源文件，默认的 Tile 里面记录着它引用的 Sprite、颜色和碰撞类型等，如果不满足默认的需求，我们还可以继承它并且重写 Tile 基类的一些重要回调方法。

如代码清单 6-4 所示，继承 Tile 类后，就可以重写 RefreshTile()、GetTileData()、GetTileAnimationData() 和 StartUp() 父类方法，来刷新 Tile、获取 Tile、获取 Tile 的动画数据，以及启动时调用。

代码清单 6-4 CustomBrush.cs 文件

```csharp
using UnityEngine;
using System.Collections;
using UnityEngine.Tilemaps;

#if UNITY_EDITOR
using UnityEditor;
#endif

public class CustomTile : Tile
{
    //需要刷新某个Tile时调用
    public override void RefreshTile(Vector3Int location, ITilemap tilemap)
    {
        base.RefreshTile(location, tilemap);
    }
    //需要获取某个TileData时调用
    public override void GetTileData(Vector3Int location, ITilemap tilemap,
        ref TileData tileData)
    {
        base.GetTileData(location, tilemap, ref tileData);
    }
    //获取Tile动画数据时调用
    public override bool GetTileAnimationData(Vector3Int position, ITilemap tilemap,
        ref TileAnimationData tileAnimationData)
    {
        return base.GetTileAnimationData(position, tilemap, ref tileAnimationData);
    }
    //启动首次调用
    public override bool StartUp(Vector3Int position, ITilemap tilemap, GameObject go)
    {
        return base.StartUp(position, tilemap, go);
```

```
}
#if UNITY_EDITOR
[MenuItem("Assets/Create/CustomTile")]
public static void CreateRoadTile()
{
    string path = EditorUtility.SaveFilePanelInProject("Save Custom Tile",
        "New Custom Tile", "Asset", "Save Custom Tile", "Assets");
    if(path == "")
        return;
    AssetDatabase.CreateAsset(ScriptableObject.CreateInstance<CustomTile>(),
        path);
}
#endif
}
```

Tile 属于 assets 资源文件，所以需要使用 `AssetDatabase.CreateAsset()` 来创建它。另外，`ScriptableObject.CreateInstance<CustomTile>()` 用于创建泛型中的数据对象。

6.3.7 更新 Tile

更新 Tile 就是调用 `tilemap.GetTile()` 和 `tilemap.SetTile()` 方法。和 2D 精灵一样，Tile 的坐标原点也是屏幕的正中心点。如图 6-26 所示，控制人物在屏幕中移动来"吃豆豆"，它的原理就是将人物的 3D 坐标转换成 Tile 的坐标信息，获取到当前所在的 Tile，最终更新它。相关代码如代码清单 6-5 所示。

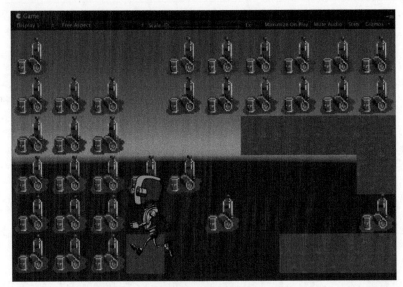

图 6-26　更新 Tile

当移动到"瓶子"位置时，调用 `tilemap.SetTile()` 方法，将"瓶子"设置成空的内容，即可"吃掉它"。

6.3 Tile 地图

代码清单 6-5　Script_06_09.cs 文件

```csharp
using UnityEngine;
using UnityEngine.Tilemaps;

public class Script_06_09 : MonoBehaviour
{
    public SpriteRenderer heroRenderer;
    public Tilemap tilemap;
    void Update()
    {
        //处理方向键
        float y = Input.GetAxis("Vertical");
        float x = Input.GetAxis("Horizontal");
        //纵向移动
        if(y > 0)
        {
            Run(Vector3.up);
        }
        else if(y < 0)
        {
            Run(Vector3.down);
        }
        //横向移动
        if(x > 0)
        {
            Run(Vector3.right, false);
        }
        else if(x < 0)
        {
            Run(Vector3.left, true);
        }

    }
    void Run(Vector3 postion,bool flipx = false)
    {
        //控制人物左右移动时镜像
        heroRenderer.flipX = flipx;
        heroRenderer.transform.position += (postion * 0.1f);

        //通过人物的坐标换算出 Tile 的位置
        Vector3Int cellpos = tilemap.WorldToCell(heroRenderer.transform.position);

        //删除 tile
        if(tilemap.GetTile(cellpos) != null) {
            tilemap.SetTile(cellpos, null);
        }

    }
}
```

在上述代码中，`GetTile()` 和 `SetTile()` 分别用于获取和设置每个 Tile。当给它们传入 null 时，表示删除这个 Tile 显示的内容。

6.4 2D 碰撞检测

Unity 2D 和 3D 的物理引擎是基于 PhysX 的，内置的碰撞检测也是基于 PhysX 的。PhysX 物理引擎是非常强大的，它可以模拟很真实的物理效果。但是作为游戏来说，太过真实的物理效果反而让它变得很假，游戏需要的是可配置性的"物理效果"，例如按帧或者时间线的方式来编辑产生类似物理的效果，所以目前大量的游戏几乎都是不使用物理引擎的。在 Unity 中可以关闭物理效果，只用它的碰撞功能，或者整体的碰撞功能都自己编写代码来完成。

6.4.1 Collider 2D

任何的碰撞现象都有两个载体，一个是发起碰撞的，另一个是接受碰撞的，所以我们首先要明确哪些物体是可以接受碰撞的。Collider 2D 并不依赖 Sprite 组件，就好比一个空气墙。如图 6-27 所示，普通的 Collider 2D 包括矩形、圆形、不规则边界、多边形和胶囊形。其中胶囊形一般用于主角，其他形状用于场景或者动态阻挡等。

图 6-27　普通 Collider 2D

Tilemap Collider 2D 专门用于 Tile 的碰撞体。如图 6-28 所示，但是它把中间不需要使用碰撞的也圈起来了，在它的基础上可以再添加一组 Composite Collider 2D 组件，这样就会自动将多余的碰撞区域去掉，像右边图显示的 Tile 一样。

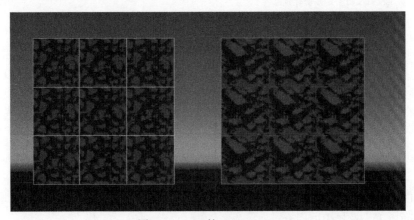

图 6-28　Tile 的 Collider 2D

6.4.2　Rigidbody 2D

Rigidbody 2D 表示 2D 刚体组件，它表示当前物体启动物理引擎。如果需要控制主角移动，并且会被上面介绍的 Collider 2D 组件阻挡住，就必须给它绑定上 Rigidbody 2D 组件。如图 6-29 所示，Body Type 一共有下面 3 个类型。

- Dynamic：表示动态刚体，完全模拟物理效果，碰到 Collider 2D 会被挡住，碰到任意 Rigidbody 2D 都会产生物理效果。它在空中会根据重力自动落下来，它的效率是最低的，仅适合给主角使用。
- Kinematic：运动学，它只能和选中 Dynamic 复选框的 Rigidbody 发生碰撞效果。如果需要碰撞事件，比如 `OnCollisionEnter2D()`，或者 Kinematic 与 Kinematic 碰撞，两者必须有一个选中 Use Full Kinematic Contacts 复选框。Kinematic 与 Static 碰撞，Kinematic 必须选中 Use Full Kinematic Contacts 复选框，否则碰撞事件也就没有了。Kinematic 适合做主角被攻击时的碰撞检测。比如像主角被别的物体击飞，发生击飞的物体可以设置 Kinematic。因为主角已经是 Dynamic 了，可以正常触发碰撞效果，如果使用 Kinematic 的话，效率比 Dynamic 要更好。
- Static：静态，只能和 Dynamic 发生碰撞效果，和 Kinematic 只能发生碰撞事件（需要保证 Kinematic 必须勾选 Use Full Kinematic Contact 复选框，它的效率是最高的）。

还需要注意的是，如果需要移动或者旋转带 Rigidbody 2D 的组件时，不能直接修改它的 `Transform.position`，而是要使用 `Rigidbody2D.position` 或者 `Rigidbody2D.rotation`。

图 6-29　Rigidbody 2D

如图 6-30 所示，使用方向键来控制角色移动，保证它不会掉下去。

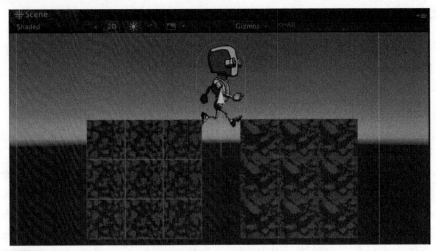

图 6-30 碰撞检测

如代码清单 6-16 所示,移动角色时不能使用 transform.position 赋值,而要采用 heroRigidbody2D.position 赋值。这样移动时,就会有碰撞效果了。

代码清单 6-16　Script_06_10.cs 文件

```
using System.Collections;
using System.Collections.Generic;
using UnityEngine;
using UnityEngine.Tilemaps;

public class Script_06_10 : MonoBehaviour
{
    public SpriteRenderer heroRenderer;
    public Rigidbody2D heroRigidbody2D;

    void Update()
    {
        //处理方向键
        if(Input.GetKey(KeyCode.W)) {
            Run(Vector2.up);
        }else if(Input.GetKey(KeyCode.S)) {
            Run(Vector2.down);
        }else if(Input.GetKey(KeyCode.A)) {
            Run(Vector2.left,true);
        }else if(Input.GetKey(KeyCode.D)) {
            Run(Vector2.right,false);
        }
    }
    void Run(Vector2 position,bool flipx = false)
    {
        //控制人物左右移动时镜像
        heroRenderer.flipX = flipx;
        //绑定 rigidbody 以后,不能再使用 transform.position 赋值
```

```
            heroRigidbody2D.position += (position * 0.1f);
        }
    }
```

由于左右移动人物图片需要翻转，所以可以设置heroRenderer.flipX来镜像图片。

6.4.3 碰撞事件

上一节中，我们讲了主角挂上 Rigidbody 2D 并且选中 Dynamic 复选框后，就可以和 Collider 2D 产生碰撞效果了。而碰撞事件和碰撞效果是两个不同的概念，碰撞事件表示 Collider 2D 被 Rigidbody 2D 碰撞后发生的事件，碰撞事件会被碰撞者和被碰撞者同时接收到。而碰撞效果好比主角是 Rigidbody 2D，空气墙是 Collider 2D，主角碰到墙以后会被墙挡住，无法继续行走。

主角碰到墙，给主角或者墙任意一方绑定脚本都可以收到事件。如果是在主角这里监听碰到什么东西，代码可以这样写。同样，如果在墙这一边想监听自己被什么东西碰到，脚本绑定在墙上就可以了，代码如下所示：

```
void OnCollisionEnter2D(Collision2D coll)
{
    Debug.LogFormat("主角开始碰到 {0} ",coll.collider.name);
}
void OnCollisionStay2D(Collision2D coll)
{
    Debug.LogFormat("主角持续碰到{0} ", coll.collider.name);
}
void OnCollisionExit2D(Collision2D coll)
{
    Debug.LogFormat("主角结束碰到 {0} ", coll.collider.name);
}
```

游戏中需要监听碰撞的事件可能比较多，并非一定要将其写在监听它的脚本中，可以将它抛出去，这样就可以在与它有关的地方统一处理。比如，将代码清单6-7所示的脚本挂在主角身上，在需要监听主角碰撞事件的地方监听即可，事件中将抛出碰撞者以及被碰撞的两个游戏对象。

代码清单 6-7　CollisionListener.cs 文件

```
using UnityEngine;
using UnityEngine.Events;

public class CollisionEvent  : UnityEvent<GameObject,GameObject>{}

public class CollisionListener : MonoBehaviour
{
    public static CollisionEvent onCollisionEnter2D = new CollisionEvent();
    public static CollisionEvent onCollisionStay2D = new CollisionEvent();
    public static CollisionEvent onCollisionExit2D = new CollisionEvent();

    //抛出事件
    void OnCollisionEnter2D(Collision2D coll)
```

```
        {
            onCollisionEnter2D.Invoke(gameObject,coll.collider.gameObject);
        }
        void OnCollisionStay2D(Collision2D coll)
        {
            onCollisionStay2D.Invoke(gameObject,coll.collider.gameObject);
        }
        void OnCollisionExit2D(Collision2D coll)
        {
            onCollisionExit2D.Invoke(gameObject,coll.collider.gameObject);
        }

}
```

游戏中大量的空气墙是不需要监听事件的,因为它仅仅起到阻挡的作用。可以考虑添加不同的 tag 来标记碰撞物体。如图 6-31 所示,在人物的左右两边各添加一堵空气墙,从而控制人物移动并且监听它的碰撞事件。

图 6-31　碰撞监听

如代码清单 6-8 所示,这里就用到了之前讲过的事件系统。由于需要监听碰撞的元素可能很多,总不能把逻辑都写在监听这里,所以需要把碰撞的事件抛出去,由外部统一来处理。

代码清单 6-8　Script_06_11.cs 文件

```
using UnityEngine;

public class Script_06_11 : MonoBehaviour
{
```

```
    public SpriteRenderer heroRenderer;
    public Rigidbody2D heroRigidbody2D;

    void Start()
    {
        CollisionListener.onCollisionEnter2D.AddListener(delegate(GameObject g1,
            GameObject g2) {
            Debug.LogFormat("{0}开始碰撞{1}",g1.name,g2.name);
        });
        CollisionListener.onCollisionStay2D.AddListener(delegate(GameObject g1,
            GameObject g2) {
            Debug.LogFormat("{0}碰撞中{1}", g1.name, g2.name);
        });
        CollisionListener.onCollisionExit2D.AddListener(delegate(GameObject g1,
            GameObject g2) {
            Debug.LogFormat("{0}结束碰撞{1}", g1.name, g2.name);
        });
    }

    void Update()
    {
        //处理方向键
        if(Input.GetKey(KeyCode.W)) {
            Run(Vector2.up);
        }else if(Input.GetKey(KeyCode.S)) {
            Run(Vector2.down);
        }else if(Input.GetKey(KeyCode.A)) {
            Run(Vector2.left,true);
        }else if(Input.GetKey(KeyCode.D)) {
            Run(Vector2.right,false);
        }
    }
    void Run(Vector2 position,bool flipx = false)
    {
        //控制人物左右移动时镜像
        heroRenderer.flipX = flipx;
        //绑定 rigidbody 以后,不能再使用 transform.position 赋值
        heroRigidbody2D.position += (position * 0.1f);
    }
}
```

其中,CollisionListener 类用来对事件进行监听,这样就可以在外部统一处理所有碰撞事件了。

6.4.4 碰撞方向

碰撞通常会有 4 个方向,跳起来脑袋碰到房顶,掉下去脚碰到地面,还有就是左右两边的碰撞了。Unity 2D 目前并没有提供方法来判断方向,但是提供了碰撞发生的坐标点,这样就可以计算碰撞方向了。如图 6-32 所示,当下面和左边同时发生碰撞时,我们将碰撞点到原点之间绘制上线。

图 6-32 碰撞方向

如代码清单 6-9 所示,使用 InverseTransformPoint()方法,可以计算两点的方向,这样就能计算出碰撞的方向了。只是由于同时碰撞的点可能有多处,例如身体正前面或者脚底板,还需要判断一下到底哪几个方向发生了碰撞。

代码清单 6-9　Script_06_12.cs 文件

```
void OnCollisionStay2D(Collision2D coll)
{
    foreach(ContactPoint2D contact in coll.contacts)
    {
        //绘制线
        Debug.DrawLine(contact.point, transform.position, Color.red);
        var direction = transform.InverseTransformPoint(contact.point);
        if(direction.x > 0f)
        {
            print("右碰撞");
        }
        if(direction.x < 0f)
        {
            print("左碰撞");
        }
        if(direction.y > 0f)
        {
            print("上碰撞");
        }
        if(direction.y < 0f)
        {
            print("下碰撞");
        }
    }
}
```

为了能更直观地看到到底哪里发生了碰撞，我们可以使用 `Debug.DrawLine()` 来绘制辅助线以查看它。

6.4.5 触发器监听

上一节中，我们讲了空气墙的碰撞，但是游戏中有些碰撞事件是不需要有碰撞效果的。比如，角色走到了一个传送点，只需要监听到触发事件就可以了。如图 6-33 所示，在 Box Collider 2D 组件中选中 Is Trigger 复选框即可。

图 6-33　触发事件

如代码清单 6-10 所示，监听方法和 CollisionListener 类类似，这里就不再赘述。

代码清单 6-10　TriggerListener.cs 文件

```
using System.Collections;
using System.Collections.Generic;
using UnityEngine;
using UnityEngine.Events;

public class TriggerEvent : UnityEvent<GameObject>{}

public class TriggerListener : MonoBehaviour
{
    public static TriggerEvent onTriggerEnter2D = new TriggerEvent();
    public static TriggerEvent onTriggerStay2D = new TriggerEvent();
    public static TriggerEvent onTriggerExit2D = new TriggerEvent();

    void OnTriggerEnter2D(Collider2D other)
    {
        onTriggerEnter2D.Invoke(gameObject);
    }
    void OnTriggerStay2D(Collider2D other)
    {
        onTriggerStay2D.Invoke(gameObject);
    }
    void OnTriggerExit2D(Collider2D other)
    {
        onTriggerExit2D.Invoke(gameObject);
    }
}
```

6.4.6 Effectors 2D

此外，Unity 还提供了一组物理效果，它可以给 Collider 2D 对象添加一种特殊的物理效果。

- **Platform Effector 2D**：一种特殊的地面，例如有些 2D 游戏有一个台子，从下面能跳上去，但是站在上面却掉不下去。
- **Surface Effector 2D**：像传输带一样带摩擦地缓慢移动。
- **Point Effector 2D**：类似炸弹一样，爆炸后可以把周围的东西向四周炸开。
- **Buoyancy Effector 2D**：模拟水中的浮力效果。
- **Area Effector 2D**：区域力。例如物体从空中掉下来，进入某个区域相互弹跳的效果。

6.4.7 优化

如果碰撞效果必须通过物理引擎，那么必须在 Rigidbody 2D 中选中 Dynamic 复选框了，这样功能虽然是最全面的，但是效率也是最低的。

另一种做法是不依赖物理引擎。换句话说，就是没有碰撞的效果（例如被墙挡住行走），仅选中运动学 Kinematic，那么只能监听到 `OnCollisionEnter2D()` 和 `OnTriggerEnter2D()` 一类的碰撞事件，无法自动处理碰撞被挡住的效果。

最后，还有一种做法，那就是文章开头提到的，完全放弃物理引擎，不使用 Rigidbody 2D 组件，碰撞效果和碰撞事件完全靠自己来计算。这么做有个好处，那就是更加灵活。不依赖物理引擎，可以极大地优化效率。其缺点就是相对更麻烦，所有算法都得靠自己来编写代码。

6.4.8 计算区域

如图 6-34 所示，获取 `SpriteRenderer` 中 4 个点的世界绝对坐标，这样就可以判断相交、重合和计算距离等。

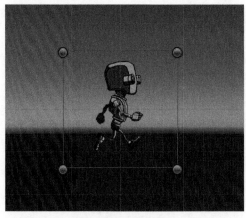

图 6-34 获取区域

如代码清单 6-11 所示，获取角色包围盒的最小点和最大点后，通过 `Debug.DrawLine()` 将它们画出来。

代码清单6-11　Script_06_14.cs文件

```
using UnityEngine;

public class Script_06_14 : MonoBehaviour
{
    public SpriteRenderer heroRenderer;

    private void Update()
    {
        Vector3 min = heroRenderer.bounds.min;
        Vector3 max = heroRenderer.bounds.max;
        //绘制4个点的连线
        Debug.DrawLine(min, new Vector3(max.x,min.y,0f),Color.red);
        Debug.DrawLine(new Vector3(max.x, min.y, 0f), max, Color.red);
        Debug.DrawLine(max, new Vector3(min.x, max.y, 0f), Color.red);
        Debug.DrawLine(new Vector3(min.x, max.y, 0f),min, Color.red);
    }
}
```

另外，原点默认在中心点，也就是 transform.position 相对整个图片的坐标。如图6-35所示，它也可以在Sprite Editor面板中编辑，很多游戏会将这个点放在脚底板的位置上。若放在正中心，如果角色高低不同，很可能就踩进地里了。

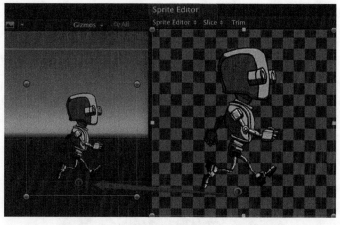

图6-35　原点

6.5　小结

本章中，我们讲了Unity 2D系统，介绍了UGUI与2D系统的区别，2D系统中摄像机组件与分辨率的自适应，SpriteRenderer，以及粒子之间的排序等。接着，介绍了2D地图，它可以使用Tile组件，可以灵活地编辑地图。此外，还学习了可编程拓展的Tile组件。最后，介绍了物理引擎，Sprite之间的碰撞效果以及碰撞事件。通过本章的学习，大家应该对2D游戏开发有个清晰的认识了。赶快开始自己的2D游戏之旅吧。

第 7 章

动画系统

Unity 的动画系统支持引擎内编辑动画,也支持外部导入 FBX 动画。由于引擎内置的动画编辑器没有提供骨骼动画的概念,所以只能编辑每一帧模型的 Transform 信息、整体的旋转、缩放和平移。假设是飘带一类的东西,它自身需要发生一些变化,此时引擎内置的编辑器就做不到了。可以用 3ds Max 来制作带骨骼信息的动画,然后将其导出 FBX 文件,最终放入 Unity 来使用。此外,Unity 还支持 FBX 网格文件的优化、动画重定向等功能。由于每个模型可能有很多动画,这样它们的切换管理就比较复杂。Unity 提供了 Animator 组件,它是可视化的状态机编辑工具,可以更方便地预览自身动画之间的切换关系,以及动画混合方式。然而游戏中可能同时会有很多模型,Unity 又提供了 TimeLine 编辑工具,它用时间线来管理模型的进度关系,像游戏中常用的过场动画、技能编辑器或者 3D 动画片等。

7.1 模型

Unity 显示模型必须给游戏对象提供 Mesh Filter 和 Mesh Renderer 组件,如图 7-1 所示,前者表示模型显示需要用的 Mesh 文件,后者通过材质的贴图和 Shader 最终将这个模型渲染出来。

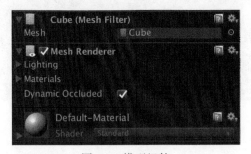

图 7-1　模型组件

7.1.1　Mesh Filter

Mesh Filter 需要绑定一个 Mesh 资源,它记录的就是模型的顶点信息。如果是外部导入 Unity 的 FBX 文件,它会自动生成 Mesh 信息,但是这个信息只能是只读的,无法二次修改它。不过可以读取原始 FBX 的 Mesh 信息,重新生成一个 Mesh 资源,这样以后就可以读写它了。

如图 7-2 所示，FBX 导入以后，建议不要选中 Import Materials 复选框，不然它会自动生成 3ds Max 中配置的材质文件。即使在 Project 视图中删除了它，该资源每次导入时，又会自动生成出来，多余的材质文件会影响到 AssetsBundle 打包的依赖关系。

图 7-2　导入材质

7.1.2　Mesh Renderer

模型的网格设置好后，就需要 Mesh Renderer 将它渲染出来了。这里需要提供一个材质，然后可以在材质上面设置贴图以及 Shader。此外，它还可以设置灯光阴影的接收信息以及烘焙的参数。

7.1.3　Prefab

游戏场景中需要很多模型，如果未来某一天想批量给某一类模型添加或修改一些参数，总不能每一个都手动地调一遍吧，所以需要引用 Prefab 的概念。制作 Prefab 的方法很简单，在 Hierarchy 视图中选择需要制作的游戏对象，然后将其直接拖入 Project 视图即可生成 Prefab。我们可以将它理解成一个快捷方式，修改原 Prefab 后，将自动影响所有引用到的地方。遗憾的是，目前 Unity 还没有提供 Prefab 嵌套的功能。

这就带来另一个问题：如果场景中的两个 Prefab，一个需要设置接收阴影，一个设置不接收阴影，那么如果把原 Prefab 改了，这岂不会把所有的 Prefab 都影响了？Prefab 自身有一个优化，如果 Hierarchy 视图中二次修改了引用的某个参数，那么 Prefab 修改后，此参数将不会被同步。如图 7-3 所示，选择 Prefab 后，可单击 Revert（还原）或者 Apply（应用到所有）按钮。另外，如果想取消 Prefab 的引用关系，可以在导航菜单栏中选择 GameObject→Break Prefab Instance 菜单。如果想彻底取消，建议直接删除。

下面我们在 Editor 中做一个小例子，来监听 Prefab 保存事件。

图 7-3　Prefab

如代码清单 7-1 所示,监听 `PrefabUtility.prefabInstanceUpdated` 事件即可知道 Prefab 何时被保存,并且在下面输出它的文件路径。

代码清单 7-1　Script_07_01.cs 文件

```
using System.Collections;
using System.Collections.Generic;
using UnityEngine;
using UnityEditor;

public class Script_07_01 {

    [InitializeOnLoadMethod]
    static void InitializeOnLoadMethod()
    {
        //监听 Prefab 保存事件
        PrefabUtility.prefabInstanceUpdated = delegate(GameObject instance) {
            Debug.LogFormat("Prefab {0} 被保存",AssetDatabase.GetAssetPath
                (PrefabUtility.GetPrefabParent(instance)));
        };
    }
}
```

7.2　动画编辑器

　　Unity 动画编辑器的原理就是通过时间线来修改组件的信息,比如修改 Transform 位置信息、修改 Renderer 组件或修改颜色信息,控制模型显示隐藏;此外,还可以在时间线上添加动画事件。编辑后的动画可以用在很多地方,比如 UI 元素、2D 元素或模型等,只要是游戏对象都可以,它需要配合 Animator 组件使用。之前已经讲过,这里就不再赘述了。

7.2.1　编辑器面板

　　使用 Window→Animation 菜单,即可打开编辑器窗口。请记住,需要提前选择一个游戏对象,并且在它身上绑定 Animator 组件。如图 7-4 所示,左上角的 New Animation 就是当前选择的动画,再点击一下可以创建新的动画。上面有个"小红点",点开后即可开启实时编辑模式。我们可以在 Scene 视图中旋转、缩放和平移当前帧模型,其中右边的 Samples 表示帧率(1 秒多少帧)。右边有两个小标志,第一个表示新添加一帧,第二个表示给当前帧添加一个事件。下面的 Add Property 按钮用于添加需要编辑的组件信息,包括模型下的所有子对象信息。在窗口最下面,还有个 Curves 按钮,它可以进入曲线编辑界面,拖动时间线就可以查看动画效果了。如果想整体播放动画,可以单击上面的播放按钮,最终动画信息将保存在 .anim 文件中。

7.2 动画编辑器　199

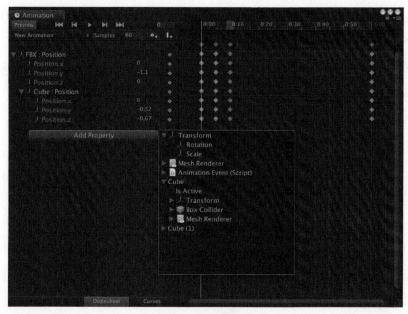

图 7-4　编辑器面板

7.2.2　在编辑器中添加事件

首先，给待编辑的对象身上绑定脚本。如果需要添加 public 方法，最多能有一个编辑参数，参数类型可以是 int、bool、string 或 GameObject 等常用类型。如图 7-5 所示，在动画编辑器的时间线上单击鼠标右键，在弹出的快捷菜单中选择 Add Animation Event 命令，即可添加一个事件，然后在 Event 面板中设置参数即可。

图 7-5　编辑器事件

如代码清单 7-2 所示，我们绑定了 `MyCustomEvent()` 方法，这里添加一个整型参数。点击 Add Animation Event 后，就可以添加事件了。

代码清单 7-2　AnimationListener.cs 文件

```
using System.Collections;
using System.Collections.Generic;
using UnityEngine;
using UnityEngine.Events;

public class AnimationEvent : UnityEvent<int>{ }
public class AnimationListener : MonoBehaviour
{
    public static AnimationEvent animationEvent = new AnimationEvent();
    public void MyCustomEvent(int intValue)
    {
        animationEvent.Invoke(intValue);
    }
}
```

`UnityEvent` 事件默认是不带参数的，如果需要参数，就要继承 `UnityEvent<T>`，其中 `T` 表示参数的数据类型，可以填入多个参数。

7.3　导入类动画

美术人员将在 3ds Max 中制作好的模型以及动画 FBX 文件导入 Unity 就可以使用了。首先，需要和美术人员约定命名规则。如图 7-6 所示，多动画文件共用同一份 Mesh 文件，动画文件以 @ 标记动画名称，命名保持一致，这样 Unity 会自动认定这是一组模型加动画。

图 7-6　文件命名

导入后，还可以对它们做很多特殊的设置，比如设置模型缩放比例，自动优化网格，自动优化动画等。导入类动画一共可分 3 种模式：Humanoid（人形重定向动画）、Generic（通用动画）和 Legacy（老版动画）。

7.3.1　人形重定向动画

人形重定向动画就是多个身形不同的人物模型也可以共用一份骨骼动画，此技术应用在游戏中会大量减少内存和包体大小，不过目前只支持人形动画。如图 7-7 所示，它会根据模型自动创建 Avatar 对象，里面记录着人物骨骼节点信息以及肌肉拉伸信息。重定向动画的原理就是将不同身形的骨骼信息套入相同的动画中来运算。前面讲过，动画需要以 @ 命名方式保持格式统一，这样点击 Update reference clips 按钮时，就会自动将模型关联到所有动画上了。选中 Optimize Game

Object 复选框,会把游戏对象下骨骼运动的节点信息删掉。由于有些节点中程序可能需要做逻辑,例如手上拿一把武器一类的,所以可以将特殊节点添加在 Extra Transforms to Expose 中,此时这些节点将会保留下来。

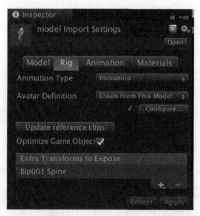

图 7-7 人形动画设置

单击 Configure 按钮,可以手动配置模型的骨骼节点以及信息。如图 7-8 所示,它会根据模型自动计算出正确的节点信息。不过我们也可以手动修改它。单击 Muscles & Settings 选项卡,可以配置骨骼的肌肉拉伸信息。

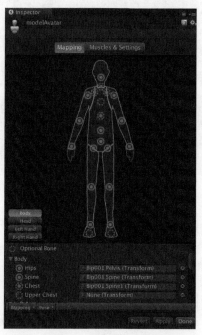

图 7-8 骨骼节点信息

可是有时候美术人员提供的模型并不完全是人形，比如武器、披风和头发等也做了人物模型中，这样播放动画时就会出问题，因为默认并没有处理非人形的骨骼。要解决这个问题，可以设置兼容播放所有骨骼。具体操作方法是：选择模型文件，切到 Animation 分页，如图 7-9 所示，在 Definition 下拉框中选择 Create From This Model 选项，在下面的 Transform 中选中剩余需要响应的骨骼节点。选中后，就表示如果这个动画被重定向，那么如果别的模型要播放这个动画，也需要这些骨骼节点信息了。最后，单击 Apply 按钮即可。

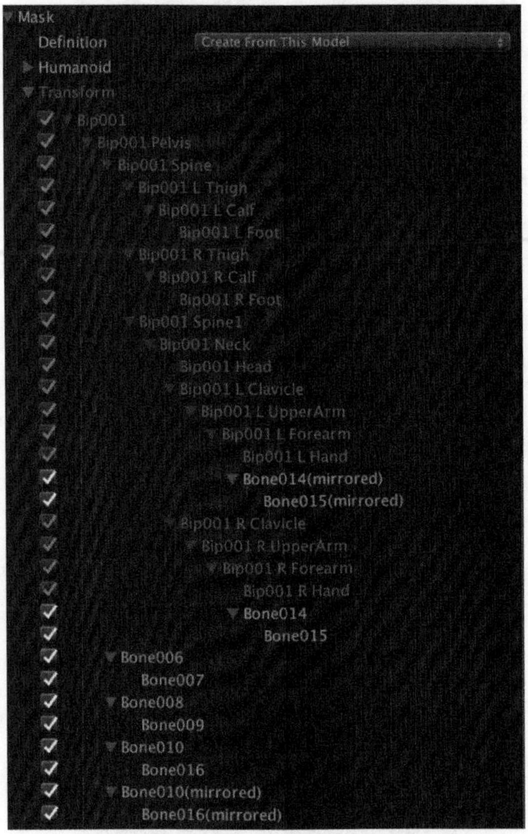

图 7-9　绑定骨骼信息

7.3.2　通用动画

通用动画就比较好理解了。它不支持重定向动画，美术人员做成什么样，拿进来就是什么样。如图 7-10 所示，在模型文件的 Avatar Definition 下拉框中选择 Create From This Model 选项，将创建 Avatar 对象。而动画文件会依赖 Avatar 对象。Root node 就是支持带位移的动画，游戏中尽可能不要使用它，因为它可能和自己的控制系统冲突。Optimize Game Object 和 Extra Transforms to Expose 与人形动画功能类似，这里不再赘述。

7.4 动画控制器

图 7-10 通用动画

7.3.3 老版动画

老版动画是 Unity 最早的动画系统，其所有功能提供的都是最原始的接口，需要开发者自行处理。其优点是开发起来比较灵活，缺点是 Unity 已经禁止对它进行更新，效率上远远不如新版动画系统了，所以建议大家以后放弃它。

7.3.4 导入类动画事件

在 Project 视图中选择动画文件，在 Animation 分页中展开 Events 标签，可以添加自定义动画事件，如图 7-11 所示。单击左上角的加号图标，可以添加一个新事件。其中 Function 就是事件回调的方法名称，Float、Int、String 和 Object 表示参数。这里回调的方法为 `MyCustomEvent()`，其代码如下：

```
public void MyCustomEvent(int intvalue)
{
    Debug.Log(intvalue);
}
```

图 7-11 导入类动画事件

7.4 动画控制器

Unity 动画控制器的原理就是状态机。传统的状态机是需要在代码里写一个很大的 `switch…case` 来处理状态，而 Unity 为我们提供了可视化的编辑工具，不需要程序员也可以编辑动画的状态。此外，它还提供了子状态机和动画混合的功能。由于状态机的原理是同一时刻只能有一个状态，所以 Unity 还提供了层的概念来将动画分成两个层来同时编辑。

7.4.1 系统状态

动画控制器默认会提供 3 个状态，我们无法删除它，并且除了 Exit 状态以外，其他状态都无法连接它。如图 7-12 所示，首先是 Entry 状态，它表示当前控制器的初始状态。右击该状态，选择 Make Transition 命令，即可连接新的状态。状态机会按照连线的状态一次切换动画。橘黄色的状态表示默认状态，如果想切换默认状态，选择另一个状态，具体操作方法是单击鼠标右键，从弹出的快捷菜单中选择 Set as Layer Default State 命令。

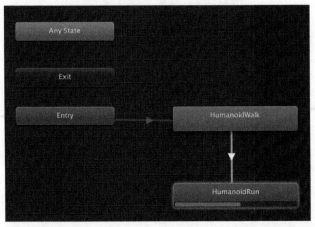

图 7-12　Entry 状态

接着是 Any State 状态，比如角色死亡一类的，需要从现有状态切换到另一个动画。如图 7-13 所示，可以从 Any State 状态连线到立刻播放的特殊状态，等它的状态处理完后，再连线回到默认状态，继续原有逻辑。

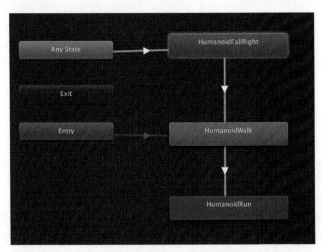

图 7-13　Any State 状态

最后是 Exit 状态。状态机可以创建子状态。如果子状态需要回到父状态 Base Layer，可以将子状态再连线到 Exit 状态，如图 7-14 所示。

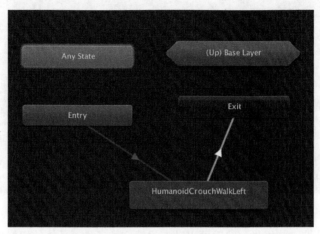

图 7-14　Exit 状态

7.4.2　切换条件

状态机中复杂的状态通过连线来确定关联关系，但是如果切换状态时发现有多条线，如何决定选择哪一条呢？这时就需要设置切换条件了。如图 7-15 所示，状态机一共支持 4 种条件：Float、Int、Bool 和 Trigger，其中前三个都是普通的数据类型，Trigger 就像 Bool 一样，设置 true 后需要立即设置 false。

图 7-15　状态条件

条件定义完毕后，就可以配置条件了。如图 7-16 所示，单击两个状态之间的连线，然后在右下角就可以添加满足的条件了，这里支持添加多个条件。状态机在同一时刻只能执行一个状态，即使两个状态的条件都满足了，也只能进入其中一个。右上角的 Solo 复选框表示即使当前别的条件达成了，也只能进入选中 Solo 的状态。Mute 复选框表示即使当前条件达成了，也不能进入选中 Mute 的状态。下面的 Has Exit Time 复选框表示不同动画切换时是否启动动画过渡，可以调节蓝色半透明区域来设置过渡的时间。

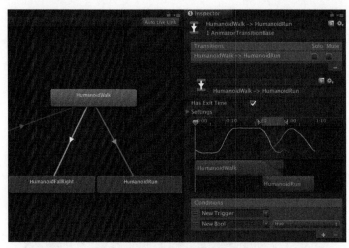

图 7-16 配置条件（另见彩插）

下述代码首先获取动画组件，然后可以动态设置它的条件，具体的切换例子在第 6 章中已经介绍过，这里就不再赘述了：

```
Animator animator = GetComponent<Animator>();
animator.SetFloat("New Float",1f);
animator.SetInteger("New Int",1);
animator.SetBool("New Bool", true);
animator.SetTrigger("New Trigger")
```

7.4.3 状态机脚本

我们可以给每个状态添加脚本来监听一些状态事件，比如状态开启、状态更新和状态退出等。另外，脚本也可以添加在子状态机上。如图 7-17 所示，选择一个状态，然后单击 Add Behaviour 按钮即可添加脚本。此外，也可以序列化常用数据，如 int、string、bool 和 object 等，然后在面板中输入参数即可。例如进入某个状态，播放一个特定的音乐或者做一些特殊的逻辑等。

图 7-17 状态脚本

NewMachineBehaviour 类的代码如下：

```csharp
using System.Collections;
using System.Collections.Generic;
using UnityEngine;

public class NewMachineBehaviour : StateMachineBehaviour {

    public string testValue;
    //进入当前状态时，调用 OnStateEnter() 方法
    override public void OnStateEnter(Animator animator, AnimatorStateInfo stateInfo,
        int layerIndex) {

    }

    //当前状态更新时，调用 OnStateUpdate() 方法。它在 OnStateEnter() 和 OnStateExit() 之间，
    //每帧都会被调用
    override public void OnStateUpdate(Animator animator, AnimatorStateInfo stateInfo,
        int layerIndex) {

    }

    //离开当前状态时，调用 OnStateExit() 方法
    override public void OnStateExit(Animator animator, AnimatorStateInfo stateInfo,
        int layerIndex) {

    }

    //OnStateMove() 方法在 Animator.OnAnimatorMove() 之后调用，这里可以处理动画根节点的位移
    override public void OnStateMove(Animator animator, AnimatorStateInfo stateInfo,
        int layerIndex) {

    }

    //OnStateIK() 方法在 Animator.OnAnimatorIK() 之后调用，这里可以处理 IK（反向运动学）动画
    override public void OnStateIK(Animator animator, AnimatorStateInfo stateInfo,
        int layerIndex) {

    }
}
```

有了这些回调事件后，就可以在各个方法中添加自己的代码了。

7.4.4 IK 动画

IK 动画的全名是 Inverse Kinematics，其意思是反向动力学，就是子骨骼节点带动父骨骼节点运动。比如跳街舞的少年用手撑着身体在地上转圈，手就是子骨骼，身体就是它的父骨骼，这样运动时手就需要带动身体来移动。如图 7-18 所示，单击 Layers 选项卡的右上角，设置勾选 IK Pass 复选框。

图 7-18 开启 IK 动画

如图 7-19 所示，在左手和右手分别绑定一个球体，移动球体来控制 IK 影响手部位的动画。

图 7-19 控制 IK 动画

如代码清单 7-3 所示，在 `OnAnimatorIK()` 方法中就可以处理 IK 动画了。这里调用了 `SetIKPosition()` 方法设置手的位置，以便带动胳膊移动。

代码清单 7-3　Script_07_03.cs 文件

```
using System.Collections;
using System.Collections.Generic;
using UnityEngine;
using UnityEditor;

public class Script_07_03 : MonoBehaviour  {

    public Animator animator;
    public Transform rightHandObj;
    public Transform leftHandObj;
    void OnAnimatorIK(int layerIndex)
    {
        if(animator)
```

```
    {
        //设置动画权重
        animator.SetIKPositionWeight(AvatarIKGoal.LeftHand,1f);
        animator.SetIKRotationWeight(AvatarIKGoal.LeftHand,1f);

        animator.SetIKPositionWeight(AvatarIKGoal.RightHand,1f);
        animator.SetIKRotationWeight(AvatarIKGoal.RightHand,1f);

        if(rightHandObj != null)
        {
            //设置右手根据目标点而旋转和移动父骨骼节点
            animator.SetIKPosition(AvatarIKGoal.RightHand,rightHandObj.position);
            animator.SetIKRotation(AvatarIKGoal.RightHand,rightHandObj.
                rotation);
        }
        if(leftHandObj != null)
        {
            //设置左手根据目标点而旋转和移动父骨骼节点
            animator.SetIKPosition(AvatarIKGoal.LeftHand,leftHandObj.position);
            animator.SetIKRotation(AvatarIKGoal.LeftHand,leftHandObj.rotation);
        }
    }
}
```

7.4.5 Root Motion

Root Motion 就是播放带位移变化的动画,并且它将影响到游戏对象的 Transform 信息。带位移的编辑可以在 3ds Max 中完成。如图 7-20 所示,首先需要选中 Apply Root Motion 复选框,这表示开启动画 Transform 位移,不选中则不会影响 Transform 信息。选中 Bake Into Pose 复选框,表示动画播放完毕后才同步位移信息,不选中表示位移随着动画同时改变。

图 7-20 位移动画

另外，在脚本中可以监听位移动画的移动更新事件。注意，位移移动事件是在 Update() 方法之后执行的。如代码清单 7-4 所示，在 OnAnimatorMove() 方法中，控制位移动画。

代码清单 7-4　Script_07_04.cs 文件

```
using System.Collections;
using System.Collections.Generic;
using UnityEngine;
using UnityEditor;

public class Script_07_04 : MonoBehaviour {

    public Animator animator;

    void OnAnimatorMove()
    {
        if(animator)
        {
            Vector3 newPosition = transform.position;
            newPosition.z += 1f * Time.deltaTime;
            transform.position = newPosition;
        }
    }
}
```

7.4.6　Avatar Mask

Avatar Mask 可以限制某些骨骼不播放动画。在 Project 视图中选择 Create→Avatar Mask 命令，即可创建它。如图 7-21 所示，如果是人形动画（Humanoid），那么可以直接设置人形遮罩骨骼，其中红色的部分表示禁止这部分骨骼的播放动作。

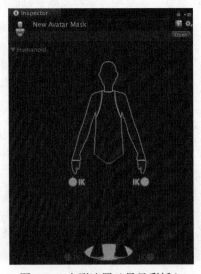

图 7-21　人形遮罩（另见彩插）

如图 7-22 所示，如果是通用动画（Generic），需要单独选中需要禁止播放动画的骨骼节点，编辑好后保存即可。

图 7-22　通用遮罩

7.4.7　层

层是用来做动画融合的，同一套骨骼上的两个动画同时播放，例如 PFS 类游戏或者篮球类游戏。下半身跑动的过程中，上半身还可以旋转投篮等。为了让上下部分的骨骼相互不影响，可以设置它们的 Avatar Mask。如图 7-23 所示，点击 Layers 面板右上角的加号，即可添加新层。可以让 Base Layer 来处理整体逻辑，而让 New Layer 专门用来做动画融合，Weight 可以设置融合的权重，Mask 就是遮罩的文件了，Blending 设置的 Override 表示直接覆盖掉其他层的动画。

图 7-23　创建层

7.4.8 Blend Tree

Blend Tree（混合树）用来做动画混合。动画混合和前面提到的动画融合是不同的概念。动画混合指两个动画切换的时候，为了避免太过生硬而混合在一起的过程，比较经典的例子就是控制角色向前跑、向左跑和向右跑，左右切换跑的时候就要用到它。Animator 提供 Blend Tree 来专门处理混合。

在 Layer 中单击鼠标右键，从弹出的快捷菜单中选择 Create State→From New Blend Tree 命令即可创建它。双击打开它，如图 7-24 所示，选择 Blend Tree 后，单击鼠标右键，从弹出的快捷菜单中选择 Add Motion 命令即可添加动画文件，这里添加了三组动画，分别是向左跑、向前跑和向右跑。接着在右上方设置 TreeValue 参数，取值范围为 –96～96。将动画文件拖到右边 Motion 面板中，取消选择 Automate Thresholds 复选框，即可手动设置动画的阈值了。最后，可以通过方向键来控制角色左右混合动画了，如图 7-25 所示。

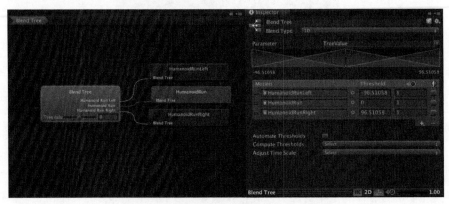

图 7-24　混合树

图 7-25　混合动画

7.4 动画控制器

如代码清单 7-5 所示，在 Update() 方法中根据左右方向设置 TreeValue 的值，控制动画左右混合。

代码清单 7-5　Script_07_05.cs 文件

```
using System.Collections;
using System.Collections.Generic;
using UnityEngine;
using UnityEditor;

public class Script_07_05 : MonoBehaviour {

    public Animator animator;

    void Update()
    {
        animator.SetFloat("TreeValue", Input.GetAxis("Horizontal") * 96.0f);
    }
}
```

7.4.9　非运行播放动画

通常，在做编辑器的时候，需要在非运行模式下也能播放动画。如图 7-26 所示，在"动画"下拉框中选择当前 Animator 组件绑定的所有动画，然后拖动进度条来调节动画的播放。

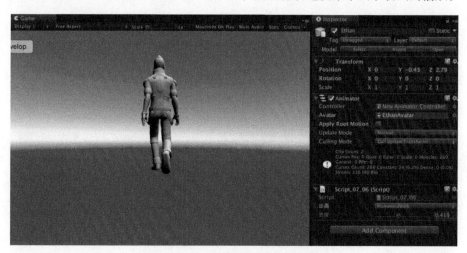

图 7-26　在非运行模式下播放动画

如代码清单 7-6 所示，SampleAnimation() 方法用于采样动画。拖动进度条设置采样事件，即可调节动画了。

代码清单 7-6　Script_07_06.cs 文件

```
using System.Collections;
using System.Collections.Generic;
```

```csharp
using UnityEngine;

#if UNITY_EDITOR
using UnityEditor;
using UnityEditor.Animations;
using System.Linq;
#endif
[RequireComponent(typeof(Animator))]
public class Script_07_06 : MonoBehaviour  {

}

#if UNITY_EDITOR
[CustomEditor(typeof(Script_07_06))]
public class ScriptEditor_07_06 : Editor
{
    private AnimationClip[] m_Clips = null;
    private Script_07_06 m_Script = null;
    void OnEnable()
    {
        m_Script = (target as Script_07_06);
        Animator animator = m_Script.gameObject.GetComponent<Animator>();
        AnimatorController controller = (AnimatorController)animator.
            runtimeAnimatorController;
        m_Clips = controller.animationClips;
    }

    private int m_SelectIndex = 0;
    private float m_SliderValue = 0;
    public override void OnInspectorGUI()
    {
        base.OnInspectorGUI();

        EditorGUI.BeginChangeCheck();
        m_SelectIndex = EditorGUILayout.Popup("动画",m_SelectIndex,m_Clips.Select
            (pkg => pkg.name).ToArray());
        m_SliderValue = EditorGUILayout.Slider("进度",m_SliderValue, 0f, 1f);
        if(EditorGUI.EndChangeCheck()) {
            AnimationClip clip = m_Clips [m_SelectIndex];
            float time = clip.length * m_SliderValue;
            clip.SampleAnimation(m_Script.gameObject, time);
        }
    }
}
#endif
```

7.4.10 Animator Override Controller

前面我们介绍了 Animator Controller 可以编辑动画之间的切换状态。在游戏中，很多模型动画的切换事件的逻辑可能都是一样的，比如游戏中的很多怪物，它们之间的区别可能就是动画文件不一样，总不能每一个怪物都编辑一套相同的 Animator Controller 控制行为吧，此时就需要使用 Animator Override Controller 了。

如图 7-27 所示，在 Controller 处绑定需要覆盖的 Controller 文件，Original 会自动列出所用到的动画文件，将它更换成新的动画文件即可。如果以后需要修改它，修改原文件，即可自动修改所有引用到的 Animator Override Controller 了。

图 7-27　Animator Override Controller

7.4.11　`RuntimeAnimatorController`

`RuntimeAnimatorController` 是用来处理 Animator Controller 动态更新的。如代码清单 7-7 所示，可以通过 `Resources.Load<RuntimeAnimatorController>()` 最终将对象赋值给 `animator`。

代码清单 7-7　Script_07_07.cs 文件

```
using System.Collections;
using System.Collections.Generic;
using UnityEngine;

[RequireComponent(typeof(Animator))]
public class Script_07_07 : MonoBehaviour  {

    public Animator animator;

    void OnGUI()
    {
        if(GUILayout.Button("<size=50>读取</size>")) {
            RuntimeAnimatorController controller =
                Resources.Load<RuntimeAnimatorController>("New Animator Controller");
            animator.runtimeAnimatorController = controller;
        }
        if(GUILayout.Button("<size=50>删除</size>")) {
            animator.runtimeAnimatorController = null;
        }
    }
}
```

7.5　TimeLine 编辑器

TimeLine 编辑器可以用来编辑游戏内剧情动画，可以添加模型、动画、声音和 Prefab 等常用组件。拖动时间线，可以编辑它们的 Transform 信息，控制动画的播放与关闭、模型的隐藏或显示。此外，还可以动态创建 Prefab。另外，它还提供了自定义脚本来拓展编辑内容，这样将大幅度提高编辑的灵活度。

7.5.1 创建 Timeline

在 Project 视图中，选择 Create→Timeline 命令即可创建 Timeline。接着，在 Hierarchy 视图中找到一个需要编辑的游戏对象。如图 7-28 所示，将刚刚创建的 Timeline 资源拖入 Playable 中，表示此对象应用了这个时间线。Update Method 表示时间线更新的方法，Play On Awake 表示是否运行就启动时间线，Wrap Mode 设置循环模式，Initial Time 表示时间线起始的时间。Playable Director 作为时间线的总控制组件，也提供了 Play()、Pause()和 Stop()等控制类方法。

图 7-28　Playable Director

接着，点击 Window→Timeline 菜单项，即可打开编辑窗口。如图 7-29 所示，单击 Add 按钮，即可在时间线上添加轨迹。

- Track Group：可将多轨迹分组，方便管理。
- Activation Track：激活轨迹。绑定游戏对象后，在时间线中进行隐藏和显示操作。
- Animation Track：动画轨迹。首先需要绑定 Animator 组件，接着在时间线中编辑动画的 Transform 信息。
- Audio Track：声音轨迹。需要绑定音频文件，在时间线中设置音频轨迹。
- Control Track：控制轨迹。需要绑定 Prefab，它可在时间线中动态实例化并且挂在某个节点下，例如特效。
- Playable Track：可拓展轨迹。可在脚本中灵活地动态扩展。

图 7-29　Track

7.5.2 Activation Track

使用 Activation，可以控制游戏对象在时间线中的隐藏和显示。如图 7-30 所示，在左边绑定一个需要控制的对象，在右边窗口中单击鼠标右键，从弹出的快捷菜单中选择 Add Activation Clip 命令，即可添加一个状态剪辑。这里可添加多个，然后编辑 Active 的区域以及位置来设置它隐藏或显示的时长。

图 7-30　Activation

如果时间线没有设置循环模式，那么它是有结束状态的。如图 7-31 所示，时间线结束后，可以设置模型的状态，其中 Active 表示激活，Inactive 表示隐藏，Revert 表示还原初始状态，Leave As Is 表示保留最后的状态。

图 7-31　结束状态

7.5.3 Animation Track

Animation Track 需要绑定 Animator。如图 7-32 所示，单击鼠标右键，从弹出的快捷菜单中选择 Add From Animation Clip 命令，可以添加动画剪辑。当然，我们也可以像 Animation 编辑器那样编辑它的位置。Add From Animation Playable Asset 表示可拓展状态资源。Add Override Track 表示还可以添加一个 Override 层，它和前面讲过的 Avatar Mask 一样，可以添加遮罩，以禁止该层上的动画。

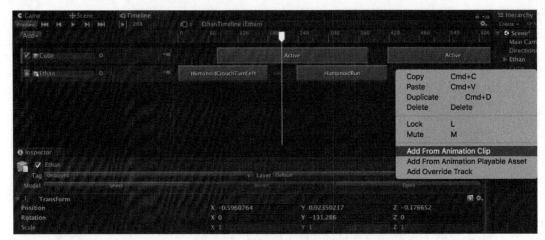

图 7-32　绑定动画

7.5.4　Audio Track

如图 7-33 所示，音频文件的做法也类似。单击鼠标右键，从弹出的快捷菜单中选择 Add From Audio Clip 命令，即可选择添加一个音频。点击音频文件，还可以设置它是否循环播放。

图 7-33　绑定音频

7.5.5　Control Track

如图 7-34 所示，Control Track 可以在时间线上动态创建 Prefab。Parent Object 表示将 Prefab 创建在这个游戏对象下面，Prefab 表示它的源文件，其他操作和别的 Track 类似，这里不再赘述。

图 7-34　控制对象

7.5.6　Playable Track

Playable Track 表示可编程型时间线，它提供了自定义时间线的数据以及更新周期。如图 7-35 所示，可以创建 Playable Behaviour 以及 Playable Asset。如图 7-36 所示，创建完 Playable Asset 资源脚本后，就可以任意添加在时间线上了。

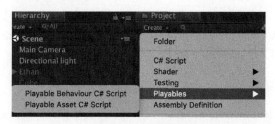

图 7-35　创建脚本

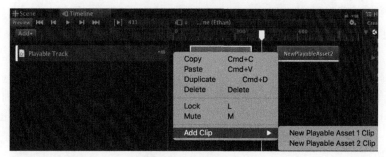

图 7-36　添加脚本

如图 7-37 所示，可以在自定义资源上序列化对象。Playable Asset 只是时间线中的一部分，那么如何来区分当前时间线是否正在执行这个资源呢？我们可以通过给它添加 Playable Behaviour 来实现。

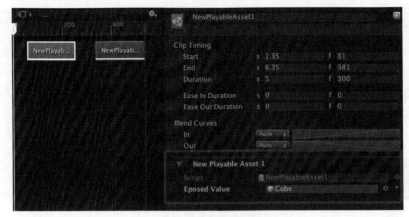

图 7-37 序列化自定义对象

如代码清单 7-8 所示，我们来实现对象与时间线的数据传递。

代码清单 7-8 NewPlayableAsset1.cs 文件

```
using System.Collections;
using System.Collections.Generic;
using UnityEngine;
using UnityEngine.Playables;

[System.Serializable]
public class NewPlayableAsset1 : PlayableAsset
{
    //序列化对象
    public ExposedReference<GameObject> eposedValue;

    public override Playable CreatePlayable(PlayableGraph graph, GameObject go)
    {
        //创建 Behaviour 脚本
        NewPlayableBehaviour1 behaviour = new NewPlayableBehaviour1();
        //向脚本中传递序列化的对象
        behaviour.eposedObjectValue = eposedValue.Resolve(graph.GetResolver());
        return ScriptPlayable<NewPlayableBehaviour1>.Create(graph, behaviour);
    }
}
```

如代码清单 7-9 所示，PlayableBehavoiur 是时间线上每个资源的生命周期，脚本可以用在多个不同的资源上。它的执行顺序是 OnGraphStart()→OnBehaviourPause()→OnBehaviourPlay()→OnBehaviourPause()→OnGraphStop()，其中 PrepareFrame() 表示每帧都会执行。

代码清单 7-9 NewPlayableBehaviour1.cs 文件

```
using System.Collections;
using System.Collections.Generic;
using UnityEngine;
```

```csharp
using UnityEngine.Playables;
public class NewPlayableBehaviour1 : PlayableBehaviour
{
    //接收序列化的对象
    public GameObject eposedObjectValue;

    //Graph 开始运行时调用
    public override void OnGraphStart(Playable playable) {
    }

    //Graph 结束运行时调用
    public override void OnGraphStop(Playable playable) {
    }

    //playable 播放时调用
    public override void OnBehaviourPlay(Playable playable, FrameData info) {
    }

    //playable 暂停时调用
    public override void OnBehaviourPause(Playable playable, FrameData info) {
    }

    //playable 每帧都会被调用
    public override void PrepareFrame(Playable playable, FrameData info) {
        //PrepareFrame()方法每帧都会执行, 可能当前还没执行到绑定 eposedObjectValue 的对象
        //所以 eposedObjectValue 会有空的情况
        if(eposedObjectValue != null) {
            //获取 PlayableDirector 对象
            PlayableDirector playableDirector = playable.GetGraph<Playable>().
                GetResolver() as PlayableDirector;
            Debug.LogFormat("playableDirector : {0} eposedObjectValue:{1}",
                playableDirector.gameObject.name,eposedObjectValue.name);
        }
    }
}
```

7.5.7 自定义 Track

Timeline 默认只支持 GameObject、Animator 和 AudioSource，但可自定义拓展 Track，这样就可以拖入自定义对象格式了，如图 7-38 所示。

图 7-38　自定义 Track

如代码清单 7-10 所示，自定义 Track 需要继承 TrackAsset 并且可以重写 CreatePlayable() 和 CreateTrackMixer() 方法。

代码清单 7-10　NewTrack.cs 文件

```csharp
using System.Collections;
using System.Collections.Generic;
using UnityEngine;
using UnityEngine.Timeline;
using UnityEngine.Playables;
//设置 Track 的颜色，即 Track 最左边的竖条颜色
[TrackColor(1f,0f,0f)]
//设置 Track 的资源对象
[TrackClipType(typeof(NewPlayableAsset1))]
//自定义绑定对象
[TrackBindingType(typeof(NewTrackType))]
public class NewTrack : TrackAsset
{
    protected override Playable CreatePlayable (UnityEngine.Playables.PlayableGraph
        graph, GameObject go, TimelineClip clip)
    {
        Debug.Log("CreatePlayable 创建");
        return base.CreatePlayable(graph, go, clip);
    }

    public override Playable CreateTrackMixer(UnityEngine.Playables.PlayableGraph
        graph, GameObject go, int inputCount)
    {
        Debug.Log("CreateTrackMixer 创建");
        return base.CreateTrackMixer(graph, go, inputCount);
    }
}
```

7.6　Playables

在 Unity 新版的动画系统中，必须创建 Animator Controller 文件（在音频系统中，也需要创建 Audio Mixer 文件）用来编辑数据。可视化的编辑确实提供了便利，但是也有一定缺陷。拿新版动画系统来说，如果游戏中的模型动画很多，每一个都需要手动用 Animator Controller 来连线的话，那就太麻烦了。截止到目前，依然有很多项目在使用老版本动画系统，其原因就是老版动画系统是纯代码控制的，不需要创建 Animator Controller 文件，控制起来更加灵活。Unity 2017 提供了 Playables 组件，这样我们就可以用纯代码来控制新版动画了。

7.6.1　播放动画

Playables 提供了工具类来直接控制播放动画。如图 7-39 所示，给脚本挂上两个不同的动画剪辑，来切换播放它。相关代码如代码清单 7-11 所示。

7.6 Playables 223

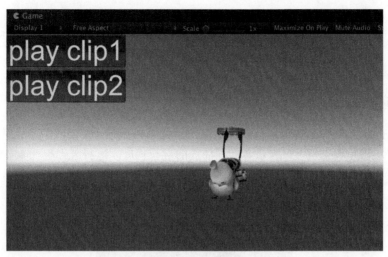

图 7-39 播放动画

需要说明的是，我们需要给 AnimationPlayableUtilities.PlayClip()传入待播放的动画剪辑。

代码清单 7-11 Script_07_10.cs 文件

```
using UnityEngine;
using UnityEngine.Playables;

public class Script_07_10 : MonoBehaviour {

    //动画
    public AnimationClip clip1;
    public AnimationClip clip2;


    PlayableGraph m_PlayableGraph;

    private void OnGUI()
    {
        if(GUILayout.Button("<size=80>play clip1</size>"))
        {
            AnimationPlayableUtilities.PlayClip(GetComponent<Animator>(),
                clip1, out m_PlayableGraph);

        }
        if(GUILayout.Button("<size=80>play clip2</size>"))
        {
            AnimationPlayableUtilities.PlayClip(GetComponent<Animator>(),
                clip2, out m_PlayableGraph);
        }
    }
    void OnDisable()
    {
```

```
            //销毁 PlayableGraph
            m_PlayableGraph.Destroy();
        }
    }
```

7.6.2 动画混合

使用 Playables 组件，也可以进行动画混合。如图 7-40 所示，设置两个动画混合播放，这里我们通过设置权重（Weight）来表示各动画播放的混合程度。运行起来后，调节它，就能看到效果。

图 7-40 混合权重

如代码清单 7-12 所示，在 Update() 中通过 SetInputWeight() 设置权重。

代码清单 7-12　Script_07_11.cs 文件

```
using UnityEngine;
using UnityEngine.Animations;
using UnityEngine.Playables;

public class Script_07_11 : MonoBehaviour {

    //动画
    public AnimationClip clip1;
    public AnimationClip clip2;
    //权重
    public float weight;
    PlayableGraph m_PlayableGraph;
    AnimationMixerPlayable m_AnimationMixerPlayable;
    private void Start()
    {
        m_AnimationMixerPlayable = AnimationPlayableUtilities.PlayMixer
            (GetComponent<Animator>(), 2, out m_PlayableGraph);
        //连接动画到 PlayableGraph
        var clipPlayable1 = AnimationClipPlayable.Create(m_PlayableGraph, clip1);
        var clipPlayable2 = AnimationClipPlayable.Create(m_PlayableGraph, clip2);
        m_PlayableGraph.Connect(clipPlayable1, 0, m_AnimationMixerPlayable, 0);
        m_PlayableGraph.Connect(clipPlayable2, 0, m_AnimationMixerPlayable, 1);
    }

    void Update()
    {
        //设置混合权重，使其保持在 0 和 1 之间
        weight = Mathf.Clamp01(weight);
```

```
        m_AnimationMixerPlayable.SetInputWeight(0, 1.0f - weight);
        m_AnimationMixerPlayable.SetInputWeight(1, weight);
}
void OnDisable()
{
        //销毁PlayableGraph
        m_PlayableGraph.Destroy();
}
}
```

7.6.3 音频混合

音频文件需要使用 `AudioSource` 来播放。如果要混合音频，需要创建 Audio Mixer 文件。使用 Playables 组件时，可以不用创建 Mixer 文件，通过脚本就可以灵活地控制混合。如图 7-41 所示，在脚本中绑定两个音频文件，然后通过调节 Weight 来控制它们的混合，相关代码如代码清单 7-23 所示。

图 7-41 混合权重

代码清单 7-13　Script_07_12.cs 文件

```
using UnityEngine;
using UnityEngine.Animations;
using UnityEngine.Audio;
using UnityEngine.Playables;

public class Script_07_12 : MonoBehaviour {

    //音频
    public AudioClip clip1;
    public AudioClip clip2;
    //权重
    public float weight;
    PlayableGraph m_PlayableGraph;
    AudioMixerPlayable m_AudioMixerPlayable;

    private void Start()
    {
        m_PlayableGraph = PlayableGraph.Create();
        m_AudioMixerPlayable = AudioMixerPlayable.Create(m_PlayableGraph, 2);
        //连接音频到PlayableGraph
        var audioClipPlayable1 = AudioClipPlayable.Create(m_PlayableGraph,
            clip1, true);
        var audioClipPlayable2 = AudioClipPlayable.Create(m_PlayableGraph,
            clip2, true);
        m_PlayableGraph.Connect(audioClipPlayable1, 0, m_AudioMixerPlayable, 0);
```

```csharp
        m_PlayableGraph.Connect(audioClipPlayable2, 0, m_AudioMixerPlayable, 1);

        //混合音频输出
        var audioPlayableOutput = AudioPlayableOutput.Create(m_PlayableGraph,
            "Audio", GetComponent<AudioSource>());
        audioPlayableOutput.SetSourcePlayable(m_AudioMixerPlayable);

        m_PlayableGraph.Play();
    }

    void Update()
    {
        //设置混合权重,保持在 0 和 1 之间
        weight = Mathf.Clamp01(weight);
        m_AudioMixerPlayable.SetInputWeight(0, 1.0f - weight);
        m_AudioMixerPlayable.SetInputWeight(1, weight);
    }

    void OnDisable()
    {
        //销毁 PlayableGraph
        m_PlayableGraph.Destroy();
    }
}
```

7.6.4 自定义脚本

使用 Playables 组件,既可以控制动画和音频,也可以创建自定义脚本。像时间线一样,我们可以在代码中更精准地控制每一帧的状态。和前面介绍的动画声音一样,自定义 Playable 组件也需要使用 Output 组件将它输出出来。

如代码清单 7-14 所示,首先使用 `ScriptPlayable<T>.Create()` 创建自定义的 PlayableBehaviour,接着使用 `SetSourcePlayable()` 进行关联即可。

代码清单 7-14　Script_07_13.cs 文件

```csharp
using UnityEngine;
using UnityEngine.Playables;

public class Script_07_13 : MonoBehaviour {

    private PlayableGraph m_PlayableGraph;

    private void Start()
    {
        m_PlayableGraph = PlayableGraph.Create();
        //创建自定义 Playable 组件
        var customPlayable = ScriptPlayable<CustomPlayableBehaviour>.Create
            (m_PlayableGraph);
        //创建自定义 Output 组件
        var customOutput = ScriptPlayableOutput.Create(m_PlayableGraph, "customOutput");
```

```csharp
        //关联
        customOutput.SetSourcePlayable(customPlayable);
        //播放
        m_PlayableGraph.Play();
    }

    void OnDisable()
    {
        //销毁 PlayableGraph
        m_PlayableGraph.Destroy();
    }
}
```

自定义脚本也可以像动画和音频那样添加混合，这样就可以在每一帧灵活地控制各种状态了。此外，它也可以监听开始或结束的触发事件。相关代码如代码清单 7-15 所示。

代码清单 7-15 CustomPlayableBehaviour.cs 文件

```csharp
using UnityEngine.Playables;
using UnityEngine;
public class CustomPlayableBehaviour : PlayableBehaviour
{
    public override void OnGraphStart(Playable playable) {
        Debug.Log("PlayableGraph 开始时触发");
    }
    public override void OnGraphStop(Playable playable) {
        Debug.Log("PlayableGraph 结束时触发");
    }

    public override void OnBehaviourPlay(Playable playable, FrameData info) {
        Debug.Log("脚本开始时触发");
    }
    public override void OnBehaviourPause(Playable playable, FrameData info) {
        Debug.Log("脚本暂停时触发");
    }

    public override void PrepareFrame(Playable playable, FrameData info) {
        //每一帧循环触发
    }
}
```

7.7 Constraint

Unity 2018 引入了一个全新概念：Constraint。之前如果想同时操作两个对象，可以将其中一个对象挂在另一个下面，成为它的子对象，操作父对象的旋转、缩放和平移会同时影响子对象，而 Constraint 可以让平级的游戏对象相互依赖。

7.7.1 Aim Constraint

Aim Constraint 用于控制游戏对象始终朝向另一个游戏对象。没有该特性之前，我们使用

`transform.LookAt(target)`方法，但 Aim Constraint 的功能更加全面。如图 7-42 所示，给 A 对象绑定 Aim Constraint 组件，在 Sources 中关联另外两个游戏对象 B 和 C，表示它会同时瞄准这两个游戏对象。Weight 用于设置瞄准的权重。下面还可以设置偏移的角度、坐标，以及冻结的轴向。运行起来后，可以看到 A 会始终朝向 B 和 C。

图 7-42　瞄准

7.7.2　Parent Constraint

此外，我们还可以给某个对象绑定父节点行为，但它们不一定保持父子关系，可以是同级关系。移动"父"对象，会自动控制子对象。如图 7-43 所示，依然可以给一个对象设置多个"父"节点。当然，也可以设置权重。另外，还可以在每个父节点对象的右边设置权重值，这里设置的 0.5 表示当"父"节点移动 1 米时，"子"节点移动 0.5 米。这个参数可以灵活控制。

图 7-43　父节点

如图 7-44 所示，Constraint 还提供了坐标、旋转和缩放等约束组件。我们可以根据需求来使用，其用法都差不多。

图 7-44 其他

7.7.3 脚本控制约束

在约束脚本中,我们也提供了面板上可操作的接口,并且在代码中可以灵活地控制它。代码清单 7-16 演示了运行时设置跟随或者取消跟随"父"节点。

代码清单 7-16　Script_07_14.cs 文件

```
using UnityEngine;
using UnityEngine.Animations;

public class Script_07_14 : MonoBehaviour {

    public ParentConstraint parentConstraint;

    private void OnGUI()
    {
        //设置跟随或者取消跟随"父"节点
        if(GUILayout.Button("constraintActive"))
        {
            parentConstraint.constraintActive = !parentConstraint.constraintActive;
        }
    }
}
```

7.8 小结

本章介绍了动画组件、动画控制组件和时间线组件。动画组件用来编辑单个动画,动画控制组件可将多个动画关联起来,我们可以通过状态机的方式来管理它们的切换事件。时间线组件更为庞大,它可以同时管理多个动画控制器、游戏对象以及音频等。借助 Timeline 的可视化图形界面,编辑起来更加方便。为了更加灵活,Timeline 编辑器还提供了脚本拓展的功能,开发者可以自由拓展它。

第 8 章

持久化数据

游戏中持久化数据一般可分为两种：第一种是静态数据，例如 Excel 数据表中由策划人员编辑的数据，其特点是运行期间程序只需要读取，不需要修改；另一种是游戏存档数据，例如记录玩家在游戏过程中的进度，其特点是运行期间既需要读取，也需要修改，并且在版本升级的时候需要考虑老数据是否需要删除或者重置。

8.1 Excel

策划人员通常都会在 Excel 中配置静态数据，例如道具表，它由主键、道具名称、描述、功能和参数等一系列数据组成。前后端使用道具主键来进行数据的通信，最终前端将主键所包含的整个数据信息展示在游戏中。

8.1.1 EPPlus

在 Windows 下，提供了很多解析 Excel 的方法。但是作为一个跨平台引擎，可能需要在多个平台都能解析 Excel，所以我们需要引用一个第三方 DLL 库 EPPlus 来处理跨平台解析 Excel。首先，需要从它的网站（https://archive.codeplex.com/?p=epplus）上将其下载下来，接着将 DLL 文件拖入 Unity 即可使用了。

8.1.2 读取 Excel

首先，我们需要创建 Excel 文件。如图 8-1 和图 8-2 所示，可以分别在不同的工作表中添加数据，接着在代码中读取这两个工作表中的所有数据。

图 8-1　工作表 1

图 8-2　工作表 2

如代码清单 8-1 所示，根据 Excel 文件的路径得到 `FileStream`，并且创建 `ExcelPackage` 对象，接着就可以用它对 Excel 进行读取了。

代码清单 8-1　Script_08_01.cs 文件

```csharp
using System.Collections;
using System.Collections.Generic;
using UnityEngine;
using UnityEditor;
using System.IO;
using OfficeOpenXml;

public class Script_08_01 {

    [MenuItem("Excel/Load Excel")]
    static void LoadExcel()
    {
        string path = Application.dataPath+ "/Excel/test.xlsx";
        //读取 Excel 文件
        using(FileStream fs = new FileStream(path, FileMode.Open, FileAccess.Read,
            FileShare.ReadWrite)) {
            using(ExcelPackage excel = new ExcelPackage(fs)) {
                ExcelWorksheets workSheets = excel.Workbook.Worksheets;
                //遍历所有工作表
                for(int i = 1; i <= workSheets.Count; i++) {
                    ExcelWorksheet workSheet = workSheets [i];
                    int colCount = workSheet.Dimension.End.Column;
                    //获取当前工作表的名字
                    Debug.LogFormat("Sheet {0}", workSheet.Name);
                    for(int row = 1, count = workSheet.Dimension.End.Row; row <= count;
                        row++) {
                        for(int col = 1; col <= colCount; col++) {
                            //读取每个单元格中的数据
                            var text = workSheet.Cells [row, col].Text ?? "";
                            Debug.LogFormat("下标:{0},{1} 内容:{2}", row, col, text);
                        }
                    }
                }
            }
        }
    }
}
```

在导航菜单栏中选择 Excel→Load Excel 命令，数据已经全部读取出来了，如图 8-3 所示。

图 8-3　读取数据

8.1.3　写入 Excel

首先,需要使用 FileInfo 来创建一个 Excel 文件,接着使用 ExcelPackage 来向 Excel 文件中写入数据,如图 8-4 所示。

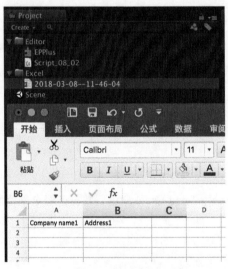

图 8-4　写入数据

如代码清单 8-2 所示,在 ExcelPackage 对象中添加 worksheet 后,即可调用 worksheet.Cells 对每个单元格的行、列赋值,最终保存即可。

代码清单 8-2　Script_08_02.cs 文件

```csharp
using System.Collections;
using System.Collections.Generic;
using UnityEngine;
using UnityEditor;
using System.IO;
using OfficeOpenXml;
using System;

public class Script_08_02 {

    [MenuItem("Excel/Write Excel")]
    static void LoadExcel()
    {
        //创建 Excel 文件
        string path = Application.dataPath+ "/Excel/"+DateTime.Now.ToString
            ("yyyy-MM-dd--hh-mm-ss")+".xlsx";
        var file = new FileInfo(path);

        using(ExcelPackage excel = new ExcelPackage(file)) {
            //向表格中写入数据
            ExcelWorksheet worksheet = excel.Workbook.Worksheets.Add("sheet1");
            worksheet.Cells[1, 1].Value = "Company name1";
            worksheet.Cells[1, 2].Value = "Address1";

            worksheet = excel.Workbook.Worksheets.Add("sheet2");
            worksheet.Cells[1, 1].Value = "Company name2";
            worksheet.Cells[1, 2].Value = "Address2";
            //保存
            excel.Save();
        }
        AssetDatabase.Refresh();
    }
}
```

保存完单元格后，为了在 Unity 中立刻看到效果，需要调用 `AssetDatabase.Refresh()` 方法进行刷新。

8.1.4　JSON

游戏运行时，我们是无法通过 EPPlus 读取 Excel 的，不过我们可以将它保存成自定义格式，例如 CSV、JSON 和 ScriptableObject 等，使用的时候将它读取进来就可以了。Unity 支持 JSON 的序列化和反序列化。需要注意的是，参与序列化的类必须在上方声明`[Serializable]`属性，并且支持类对象的相互嵌套。我们可以使用 `JsonUtility.ToJson()` 以及 `JsonUtility.FromJson<T>()` 来进行序列化以及反序列化。比较遗憾的是，它并不支持字典类型的序列化。如图 8-5 所示，将数据对象转成 JSON 字符串，再从 JSON 字符串还原数据对象，并且将数据输出。相关代码如代码清单 8-3 所示。

图 8-5　JSON

代码清单 8-3　Script_08_03.cs 文件

```
using System.Collections;
using System.Collections.Generic;
using UnityEngine;
using UnityEditor;
using System;

public class Script_08_03  {

    [MenuItem("Excel/Json")]
    static void LoadJson()
    {
        Data data = new Data();
        data.name = "Data";
        data.subData.Add(new SubData(){ intValue = 1, boolValue = true, floatValue =
            0.1f, stringValue = "one" });
        data.subData.Add(new SubData(){ intValue = 2, boolValue = true, floatValue =
            0.1f, stringValue = "two" });

        string json =  JsonUtility.ToJson(data);
        Debug.Log(json);
        data = JsonUtility.FromJson<Data>(json);
        Debug.LogFormat("name = {0}", data.name);
        foreach(var item in data.subData) {
            Debug.LogFormat("intValue = {0} boolValue = {0} floatValue = {0}
                stringValue = {0}",
                    item.intValue, item.boolValue, item.floatValue, item.stringValue);
        }
    }

    [Serializable]
    public class Data
    {
        public string name;
        public List<SubData> subData =new List<SubData>();
    }

    [Serializable]
```

```csharp
    public class SubData
    {
        public int intValue;
        public bool boolValue;
        public float floatValue;
        public string stringValue;
    }
}
```

8.1.5 JSON 支持字典

Unity 的 JSON 是不支持字典的，不过可以继承 `ISerializationCallbackReceiver` 接口，间接地实现字典序列化，相关代码如代码清单 8-4 所示。

代码清单 8-4 Script_08_04.cs 文件

```csharp
using System.Collections;
using System.Collections.Generic;
using UnityEngine;
using UnityEditor;

public class Script_08_04
{

    [MenuItem("Excel/Load Dictionary")]
    static void SerializableDictionary()
    {

        SerializableDictionary<int,string> serializableDictionary =
            new SerializableDictionary<int,string>();
        serializableDictionary [100] = "雨松 momo";
        serializableDictionary [200] = "好好学习";
        serializableDictionary [300] = "天天向上";
        string json = JsonUtility.ToJson(serializableDictionary);
        Debug.Log(json);

        serializableDictionary = JsonUtility.FromJson<SerializableDictionary<int,
            string>>(json);
        Debug.Log(serializableDictionary [100]);

    }

}
```

如代码清单 8-5 所示，序列化两个 List 元素来保存键和值，接着将 C# 的泛型传入，这样键和值就更加灵活了，在 `OnBeforeSerialize()` 和 `OnAfterDeserialize()` 进行序列化和反序列化赋值操作。

代码清单 8-5 SerializableDictionary.cs 文件

```csharp
using System.Collections;
using System.Collections.Generic;
using UnityEngine;
```

```csharp
public class SerializableDictionary<K, V> : ISerializationCallbackReceiver
{
    [SerializeField]
    private List<K> m_keys;
    [SerializeField]
    private List<V> m_values;

    private Dictionary<K, V> m_Dictionary = new Dictionary<K, V>();

    public V this[K key]
    {
        get{
            if(!m_Dictionary.ContainsKey(key))
                return default(V);
            return m_Dictionary[key];
        }
        set{
            m_Dictionary [key] = value;
        }
    }

    public void OnAfterDeserialize() {
        int length = m_keys.Count;
        m_Dictionary = new Dictionary<K, V>();
        for(int i = 0; i < length; i++) {
            m_Dictionary[m_keys[i]] = m_values[i];
        }
        m_keys = null;
        m_values = null;
    }

    public void OnBeforeSerialize() {
        m_keys = new List<K>();
        m_values = new List<V>();

        foreach(var item in m_Dictionary) {
            m_keys.Add(item.Key);
            m_values.Add(item.Value);
        }
    }
}
```

8.2 文件读取与写入

游戏中有很多数据需要在运行期间读取或者写入，最典型的就是游戏存档功能。Unity 自己也提供了一套存档的 API，但是功能比较单一，只支持保存 int、float 和 string 这三种类型。不过 C# 支持文件的读写，我们可以灵活地扩展它。

8.2.1 PlayerPrefs

PlayerPrefs 是 Unity 自带的存档方法,它的优点是使用起来非常方便。引擎已经封装好 GetKey 以及 SetKey 的方法,并且还做保存数据的优化。由于保存数据可能是个耗时操作,频繁地保存可能会带来卡顿,所以 Unity 默认会在应用程序将切入后台时统一保存文件,开发者也可以强制调用 PlayerPrefs.Save() 来保存。

然而它的缺点就是,编辑模式下查看存档非常不方便,macOS 的存档在 ~/Library/Preferences folder 目录下,Windows 的存档在 HKCU\Software\[company name]\[product name]注册表中。如代码清单 8-6 所示,我们使用 PlayerPrefs 对数据进行保存和读取操作。

代码清单 8-6　Script_08_05.cs 文件

```
using System.Collections;
using System.Collections.Generic;
using UnityEngine;

public class Script_08_05 : MonoBehaviour
{
    void Start()
    {
        PlayerPrefs.SetInt("MyInt", 100);
        PlayerPrefs.SetFloat("MyFloat", 200f);
        PlayerPrefs.SetString("MyString", "雨松MOMO");

        Debug.Log(PlayerPrefs.GetInt("MyInt", 0));
        Debug.Log(PlayerPrefs.GetFloat("MyFloat", 0f));
        Debug.Log(PlayerPrefs.GetString("MyString", "没有返回默认值"));

        //判断是否有某个键
        if(PlayerPrefs.HasKey("MyInt")) {

        }
        //删除某个键
        PlayerPrefs.DeleteKey("MyInt");

        //删除所有键
        PlayerPrefs.DeleteAll();

        //强制保存数据
        PlayerPrefs.Save();
    }
}
```

8.2.2 EditorPrefs

在编辑器模式下,Unity 也提供了一组存档功能,它不需要考虑运行时的效率,所有没有采用 PlayerPrefs 优化的方式，而是立即就保存了，其使用方法和 PlayerPrefs 类似，这里就不再赘述。相关代码如代码清单 8-7 所示。

代码清单 8-7   Script_08_06.cs 文件

```
using System.Collections;
using System.Collections.Generic;
using UnityEngine;
using UnityEditor;

public class Script_08_06
{
    [MenuItem("Tools/Save")]
    static void Save()
    {
        EditorPrefs.SetInt("MyInt", 100);
        EditorPrefs.SetFloat("MyFloat", 200f);
        EditorPrefs.SetString("MyString", "雨松MOMO");

        Debug.Log(EditorPrefs.GetInt("MyInt", 0));
        Debug.Log(EditorPrefs.GetFloat("MyFloat", 0f));
        Debug.Log(EditorPrefs.GetString("MyString", "没有返回默认值"));

        //判断是否有某个键
        if(EditorPrefs.HasKey("MyInt")) {

        }
        //删除某个键
        EditorPrefs.DeleteKey("MyInt");

        //删除所有键
        EditorPrefs.DeleteAll();
    }

}
```

8.2.3 PlayerPrefs 保存复杂结构

PlayerPrefs 可以保存字符串，结合 JSON 的序列化和反序列功能，它就可以保存各种复杂的数据结构了。另外，保存存档取决于硬件当时的条件，完全有保存不上的情况，所以可以通过 try...catch 来捕获保存时的错误异常。

如代码清单 8-8 所示，使用 JsonUtility.ToJson()方法将对象保存成 JSON 字符串，读取的时候再使用 JsonUtility.FromJson 将 JSON 字符串还原为类对象。

代码清单 8-8   Script_08_07.cs 文件

```
using System.Collections;
using System.Collections.Generic;
using UnityEngine;
```

```
public class Script_08_07 : MonoBehaviour
{
    void Start()
    {
        //保存游戏存档
        Record record = new Record();
        record.stringValue = "雨松MOMO";
        record.intValue = 200;
        record.names = new List<string>(){ "test1", "test2" };
        string json = JsonUtility.ToJson(record);
        //可以使用try…catch来捕获异常
        try
        {
            PlayerPrefs.SetString("record", json);
        }
        catch(System.Exception err)
        {
            Debug.Log("Got: " + err);
        }
        //读取存档
        record = JsonUtility.FromJson<Record>(PlayerPrefs.GetString("record"));
        Debug.LogFormat("stringValue = {0} intValue={1}", record.stringValue,
            record.intValue);
    }

    //存档对象
    [System.Serializable]
    public class Record
    {
        public string stringValue;
        public int intValue;
        public List<string> names;
    }

}
```

需要注意的是，凡是参与 JSON 序列化的对象都需要标记 `[System.Serializable]` 对象。

8.2.4 TextAsset

TextAsset 是 Unity 提供的一个文本对象，它可以通过 `Resources.Load()` 或者 Asset Bundle 来读取数据，其中数据是 `string` 格式的。当然，我们也可以按 `byte[]` 读取。它支持读取的文本格式包括.txt、.html、.htm、.xml、.bytes、.json、.csv、.yaml 和.fnt。如代码清单 8-9 所示，我们来读取 Resources 目录下的 MyText 文本。

代码清单 8-9　Script_08_08.cs 文件

```
using System.Collections;
using System.Collections.Generic;
using UnityEngine;
```

```csharp
public class Script_08_08 : MonoBehaviour
{
    void Start()
    {
        Debug.Log(Resources.Load<TextAsset>("MyText").text);
    }
}
```

8.2.5 读写文本

Unity 可以利用 C# 的 `File` 类来读写文本，此时只需要提供一个目录即可。这里需要注意的是，编辑器模式下读写文本是很方便的，但是一旦打包发布，Assets/目录都不存在了，运行时是无法读取它目录下的文本的。

如代码清单 8-10 所示，通过 `File.WriteAllText()` 和 `File.ReadAllText()` 来对文本进行读取和写入。

代码清单 8-10　Script_08_09.cs 文件

```csharp
using System.Collections;
using System.Collections.Generic;
using UnityEngine;
using UnityEditor;
using System.IO;
using System.Text;

public class Script_08_09
{
    [MenuItem("Tool/File")]
    static void Start()
    {
        string path = Path.Combine(Application.dataPath, "test.txt");
        //如果文件存在，就删除它
        if(File.Exists(path)) {
            File.Delete(path);
        }

        //写入文件
        StringBuilder sb = new StringBuilder();
        sb.AppendFormat ("第一行:{0}", 100).AppendLine();
        sb.AppendFormat ("第二行:{0}", 200).AppendLine();
        File.WriteAllText(path,sb.ToString());

        //读取文件
        Debug.Log(File.ReadAllText(path));
    }
}
```

8.2.6 运行期读写文本

在游戏运行期间，只有 Resources 和 StreamingAssets 目录具有读取权限，其中 Resources 用来读取游戏资源，而 StreamingAssets 可使用 File 类来读取文件（除了个别平台外），但都是只读的，并不能写。只有 Application.persistentDataPath 目录是可读、可写的。如图 8-6 所示，我们分别来读写这 3 个目录，最终需要打一个 Mac 包来运行测试，相关代码如代码清单 8-11 所示。

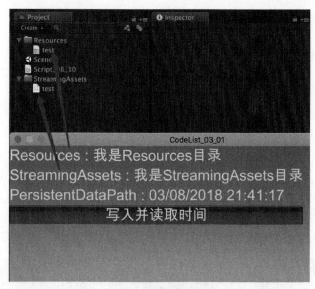

图 8-6　读写文件

代码清单 8-11　Script_08_10.cs 文件

```
using System.Collections;
using System.Collections.Generic;
using UnityEngine;
using System.IO;
using System.Text;
using System;

public class Script_08_10 : MonoBehaviour
{
    //可读不可写
    string m_ResourcesTxt = string.Empty;
    //可读不可写
    string m_StreamingAssetsTxt = string.Empty;
    //可读可写
    string m_PersistentDataTxt = string.Empty;

    void Start()
    {
        m_ResourcesTxt = Resources.Load<TextAsset>("test").text;
        m_StreamingAssetsTxt = File.ReadAllText(System.IO.Path.Combine(Application.
            streamingAssetsPath, "test.txt"));
    }

    void OnGUI()
```

```
    {
        GUILayout.Label(string.Format("<size=50>Resources : {0}</size>",
            m_ResourcesTxt));
        GUILayout.Label(string.Format("<size=50>StreamingAssets : {0}</size>",
            m_StreamingAssetsTxt));
        GUILayout.Label(string.Format("<size=50>PersistentDataPath : {0}</size>",
            m_PersistentDataTxt));

        if(GUILayout.Button("<size=50>写入并读取时间</size>")) {
            string path = Path.Combine(Application.persistentDataPath, "test.txt");
            File.WriteAllText(path,DateTime.Now.ToString());
            m_PersistentDataTxt = File.ReadAllText(path);
        }
    }
}
```

8.2.7 PersistentDataPath 目录

PersistentDataPath 目录本身并没有什么问题，但是如果平常开发中也在这个目录下进行读写操作的话，就会比较麻烦，因为它在 Windows 以及 Mac 下的目录是很难找的。例如，开发过程中需要验证一下保存的文件是否正确，我们随时都需要很快地找到它，如图 8-7 所示。

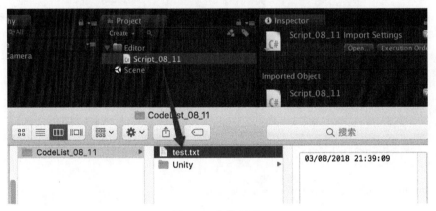

图 8-7 定位目录

如代码清单 8-12 所示，调用 `EditorUtility.RevealInFinder()` 方法，就可以立即定位到指定目录。

代码清单 8-12　Script_08_11.cs 文件

```
using System.Collections;
using UnityEngine;
using UnityEditor;

public class Script_08_11 : MonoBehaviour
{
    [MenuItem("Assets/Open PersistentDataPath")]
```

```
    static void Open()
    {
        EditorUtility.RevealInFinder(Application.persistentDataPath);
    }
}
```

8.2.8 游戏存档

我们需要一个不影响开发的存档,并且查看要非常方便。如图 8-8 所示,可以自己写一个存档类,在编辑模式下将存档保存在 Assets 同级目录下,这样查看存档内容就方便多了。

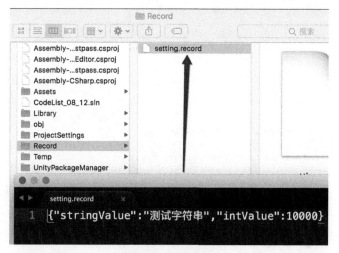

图 8-8 游戏存档

如代码清单 8-13 所示,在编辑模式下将数据保存于 Application.dataPath 目录,而在真实环境下将数据保存在 Application.persistentDataPath 下。只有调用 Save() 方法时,数据才会被强制写入。

代码清单 8-13 RecordUtil.cs 文件

```
using System.Collections;
using System.Collections.Generic;
using UnityEngine;
using System.IO;
using System;
using System.Text;

public class RecordUtil {

    //游戏存档保存的根目录
    static string RecordRootPath
    {
        get{
            #if(UNITY_EDITOR || UNITY_STANDALONE)
                return Application.dataPath + "/../Record/";
```

```csharp
        #else
            return Application.persistentDataPath + "/Record/";
        #endif
    }
}

//游戏存档
static Dictionary<string,string> recordDic = new Dictionary<string, string>();
//标记某个游戏存档是否需要重新写入
static List<string> recordDirty =new List<string>();
//标记某个游戏存档是否需要删除
static List<string> deleteDirty =new List<string>();
//表示某个游戏存档读取时需要重新从文件中读取
static List<string> readDirty =new List<string>();

static private readonly UTF8Encoding UTF8 = new UTF8Encoding(false);

static RecordUtil()
{
    readDirty.Clear();

    if(Directory.Exists(RecordRootPath)) {
        foreach(string file in Directory.GetFiles(RecordRootPath,"*.record",
            SearchOption.TopDirectoryOnly)) {

            string name = Path.GetFileNameWithoutExtension(file);
            if(!readDirty.Contains(name)) {
                readDirty.Add(name);
                Get(name);
            }
        }
    }
}




//强制写入文件
public static void Save()
{
    foreach(string key in deleteDirty) {
        try {
            string path = Path.Combine(RecordRootPath, key + ".record");
            if(recordDirty.Contains(key)){
                recordDirty.Remove(key);
            }
            if(File.Exists(path)){
                File.Delete(path);
            }
        } catch(Exception ex) {
            Debug.LogError(ex.Message);
        }
    }
    deleteDirty.Clear();
```

```csharp
        foreach(string key in recordDirty) {
            string value;
            if(recordDic.TryGetValue(key, out value)) {
                if(!readDirty.Contains(key)) {
                    readDirty.Add(key);
                }
                string path = Path.Combine(RecordRootPath, key + ".record");
                recordDic [key] = value;
                try {
                    Directory.CreateDirectory(Path.GetDirectoryName(path));
                    File.WriteAllText(path, value, UTF8);
                } catch(Exception ex) {
                    Debug.LogError(ex.Message);
                }

            }
        }
        recordDirty.Clear();
    }



    public static void Set(string key,string value)
    {
        recordDic [key] = value;
        if(!recordDirty.Contains(key)) {
            recordDirty.Add(key);
        }
        #if UNITY_EDITOR || UNITY_STANDALONE
        Save();
        #endif
    }

    public static string Get(string key)
    {
        return Get(key, string.Empty);
    }

    public static string Get(string key,string defaultValue)
    {
        if(readDirty.Contains(key)) {
            string path = Path.Combine(RecordRootPath, key + ".record");
            try {
                string readStr = File.ReadAllText(path, UTF8);
                recordDic [key] = readStr;
            } catch(Exception ex) {
                Debug.LogError(ex.Message);
            }

            readDirty.Remove(key);
        }

        string value;
```

```
        if(recordDic.TryGetValue(key, out value)) {
            return value;
        } else {
            return defaultValue;
        }
    }

    public static void Delete(string key)
    {
        if(recordDic.ContainsKey(key)) {
            recordDic.Remove(key);
        }
        if(!deleteDirty.Contains(key)) {
            deleteDirty.Add(key);
        }

        #if UNITY_EDITOR || UNITY_STANDALONE
        Save();
        #endif
    }
}
```

在编辑模式下,数据发生改变时,会立刻写入文件并保存,而真实环境下出于性能考虑,可以在某个特定的时间点保存数据。

如代码清单 8-14 所示,由于是自己写的存档类,处理起来会更加灵活。当然,我们也可以仿照 PlayerPrefs 在应用程序即将进入后台时保存。当调用 OnApplicationPause()时,表示应用进入后台后再保存数据。

代码清单 8-14　Script_08_12.cs 文件

```
using System.Collections;
using UnityEngine;
using UnityEditor;

public class Script_08_12 : MonoBehaviour
{
    void Start()
    {
        Setting setting = new Setting();
        setting.stringValue = "测试字符串";
        setting.intValue = 10000;

        RecordUtil.Set("setting",JsonUtility.ToJson(setting));


    }

    private Setting m_Setting = null;
    void OnGUI()
```

```
        {
            if(GUILayout.Button("<size=50>获取存档</size>")) {
                m_Setting = JsonUtility.FromJson<Setting>(RecordUtil.Get("setting"));
            }
            if(m_Setting != null) {
                GUILayout.Label(string.Format("<size=50> {0},{1} </size>", m_Setting.
                    intValue, m_Setting.stringValue));
            }
        }

        void OnApplicationPause(bool pauseStatus)
        {
            //当游戏即将进入后台时,保存存档
            if(pauseStatus) {
                RecordUtil.Save();
            }
        }


        [System.Serializable]
        class Setting
        {
            public string stringValue;
            public int intValue;
        }

    }
```

默认情况下,应用进入后台后才会保存数据。如果应用进入后台之前发生了闪退现象,那么数据就无法保存了,所以某些非常重要的数据需要强制调用 Save() 方法。

8.3 XML

XML 在开发中使用也很频繁,此时要以标签的形式来组织数据结构。C# 提供了创建、解析、修改和查询等方法,可以很方便地操作它。

8.3.1 创建 XML

操作 XML 时,需要用到 System.Xml 命名空间。如图 8-9 所示,我们可以在运行时动态创建 XML 字符串,并且在节点下添加数据。

图 8-9 创建 XML 字符串

如代码清单 8-15 所示,首先需要引用 System.Xml 命名空间,接着创建 XmlDocument 对象,然后就可以给 XML 节点添加数据了。

代码清单 8-15　Script_08_13.cs 文件

```csharp
using UnityEngine;
using System.IO;
using System.Xml;

public class Script_08_13 : MonoBehaviour
{
    void Start()
    {
        //创建 XmlDocument
        XmlDocument xmlDoc = new XmlDocument();
        XmlDeclaration xmlDeclaration = xmlDoc.CreateXmlDeclaration("1.0",
            "UTF-8", null);
        xmlDoc.AppendChild(xmlDeclaration);

        //在节点中写入数据
        XmlElement root = xmlDoc.CreateElement("XmlRoot");
        xmlDoc.AppendChild(root);
        XmlElement group = xmlDoc.CreateElement("Group");
        group.SetAttribute("username", "雨松momo");
        group.SetAttribute("password", "123456");
        root.AppendChild(group);

        //读取节点并输出 XML 字符串
        using(StringWriter stringwriter = new StringWriter())
        {
            using(XmlTextWriter xmlTextWriter = new XmlTextWriter(stringwriter))
            {
                xmlDoc.WriteTo(xmlTextWriter);
                xmlTextWriter.Flush();
                Debug.Log(stringwriter.ToString());
            }
        }
    }
}
```

8.3.2　读取与修改

XML 可作为字符串来传递。如图 8-10 所示,我们可以动态读取 XML 字符串中的内容,并且修改它的内容,以重新生成新的 XML 字符串。

图 8-10　读取 XML 字符串

如代码清单 8-16 所示，创建 XmlDocument 对象后，需要读取 XML 文件，通过循环可以遍历所有子节点对它们进行修改。

代码清单 8-16　Script_08_14.cs 文件

```csharp
using UnityEngine;
using System.IO;
using System.Xml;

public class Script_08_14 : MonoBehaviour
{
    void Start()
    {
        //xml 字符串
        string xml ="<?xml version=\"1.0\" encoding=\"UTF-8\"?><XmlRoot><Group 
            username=\"雨松 momo\" password=\"123456\" /></XmlRoot>";

        //读取字符串 xml
        XmlDocument xmlDoc = new XmlDocument();
        xmlDoc.LoadXml(xml);
        //遍历节点
        XmlNode nodes = xmlDoc.SelectSingleNode("XmlRoot");
        foreach(XmlNode node in nodes.ChildNodes)
        {
            string username = node.Attributes["username"].Value;
            string password = node.Attributes["password"].Value;
            Debug.LogFormat("username={0} password={1}", username, password);
            //修改其中一条数据
            node.Attributes["password"].Value = "88888888";
        }

        //读取节点并输出 XML 字符串
        using(StringWriter stringwriter = new StringWriter())
        {
            using(XmlTextWriter xmlTextWriter = new XmlTextWriter(stringwriter))
            {
                xmlDoc.WriteTo(xmlTextWriter);
                xmlTextWriter.Flush();
                Debug.Log(stringwriter.ToString());
            }
        }
    }
}
```

8.3.3　XML 文件

XmlDocument 类也提供了从文件中读取 XML，或者将 XML 写入本地路径的方法。如图 8-11 所示，将 XML 写入本地文件，读取后再输出节点中的内容。

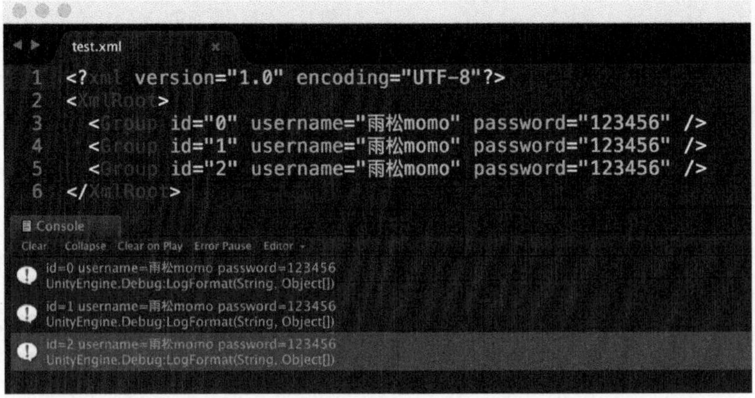

250 第 8 章 持久化数据

```
                                test.xml
                         1   <?xml version="1.0" encoding="UTF-8"?>
                         2   <XmlRoot>
                         3       <Group id="0" username="雨松momo" password="123456" />
                         4       <Group id="1" username="雨松momo" password="123456" />
                         5       <Group id="2" username="雨松momo" password="123456" />
                         6   </XmlRoot>

    Console
    Clear   Collapse   Clear on Play   Error Pause   Editor
      id=0 username=雨松momo password=123456
      UnityEngine.Debug:LogFormat(String, Object[])
      id=1 username=雨松momo password=123456
      UnityEngine.Debug:LogFormat(String, Object[])
      id=2 username=雨松momo password=123456
      UnityEngine.Debug:LogFormat(String, Object[])
```

图 8-11 XML 文件的读取与写入

如代码清单 8-17 所示，我们对 XML 文件进行了读取与写入。

代码清单 8-17　Script_08_15.cs 文件

```csharp
using UnityEngine;
using System.IO;
using System.Xml;
using UnityEditor;

public class Script_08_15
{
    [MenuItem("XML/WriteXml")]
    static void WriteXml()
    {
        string xmlPath = Path.Combine(Application.dataPath, "test.xml");
        //如果 XML 文件已经存在，就删除它
        if(File.Exists(xmlPath))
        {
            File.Delete(xmlPath);
        }
        //创建 XmlDocument
        XmlDocument xmlDoc = new XmlDocument();
        XmlDeclaration xmlDeclaration = xmlDoc.CreateXmlDeclaration("1.0", "UTF-8",
            null);
        xmlDoc.AppendChild(xmlDeclaration);

        //在节点中写入数据
        XmlElement root = xmlDoc.CreateElement("XmlRoot");
        xmlDoc.AppendChild(root);

        //循环写入 3 条数据
        for(int i = 0; i < 3; i++)
        {
            XmlElement group = xmlDoc.CreateElement("Group");
            group.SetAttribute("id",i.ToString());
            group.SetAttribute("username", "雨松 momo");
            group.SetAttribute("password", "123456");
```

```
        root.AppendChild(group);
    }
    //写入文件
    xmlDoc.Save(xmlPath);
    AssetDatabase.Refresh();
}

[MenuItem("XML/LoadXml")]
static void LoadXml()
{
    string xmlPath = Path.Combine(Application.dataPath, "test.xml");
    //XML 文件只有存在，才能读取
    if(File.Exists(xmlPath))
    {
        XmlDocument xmlDoc = new XmlDocument();
        xmlDoc.Load(xmlPath);
        //遍历节点
        XmlNode nodes = xmlDoc.SelectSingleNode("XmlRoot");
        foreach(XmlNode node in nodes.ChildNodes)
        {
            string id = node.Attributes["id"].Value;
            string username = node.Attributes["username"].Value;
            string password = node.Attributes["password"].Value;
            Debug.LogFormat("id={0} username={1} password={2}", id,username,
                password);
        }
    }
}
```

8.4　YAML

前面我们介绍了 JSON 和 XML 其实已经在大量使用了，当数据多了以后，JSON 有个最大的问题，那就是可读性很差。XML 比 JSON 的可读性会好一些。无论 JSON 还是 XML，编辑都会很麻烦，它们的数据格式要求很严格，少写括号或者逗号都不行。

Unity 就没有使用 JSON 或者 XML 来描述结构，采取的是 YAML 格式。如图 8-12 所示，它的预览性以及编辑性都非常好，数据与变量通过冒号来连接。例如，游戏中一些服务器列表的配置，或者调试性的开关，不太方便配置在表格中的数据，或者修改比较频繁的数据都可以使用 YAML 来配置，随时用随时改。

图 8-12　YAML 格式

8.4.1 YamlDotNet

YAML 提供了 .NET 的类库，即 YamlDotNet。Unity 中直接提供了插件，它可以在 Asset Store 中免费下载到。如图 8-13 所示，YamlDotNet 支持 PC 和移动端，下载后导入工程就可以使用了。

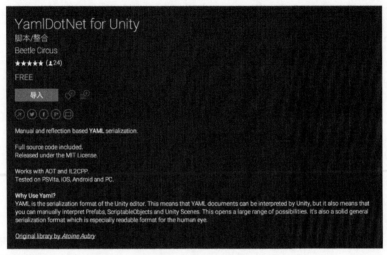

图 8-13　YamlDotNet

8.4.2 序列化和反序列化

YamlDotNet 提供了运行时序列化和反序列化的接口。这里需要注意的是，对于参与序列化的类中的变量，其属性必须设置成 get 或者 set，不然无法序列化。如图 8-14 所示，在程序运行中，可以序列化和反序列化数据。

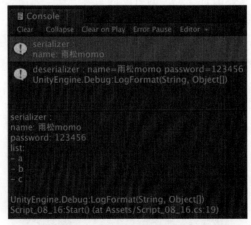

图 8-14　序列化和反序列化

如代码清单 8-18 所示，使用 Serialize() 和 Deserialize() 方法就可以进行序列化和反序列化操作。

代码清单 8-18　Script_08_16.cs 文件

```csharp
using System.Collections.Generic;
using UnityEngine;
using YamlDotNet.Serialization;

public class Script_08_16 : MonoBehaviour
{
    private void Start()
    {
        //创建对象
        Data data = new Data();
        data.name = "雨松 momo";
        data.password = "123456";
        data.list = new List<string>(){"a","b","c"};

        //序列化 YAML 字符串
        Serializer serializer = new Serializer();
        string yaml = serializer.Serialize(data);
        Debug.LogFormat("serializer : \n{0}",yaml);

        //反序列化成类对象
        Deserializer deserializer = new Deserializer();
        Data data1 = deserializer.Deserialize<Data>(yaml);
        Debug.LogFormat("deserializer : name={0} password={1}",
            data1.name,data1.password);
    }

    class Data
    {
        public string name { get; set; }
        public string password { get; set; }
        public List<string> list { get; set; }
    }
}
```

8.4.3　读取配置

在游戏中，一些服务器列表或者一些临时调试的配置信息，可能需要频繁地添加、删除和修改等。由于这些测试数据都是临时性的，就不太适合配置在 Excel 表格中。如图 8-15 所示，当包打出来以后，直接修改配置文件就可以立即生效到游戏中。在 YAML 中，可以使用"#"符号来表示注释部分。

254 | 第 8 章 持久化数据

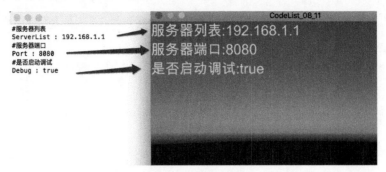

图 8-15 YAML 配置

如代码清单 8-19 所示,运行中通过 YamlStream() 来读取数据。

代码清单 8-19 Script_08_17.cs 文件

```csharp
using System.Collections.Generic;
using System.IO;
using UnityEngine;
using YamlDotNet.RepresentationModel;

public class Script_08_17 : MonoBehaviour
{
    private IDictionary<YamlNode, YamlNode> m_MappingData;
    private void Start()
    {
        //读取 YAML 字符串
        string document = File.ReadAllText(Path.Combine(Application.streamingAssetsPath,
            "yaml.txt"));
        var input = new StringReader(document);
        var yaml = new YamlStream();
        yaml.Load(input);

        //读取 root 节点
        var mapping = 
            (YamlMappingNode)yaml.Documents[0].RootNode;

        m_MappingData = mapping.Children;
    }


    private void OnGUI()
    {
        GUILayout.Label(string.Format("<size=50>服务器列表:{0}</size>",
            m_MappingData["ServerList"]));
        GUILayout.Label(string.Format("<size=50>服务器端口:{0}</size>",
            m_MappingData["Port"]));
        GUILayout.Label(string.Format("<size=50>是否启动调试:{0}</size>",
            m_MappingData["Debug"]));
    }

}
```

8.5 小结

本章中，我们学习了游戏存档。存档可分为静态存档和动态存档。静态存档时，在游戏运行过程中，对它只能读取不能写入。这好比 Excel 表格数据，最终表格数据可在编辑模式下利用 EPPlus 转成程序可读文件类型。动态存档应用得就更多了，它在玩游戏的过程中记录玩家游戏的进度，或者一些设置选项，Unity 提供了 `PlayerPrefs` 类来处理存档的读与写，我们也可以利用 C# 的 `File` 类来自行保存存档。

第 9 章

静态对象

静态对象是 Unity 提供的一个属性,它可以附加在游戏对象或者 Prefab 上。它的原理是限制物体在运行中不能发生位移变化,预先生成一些辅助的数据,从而达成一种用内存换时间的优化方式。静态元素的种类很多。如图 9-1 所示,选择任意游戏对象,单击右上角的 Static 下拉框,即可设置该对象的静态元素了,具体如下。

- Lightmap Static:用来表示接受烘焙光照计算,可烘焙光照贴图。
- Occluder Static:表示自身是否可以遮挡其他元素。
- Batching Static:表示支持静态合批。
- Navigation Static:表示可烘焙寻路网格。
- Occludee Static:表示自身可以被遮挡剔除掉。
- Off Mesh Link Generation:寻路连接不同区域的点,就像角色从山顶跳下来。
- Reflection Probe Static:反射探头,就像玻璃反射一样的镜面效果。

图 9-1 静态对象

9.1 Lightmap

Lightmap 技术的原理是将场景中的灯光与物体产生的光照与阴影信息烘焙在一张或者多张 Lightmap 贴图中,这些物体将不再参与实时光照计算,从而减少了大量的性能开销。它的缺点就是参与烘焙计算的对象在游戏过程中不能发生移动,所以游戏中通常会将物体分成两类:一类是可发生位移变化的,它们使用实时光照计算;另一类是不可发生位移变化的,它们采取预先烘焙 Lightmap。

9.1.1 设置烘焙贴图

首先,需要在场景中选中需要参与烘焙计算的游戏对象,如图 9-2 所示。接着,下面就会出现烘焙信息参数了，我们可以单独调整某一个对象。

图 9-2　设置 Lightmap 参数

当烘焙对象都设置完毕后，在导航菜单栏中选择 Window→Lighting→Setting 命令，即可打开烘焙面板，如图 9-3 所示。同样，设置完烘焙参数后，单击右下角的 Bake Reflection Probes 命令，即可开始烘焙。如果选中了左边的 Auto Generate 复选框，将会自动烘焙。但是如果场景中元素很多，可能会造成卡顿，因此不建议开启。

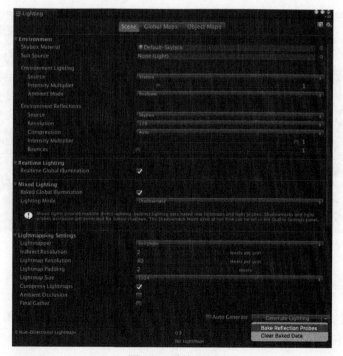

图 9-3　烘焙面板

9.1.2 实时光和焙贴光共存

在游戏中，少部分物体确实需要实时光，例如控制主角移动时，需要动态地产生光照和阴影信息。如图 9-4 所示，可以在 Mode 中设置灯光的属性，其中 Realtime 表示实时光，Mixed 表示实时光和烘焙光的混合模式，Baked 表示仅烘焙光。所以，游戏中更多的会使用 Mixed 模式。

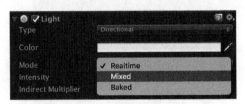

图 9-4　设置灯光的属性

9.1.3 灯光管理

游戏做到后期，光源是非常多的，如何管理就是个问题。新版的 Unity 提供了管理光源的菜单，在导航菜单栏中选择 Window→Lighting→Light Explorer 命令即可，如图 9-5 所示。我们可以快速设置灯光开关状态、灯光类型和模式等，并且点击其中一个光源，即可快速在 Scene 视图中找到它，使用起来确实很方便。

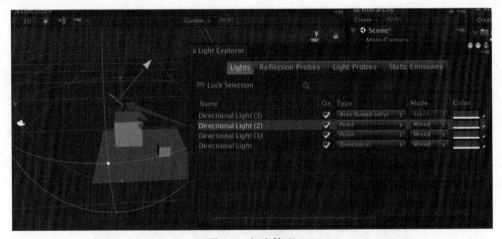

图 9-5　灯光管理

9.1.4 运行时更换烘焙贴图

如果游戏中有一个白天场景和夜晚场景，那么就需要烘焙出多张烘焙贴图了。在程序中，可以动态更换白天和夜晚的烘焙贴图，如图 9-6 所示。

9.1 Lightmap

图 9-6 更换烘焙贴图

如代码清单 9-1 所示，首先创建 `LightmapData` 对象，最终将需要更换的烘焙贴图放入 `LightmapSettings.lightmaps` 中即可。

代码清单 9-1 Script_09_01.cs 文件

```csharp
using UnityEngine;

public class Script_09_01 : MonoBehaviour
{
    //烘焙贴图1
    public Texture2D lightmap1;
    //烘焙贴图2
    public Texture2D lightmap2;

    void OnGUI()
    {
        if(GUILayout.Button("<size=50>lightmap1</size>"))
        {
            LightmapData data = new LightmapData();
            data.lightmapColor = lightmap1;
            LightmapSettings.lightmaps = new LightmapData[1]{data};
        }

        if(GUILayout.Button("<size=50>lightmap2</size>"))
        {
            LightmapData data = new LightmapData();
            data.lightmapColor = lightmap2;
            LightmapSettings.lightmaps = new LightmapData[1]{data};
        }
    }
}
```

在上述代码中，我们通过点击按钮来动态切换烘焙贴图，例如切换白天与夜晚的效果。

9.1.5 动态更换游戏对象

光照和阴影信息都记录在烘焙贴图上，但是如果需要动态地加载 Prefab，就没有烘焙信息了，此时可以给它绑定一个脚本，在生成 Prefab 的同时将烘焙信息写入这个脚本中，以便在实例化 Prefab 时再将信息写入。

如图 9-7 所示，当场景烘焙完后，选择任意游戏对象，然后在菜单中选择 Light→ToPrefab 命令，接着在代码中智能判断这个对象是否已经生成 Prefab，如果没有生成，则创建新的，最终将烘焙信息序列化在 PrefabLightmap 脚本中。当以后这个 Prefab 实例化进场景时，将保存的烘焙预制信息重新赋值给它即可。

图 9-7 保存 Prefab

如图 9-8 所示，单击 Load 按钮后，在代码中实例化 Prefab，相关代码如代码清单 9-2 所示。

代码清单 9-2　Script_09_02.cs 文件

```
using UnityEngine;

public class Script_09_02 : MonoBehaviour
{
    public GameObject prefab;

    void OnGUI()
    {
        if(GUILayout.Button("<size=50>Load</size>"))
        {
            GameObject.Instantiate<GameObject>(prefab);
        }
    }
}
```

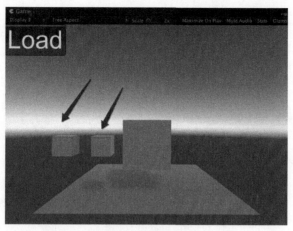

图 9-8　实例化 Prefab

如代码清单 9-3 所示，Prefab 对象绑定了 PrefabLightmap 脚本，所以在使用 Awake() 的时候，可以将之前保存的烘焙信息重新赋值给它。

代码清单 9-3　PrefabLightmap.cs 文件

```
using System.Collections;
using System.Collections.Generic;
using UnityEngine;
#if UNITY_EDITOR
using UnityEditor;
#endif

public class PrefabLightmap : MonoBehaviour {

    public int lightmapIndex;
    public Vector4 lightmapScaleOffset;

    void Awake()
    {
        //Prefab 实例化后赋值
        Renderer renderer = GetComponent<Renderer>();
        if(renderer) {
            renderer.lightmapIndex = lightmapIndex;
            renderer.lightmapScaleOffset = lightmapScaleOffset;
        }
    }
#if UNITY_EDITOR
[MenuItem("GameObject/Light/ToPrefab")]
    static void ToPrefab()
    {
        //确保选择 Hierarchy 视图下的一个游戏对象
        if(Selection.activeTransform) {
            Renderer renderer = Selection.activeTransform.GetComponent<Renderer>();
            //确保有 renderer 组件
            if(renderer) {
```

```
            PrefabLightmap prefabLightmap = Selection.activeTransform.
                GetComponent<PrefabLightmap>();
            if(!prefabLightmap) {
                prefabLightmap = Selection.activeTransform.gameObject.
                    AddComponent<PrefabLightmap>();
            }
            prefabLightmap.lightmapIndex = renderer.lightmapIndex;
            prefabLightmap.lightmapScaleOffset = renderer.lightmapScaleOffset;

            Object prefab = PrefabUtility.GetPrefabParent(renderer.gameObject);
            //如果有 Prefab 文件，就更新；没有就创建新的
            if(prefab) {
                PrefabUtility.ReplacePrefab(Selection.activeTransform.gameObject,
                    prefab, ReplacePrefabOptions.ConnectToPrefab);
            } else {
                PrefabUtility.CreatePrefab(string.Format ("Assets/Resources/{0}.
                    prefab", Selection.activeTransform.name), Selection.
                    activeTransform.gameObject, ReplacePrefabOptions.
                    ConnectToPrefab);
            }
        }
    }
    #endif
}
```

通过上述代码可以看到，更换烘焙贴图实际上就是设置正确的 `lightmapIndex` 和 `lightmapScaleOffset`。

9.1.6 复制游戏对象

光照和阴影信息场景烘焙完毕后，如果直接按 Command+D 快捷键来复制游戏对象，烘焙信息就是不对的，如图 9-9 所示，必须要重新烘焙才行。如果不想重新烘焙，可以自己拓展一个菜单，定义一个新的快捷键 Command+Shift+D 来执行复制游戏对象的操作，并且动态设置烘焙信息给它。如图 9-10 所示，新复制出来的对象中烘焙信息就正确了。需要注意的是，我们只能复制物体身上的光照烘焙信息，物体产生的阴影是无法复制的。

图 9-9　普通复制

图 9-10　自定义复制

如代码清单 9-4 所示，复制游戏对象的同时，将 lightmapIndex 和 lightmapScaleOffset 信息赋值给新对象即可。

代码清单 9-4　Script_09_03.cs 文件

```
using UnityEngine;
using UnityEditor;
using System.Collections.Generic;

public class Script_09_03 : MonoBehaviour
{

    [MenuItem("Tool/DuplicateGameObject %#d")]
    static void DuplicateGameObject()
    {
        if(Selection.activeTransform)
        {
            Dictionary<string, Renderer> save = new Dictionary<string, Renderer>();

            //根据相对路径保存Renderer信息
            foreach(var renderer in Selection.activeTransform.GetComponentsInChildren
                <Renderer>()) {
                string path = AnimationUtility.CalculateTransformPath(renderer.
                    transform, Selection.activeTransform);
                save [path] = renderer;
            }
            //执行复制
            EditorApplication.ExecuteMenuItem("Edit/Duplicate");
            //还原烘焙信息
            foreach(var renderer in Selection.activeTransform.GetComponentsInChildren
                <Renderer>()) {
                string path = AnimationUtility.CalculateTransformPath(renderer.
                    transform, Selection.activeTransform);
                if(save.ContainsKey(path)) {
                    renderer.lightmapIndex = save [path].lightmapIndex;
                    renderer.lightmapScaleOffset = save [path].lightmapScaleOffset;
                }
            }
        }
    }
}
```

9.2　遮挡剔除

　　游戏中的元素非常多，但是摄像机能看到的内容是有限的，并且有些元素会被另外一些元素挡住，例如城墙一类的，城墙后面的元素就会被它挡住。如果不处理的话，这些元素也会带来一定的开销，此时可以使用遮挡剔除技术来剔除掉这些被挡住的元素。如图 9-11 所示，只有摄像机能看到的内容才会被动态保留下来。

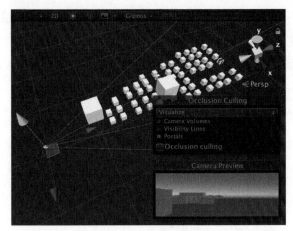

图 9-11　遮挡剔除

9.2.1　遮挡与被遮挡

遮挡关系是由遮挡物与被遮挡物构成的，例如一面墙后面放了很多元素，那么墙属于遮挡物，元素就属于被遮挡物。按照遮挡剔除的原理，墙后面的元素会被剔除掉，这样就会有一个新问题：如果墙是一面透明的墙，显示时它就不会挡住后面的元素了。因此，我们需要设置元素的遮挡与被遮挡关系了。

首先，在场景中将需要参与遮挡以及被遮挡的游戏对象中，选中 Occluder Static 和 Occludee Static 标记；接着在导航菜单栏中选择 Window→Occlusion Culling 命令，打开烘焙面板，如图 9-12 所示。我们可以在这里设置最小的遮挡距离、最小的遮挡空隙以及背面的阈值。最后，单击 Bake 按钮，即可烘焙当前场景。烘焙结束后，Unity 会自动在场景所在的位置创建一个同名的文件夹，并且往其中放入 OcclusionCullingData.asset 文件。

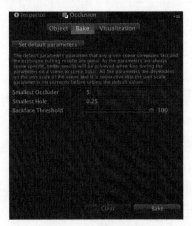

图 9-12　遮挡烘焙

运行游戏后，移动摄像机的位置，当墙完全挡住背景的元素时，将自动剔除背景墙后面的元素，如图 9-13 所示。

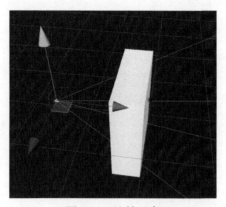

图 9-13　遮挡元素

如果这面墙是透明的，那么当背景元素被剔除时，显示就有问题了，此时墙后面的元素可以取消选择 Occludee Static 标志。如图 9-14 所示，这样无论如何移动摄像机，墙后面的元素都不会被剔除掉。如果墙后的元素同样也是一面墙，并且还需要剔除后面的元素，它自身只需要选择 Occluder Static 标志即可。

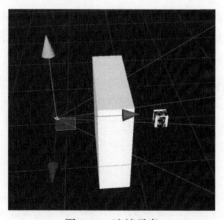

图 9-14　遮挡元素

9.2.2　遮挡与被遮挡事件

当发生遮挡剔除时，Unity 会自动调用 `gameObject.SetActive(false)` 方法，这样整个对象的渲染就会被暂停，直到它重新被启动。如代码清单 9-5 所示，可以监听 `OnBecameInvisible()` 和 `OnBecameVisible()` 方法来处理即将隐藏或显示的逻辑。

代码清单 9-5　OcclusionListener.cs 文件

```
using System.Collections;
using System.Collections.Generic;
using UnityEngine;
using UnityEngine.Events;

public class OcclusionEvent : UnityEvent<GameObject>{}


public class OcclusionListener : MonoBehaviour
{
    public static OcclusionEvent onInvisible = new OcclusionEvent();
    public static OcclusionEvent onVisible = new OcclusionEvent();

    //隐藏状态
    void OnBecameInvisible()
    {
        onInvisible.Invoke(gameObject);
    }
    //显示状态
    void OnBecameVisible()
    {
        onVisible.Invoke(gameObject);
    }

}
```

如代码清单 9-6 所示，我们在代码初始化的地方，自动给所有的 Renderer 组件挂上脚本，统一监听它的剔除以及显示的方法。如果想主动判断某个对象是否在摄像机显示区域内，也可以调用 Renderer.isVisible() 方法。

代码清单 9-6　Script_09_04.cs 文件

```
using UnityEngine;

public class Script_09_04 : MonoBehaviour
{

    void Start()
    {
        foreach(var item in GameObject.FindObjectsOfType<Renderer>()) {
            item.gameObject.AddComponent<OcclusionListener>();
        }

        OcclusionListener.onInvisible.AddListener(delegate(GameObject gameObject) {
            Debug.LogFormat("gameobject {0} 隐藏",gameObject);
        });
        OcclusionListener.onVisible.AddListener(delegate(GameObject gameObject) {
            Debug.LogFormat("gameobject {0} 显示",gameObject);
        });
    }
}
```

9.2.3 动态剔除

在游戏对象中，一旦勾选 Occluder Static 或 Occludee Static 标记，运行期间就无法修改它们的 Transform 信息了。如图 9-15 所示，可以在 Mesh Renerer 组件中勾选 Dynamic Occluded 复选框，表示它将被动态剔除掉。

图 9-15 动态剔除设置

运行游戏后，在 Scene 视图中将 Mesh Renderer 移出摄像机的显示区域，它立刻就被剔除掉了，如图 9-16 所示。注意它只会剔除掉渲染，Update 还是会更新。默认情况下，建议选中 Dynamic Occluded 复选框。

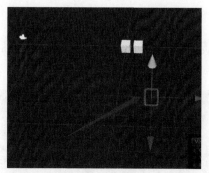

图 9-16 动态剔除

9.2.4 自定义遮挡剔除

遮挡剔除虽然很方便，但也未必是好事。如果参与烘焙的元素多了，每次移动摄像机时，遮挡剔除会产生大量的计算，尤其移动平台更为明显。

其实，我们可以自己来实现遮挡剔除。比如，可以将场景上的元素按位置来划分成若干个格子，每个格子里面就是场景中的游戏对象了，如图 9-17 所示。无论游戏场景有多大，玩家同一时刻关心的只有 1~9 这些区域中的元素。当角色向左上方移动并超出当前格子的位置时，那么红色区域表示需要新加载的，黄色区域表示需要保留的，蓝色区域表示需要释放的。

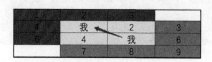

图 9-17 移动（另见彩插）

这可以保证最小化地管理所有游戏对象，而且这么做还有个好处：当需要在主角范围内查找最近单元时，参与判断的对象如果很多，就会带来 `for` 循环判断的开销，但是由于我们只保留格子范围内的元素，判断就会非常快了。至于遮挡剔除，由于需要管理的对象已经很少了，遮挡剔除的优化几乎可以忽略，所以在移动摄像机的时候，就不会再带来额外的开销了。

另外，还需要处理的是近景、远景的元素。比如近处的房子、树和草可能需要很快剔除掉，但是远处的城楼、山和云彩等不需要剔除，此时可以和美术人员来做个约定，将特殊的元素过滤掉即可。

9.3 Batching（静态合批）

美术人员做的模型是一个一个独立的，它们的材质以及贴图很有可能都是完全一样的，但是由于模型不一样，放入 Unity 时就会多占很多的 DrawCall，所以 Unity 提供了一个属性来做静态合批，可以将它们合并在一个 Mesh 里。

9.3.1 设置静态合批

静态合批首先需要在 Player Settings 页面中勾选 Static Batching 和 Dynamic Batching 复选框（表示动态合并批次），如图 9-18 所示。

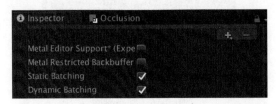

图 9-18　启动合批

接着，在游戏场景中选择需要合批的游戏对象，并选中 Batching Static 标记，然后运行游戏。如图 9-19 所示，Mesh Filter 会自动生成一个新的 Mesh，这样如果有相同的材质、Shader 并且参数一致的话，就会合并 DrawCall。

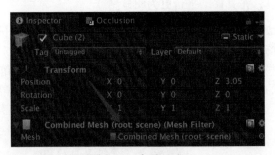

图 9-19　合并批次

9.3.2 脚本静态合批

自动的静态合批用起来很方便，但是也有隐患。假如场景非常庞大，那么合并出来的 Mesh 就会非常大。运行游戏后，只要其中有一小部分出现在摄像机内，那么整个 Mesh 都需要参与渲染。另外，静态合批的最大顶点数是 65535，如果顶点数超过了它，Unity 就会自动合并出多个 Mesh。我们可以利用脚本来动态设置需要合批在一起的游戏对象。注意，如果使用脚本合并，游戏对象不需要选中 Static 标记。

如代码清单 9-7 所示，只需要调用 StaticBatchingUtility.Combine() 方法即可动态设置合批。

代码清单 9-7　Script_09_05.cs 文件

```
using UnityEngine;

public class Script_09_05 : MonoBehaviour
{
    public GameObject[] datas;

    void Start() {
        StaticBatchingUtility.Combine(datas, gameObject);
    }
}
```

这段代码的含义就是将数组中的游戏对象合并在同一个 Root 节点下，也就是第二个参数指定的。另外，运行游戏后，合并过的 Mesh 对象是不可以发生位移的，但是可以移动它指定的 Root 节点。如图 9-20 所示，Root 游戏对象可以在运行时任意修改位置。

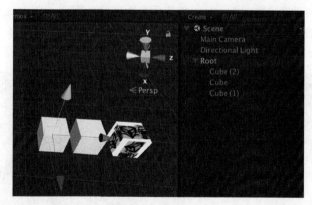

图 9-20　移动 Root

9.3.3 动态合批

动态合批是全自动的，我们不需要做任何事情。但它是有要求的，Mesh 的顶点数量需要小于 300。如果 Shader 中使用了顶点位置、法线、UV0、UV1 和切线，Mesh 的顶点数必须小于 180。

可能会有朋友问：这么小限制的动态合批适用于哪里呢？其实在粒子特效中它发挥了很大的优势。由于每个特效喷射出来以后都是 Mesh，如果不开启动态合批，DrawCall 就会非常大。

9.3.4 静态合批的隐患

静态合批的原理就是自动生成 Mesh，但是不同 Mesh 保存的信息可能是不同的。例如 Mesh 中可能会保存 color 和 tangent，但是大部分 Mesh 都是不需要这个信息的，如果静态合批中有一个 Mesh 包含了这个信息，那么合并以后整个 Mesh 都会带上它，这样无疑会增加一些额外的开销。据我的经验，更多的时候是由于美术人员在导出 FBX 时，操作不当导致添加了没用的 color 或 tangent 信息，所以可以利用 FBX 官网提供的 FBX 接口，自己写一个 Python 脚本来删除它们。

9.4 寻路网格

寻路就是提供一个目标点，根据障碍物自动计算出一条最优的路径，Unity 寻路使用的是 A* 算法。寻路可分为动态寻路以及静态寻路两种。动态寻路就是障碍物的位置可以动态修改，而静态寻路表示障碍物永远都不会发生改变。由此可见，静态寻路的效率会更高。

9.4.1 设置寻路

参与寻路计算的游戏对象需要选中 Navigation Static 复选框，接着在导航菜单栏中选择 Window→ Navigation 命令，打开寻路烘焙面板，如图 9-21 所示。这里还需要设置控制角色寻路的一些基本信息，其中 Agent Radius 表示角色胶囊体的半径，Agent Height 表示胶囊体的高度，Max Slope 表示爬坡的最高坡度，Step Height 表示每次爬楼的高度。Generated Off Mesh Links 用于设置角色落下或者跳起来没有在连接在一起的两个点的高度和距离，例如角色可以跳过一条水沟。最后，单击 Bake 按钮即可。

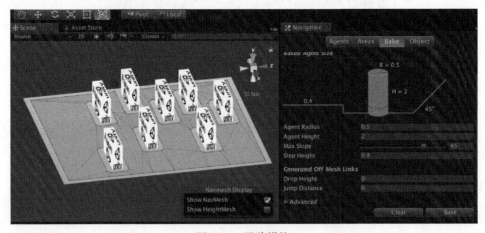

图 9-21　寻路烘焙

如图9-22所示,我们来做一个简单的寻路。点击地面,让"方块"越过障碍物自动走过去,点击屏幕时,需要使用射线计算出点在地面的位置,接着就可以控制"方块"寻路过去了。

图9-22 点击寻路

如代码清单9-8所示,控制角色寻路时,需要调用navMeshAgent.SetDestination()来设置它的位移。

代码清单9-8 Script_09_06.cs文件

```
using UnityEngine;
using UnityEngine.AI;

public class Script_09_06 : MonoBehaviour
{
    public  NavMeshAgent navMeshAgent;

    public void Update()
    {
        if(Input.GetMouseButton(0)) {
            Ray ray = Camera.main.ScreenPointToRay(Input.mousePosition);;
            //穿透所有Mesh直到找到地面
            RaycastHit[] hits = Physics.RaycastAll(ray);
            foreach(var hit in hits) {
                string name = hit.collider.gameObject.name;
                if(name == "Plane") {
                    //移动方块
                    navMeshAgent.SetDestination(hit.point);
                }
            }
        }
    }
}
```

9.4.2 连接两点

寻路必须保证两点是能走过去的，但有时候设计上不一定是走过去，例如跳过去、掉下去或空中飞过去等。寻路专门提供了一个 Off Mesh Link 组件来处理两点之间的连接，如图 9-23 所示。

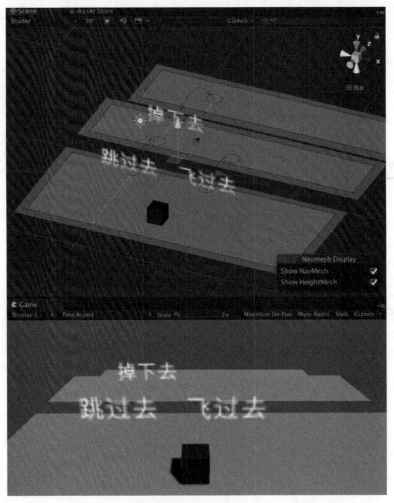

图 9-23　连接两点

9.4.3 获取寻路路径

有时候，需要在寻路之前判断一下目标点是否合法，或者寻路的路径是否合法，此时就要提前获取寻路的完整路径了，如图 9-24 所示。在代码中，我们可以使用 `NavMesh.CalculatePath()` 方法来提前计算出目标点的路径。

图 9-24 计算路径

如代码清单 9-9 所示,调用 NavMesh.CalculatePath()方法提前计算寻路路径,接着通过 Debug.DrawLine()方法将路径绘制在 Scene 中查看。

代码清单 9-9　Script_09_08.cs 文件

```
using UnityEngine;
using UnityEngine.AI;

public class Script_09_08 : MonoBehaviour
{
    public   NavMeshAgent navMeshAgent;
    public Transform target;
    private NavMeshPath m_Path = null;
    void Start()
    {
        m_Path = new NavMeshPath();
        //计算路径
        NavMesh.CalculatePath(transform.position, target.position,
            NavMesh.AllAreas, m_Path);

    }

    void Update() {
        //绘制路径
        for(int i = 0; i < m_Path.corners.Length-1; i++)
            Debug.DrawLine(m_Path.corners[i], m_Path.corners[i+1], Color.red);

    }
}
```

9.4.4 动态阻挡

在 Unity 的寻路中，很多元素是需要支持动态阻挡的，例如一堵空气墙，玩家在经历某种特殊事件之前是不能走过去的。如图 9-25 所示，给需要动态阻挡的游戏对象添加 Nav Mesh Obstacle 组件。设置游戏对象的隐藏或显示，即可控制是否发生动态阻挡。

这里需要介绍一下 Carve 属性，一旦选中它，表示这个对象支持动态烘焙。其中 Move Threshold 表示移动多长的距离后启动动态烘焙，Time To Stationary 表示元素停止运动后多久标记为静止状态，Carve Only Stationary 表示元素是否需要移动，例如空气墙，只有开启或关闭两个状态。

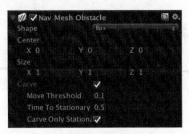

图 9-25　动态阻挡

9.4.5 导出寻路网格信息

Unity 的寻路是能满足客户端的，但是如果是网络游戏，服务器需要控制怪物寻找主角，此时就需要将寻路的网格信息导出来。如图 9-26 所示，可以利用发射线的方式来检测到当前地面是否可以行走，接着导出一个二维数组，其中 0 表示不可走，1 表示可走，也就是图中红色和蓝色的射线区域。

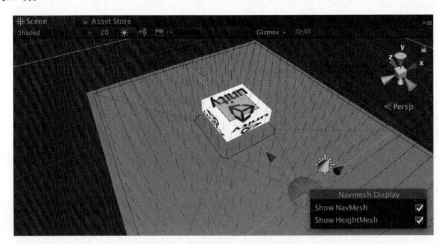

图 9-26　计算阻挡（另见彩插）

9.4 寻路网格

如图 9-27 所示，首先需要设置 X 坐标格子的数量、Y 坐标格子的数量以及每个格子的大小，接着利用 Gizmos 绘制射线来查看效果。

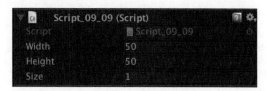

图 9-27 设置格子

如代码清单 9-10 所示，绑定脚本后，在 Scene 视图中单击一下该对象，即可渲染射线区域并自动生成网格信息文本。

代码清单 9-10　Script_09_09.cs 文件

```
#if UNITY_EDITOR
using UnityEngine;
using UnityEngine.AI;
using System.IO;
using UnityEditor;
using System.Text;

public class Script_09_09 : MonoBehaviour
{
    //X 坐标格子的数量
    public int width;
    //Y 坐标格子的数量
    public int height;
    //每个格子的大小
    public int size;

    void OnDrawGizmosSelected()
    {
        //确保当前场景烘焙过
        if(NavMesh.CalculateTriangulation().indices.Length > 0) {
            //获取场景名
            string scenePath = UnityEditor.SceneManagement.EditorSceneManager.
                GetSceneAt(0).path;
            string sceneName = System.IO.Path.GetFileName(scenePath);
            string filePath = Path.ChangeExtension(Path.Combine(Application.
                dataPath, sceneName),"txt");
            if(File.Exists(filePath)) {
                File.Delete(filePath);
            }
            //准备写入数据
            StringBuilder sb = new StringBuilder();
            sb.AppendFormat("scene={0}", sceneName).AppendLine();
            sb.AppendFormat("width={0}", width).AppendLine();
            sb.AppendFormat("height={0}", height).AppendLine();
            sb.AppendFormat("size={0}", size).AppendLine();
```

```csharp
            sb.Append("data={").AppendLine();

            Gizmos.color = Color.yellow;
            Gizmos.DrawSphere(transform.position, 1);

            float widthHalf = (float)width / 2f;
            float heightHalf = (float)height / 2f;
            float sizeHalf = (float)size / 2f;
            //从左到右、从下到上一次性写入每个格子的数据
            for(int i = 0; i < height; i++) {
                sb.Append("\t{");
                Vector3 startPos = new Vector3(-widthHalf + sizeHalf, 0,
                    -heightHalf + (i * size) + sizeHalf);
                for(int j = 0; j < width; j++) {
                    Vector3 source = startPos + Vector3.right * size * j;
                    NavMeshHit hit;
                    Color color = Color.red;
                    int a = 0;
                    //检测当前格子是否可以行走
                    if(NavMesh.SamplePosition(source, out hit, 0.2f, NavMesh.AllAreas)) {
                        color = Color.blue;
                        a = 1;
                    }
                    sb.AppendFormat(j > 0?",{0}":"{0}", a);
                    Debug.DrawRay(source, Vector3.up, color);
                }
                sb.Append("}").AppendLine();
            }
            sb.Append("}").AppendLine();
            //绘制格子的总区域
            Gizmos.DrawLine(new Vector3(-widthHalf, 0, -heightHalf),
                new Vector3(widthHalf, 0, -heightHalf));
            Gizmos.DrawLine(new Vector3(widthHalf, 0, -heightHalf),
                new Vector3(widthHalf, 0, heightHalf));
            Gizmos.DrawLine(new Vector3(widthHalf, 0, heightHalf),
                new Vector3(-widthHalf, 0, heightHalf));
            Gizmos.DrawLine(new Vector3(-widthHalf, 0, heightHalf),
                new Vector3(-widthHalf, 0, -heightHalf));

            //写入文件
            File.WriteAllText(filePath, sb.ToString());
        }
    }
}

#endif
```

如图 9-28 所示，行走区域的二维数组已生成完毕，数据的排序是从左到右、从下到上，服务端拿到这个数据后，即可按照此格式来解析了。

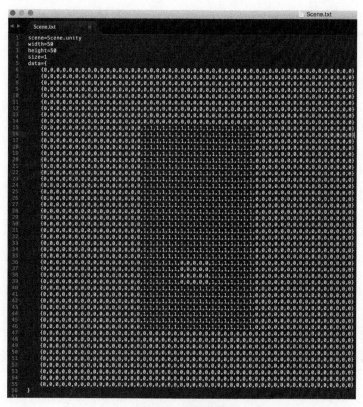

图 9-28　网格信息

9.5　反射探头

反射探头需要设置参与反射的区域，根据区域内的元素生成 cubemap，从而使周围的物体得到精准的环境反射效果。首先，需要给具有反射效果的物体绑定 Reflection Probe 组件。如图 9-29 所示，给球体绑定 Reflection Probe 组件后，就可以看到一个参与反射的矩形区域。

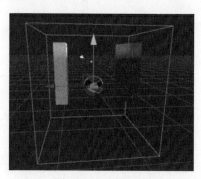

图 9-29　反射区域

接着，给球体设置 Standard（Specular setup）着色器，让球体支持镜面反射。如图 9-30 所示，球体左右的两面墙已经反射进球体中。

图 9-30　反射效果（另见彩插）

运行时实时生成 cubemap 很低效，所以可以设置 Reflection Probe 为 Baked（烘焙）模式，同时给参与反射的游戏对象设置 Reflection Probe Static 静态属性，最后正常烘焙场景即可。

9.6　小结

本章学习了 Unity 所有的静态元素，包括烘焙贴图、遮挡剔除、静态合批和寻路网格。总体来说，静态元素就是利用物体不能发生位置上的变化而产生的一种优化方式。例如，烘焙贴图将元素的光照信息烘焙在一张图上，这样就避免了实时光带来的开销；遮挡剔除也是将场景中的元素进行烘焙，动态计算遮挡者以及被遮挡者剔除掉不需要的元素；合并批次可将多个 Mesh 文件合并在一起从而减少 DrawCall；寻路网格也是烘焙出静态遮挡信息，从而使用 A* 算法来寻路。

第 10 章

多 媒 体

在游戏中，都需要播放声音与视频，其中声音又可以分为音乐和音效两种。音乐就和背景音乐一样，一直在循环播放；而音效用得就更多了，例如释放技能、被攻击所带来的声音，而且音效很可能同时需要播放好几组。为了满足更加逼真的效果，Unity 还提供了强大的混音模式以及 3D 音频模式，这就好比远处和近处都有怪物在发生攻击的音效，听起来需要有 3D 的层次感。另外，Unity 还提供了 Audio Mixer 组件，它可以很方便地在编辑模式下组合播放各种声音以及混合播放多音频。新版的视频功能用起来非常方便，既可以将它输出在摄像机的前面或后面，也可以很容易地将视频放在 UI 以及场景中间，还可以将视频材质输出并自定义渲染在贴图上。

10.1 音频

音频由多个 Audio Source 组件和一个 Audio Listener 组件组成，其中 Audio Listener 负责监听所有的 Audio Source，最终通过设备的扬声器播放出来。如果是 3D 音效，Unity 会自动判断 Audio Listener 与音频的距离，从而增强或者减少音量。另外，同一个场景只能有一个 Audio Listener 组件启用，它会默认添加在主摄像机上。

10.1.1 音频文件

首先，Unity 支持的音频文件是非常丰富的，它们包括 .mp3、.ogg、.wav、.aiff、.aif、.mod、.it、.s3m 和 .xm，最常用的就是 .mp3 和 .ogg 声音文件了。首先，我们将一个 .mp3 文件拖入 Project 视图中。我们可以单独设置声音文件是否在后台加载，以及声音文件的压缩格式等。这些设置并不是修改了声音文件本身，而是 Unity 在导入时会自动生成一个新的音频文件，它与原始文件只有简单的引用关系，将来打包发布后，其实使用的是新的音频，而使用者是毫无感知的。如图 10-1 所示，在右下角处，可以预览播放当前选择的声音文件，并且设置是否循环播放。

图 10-1　导入音频

10.1.2　Audio Source

给任意游戏对象绑定 Audio Source 组件，即可播放声音。如图 10-2 所示，将音频文件绑定在 AudioClip 处，再选中 Play On Awake 和 Loop 复选框，直接运行游戏，就可以循环播放声音了。

图 10-2　Audio Source

10.1.3　3D 音频

3D 音频就是根据声源与主角的距离自动增加以及减弱的音频。由于 Audio Listener 是绑定在主摄像机上的，所以控制摄像机的远近即可满足 3D 音频的条件。不过默认情况下，Audio Source 是 2D 音频。如图 10-3 所示，调节 Spatial Blend，其中 0 表示 2D 音频、1 表示 3D 音频、0 和 1 之间表示 2D 和 3D 之间的插值音频。

另外，Priority 表示声音的优先级。由于同时播放的音频是有最大上限的，一旦超过上限，会自动关闭一个优先级最高的音频，所以背景音乐一类的就适合设置成 0。Volume 表示声音的音量，Pitch 表示音频的播放速度，Stereo Pan 表示左声道和右声道占比，Reverb Zone Mix 用于设置回音混合。

图 10-3　3D 音频

开启 3D 音频后，就可以调节最小距离以及最大距离了。如图 10-4 所示，这两个球形区域表示 3D 声音最大区域和最小区域，点击周围的蓝色小方块，拖动鼠标即可调节它的区域。如果角色在最小音频区域内听到的音量最大，在最小区域与最大区域之间听到的音量递减，则当角色超出最大区域时，会保持递减后最小的音量。最后播放游戏后，直接移动场景中的摄像机，即可听到效果。

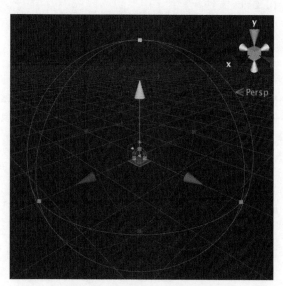

图 10-4　3D 距离（另见彩插）

10.1.4　代码控制播放

在代码中获取 Audio Source 组件，就可以动态控制音频了。如图 10-5 所示，我们可以切换播放两个音频。由于音频组件没有提供播放音频结束的回调，所以可以先获取音频的播放时间，接着通过添加定时器来等待音频播放结束，然后再处理结束后的事件。

第 10 章 多媒体

图 10-5 代码控制音频播放

如代码清单 10-1 所示，clip.length 用于获取音频的长度，yield return new WaitForSeconds() 等音频播放完毕后抛出事件。

代码清单 10-1　Script_10_01.cs 文件

```csharp
using UnityEngine;
using System.Collections;
using UnityEngine.Events;

public class Script_10_01 : MonoBehaviour
{
    public AudioClip clip1;
    public AudioClip clip2;

    public AudioSource source;

    void OnGUI()
    {

        if(GUILayout.Button("<size=50>播放音频1</size>")) {
            PlayAudioClip(clip1,delegate(AudioClip clip) {
                Debug.LogFormat("音频：{0}播放结束",clip.name);
            });

        }
        if(GUILayout.Button("<size=50>播放音频2</size>")) {
            PlayAudioClip(clip2,delegate(AudioClip clip) {
                Debug.LogFormat("音频：{0}播放结束",clip.name);
            });
        }
    }

    private Coroutine m_Coroutine = null;
    void PlayAudioClip(AudioClip clip,UnityAction<AudioClip> callback)
```

```
    {
        StopAudioClip();
        source.clip = clip;
        source.Play();
        m_Coroutine = StartCoroutine(AduioClipCallback(clip, callback));
    }

    void StopAudioClip()
    {
        if(m_Coroutine != null) {
            StopCoroutine(m_Coroutine);
            m_Coroutine = null;
        }
        source.Stop();
    }

    private IEnumerator AduioClipCallback(AudioClip clip,UnityAction<AudioClip>
        callback)
    {
        yield return new WaitForSeconds(clip.length);
        callback(clip);
    }
}
```

10.1.5 混音区

真实的游戏场景可能很复杂，比如热闹的市场、黑暗的山洞和安静的卧室，所以声音需要结合实际场景来混音，才会更加逼真。如图 10-6 所示，可创建一个 Audio Reverb Zone 组件，其中混音区同样提供了最大区域以及最小区域。在 Reverb Preset 中，可以选择混音的场景，其中常见的场景混音参数已经提供好了。如果选择 User，表示开启用户自定义模式，这样就可以灵活调节每项参数了。

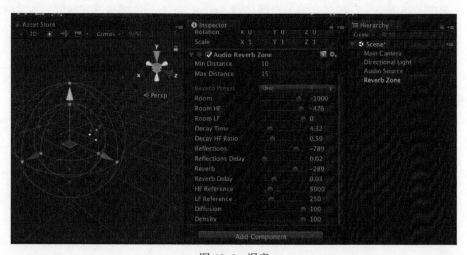

图 10-6　混音

10.1.6 Audio Mixer

音效混合器就是同时将多个音频文件混合播放，它可以任意调节混合的参数。首先，在场景中创建多个 Audio Source，其中必须有一个主音频 Master。如图 10-7 所示，添加 Mixers 文件并且创建了两个组，接着在 Audio Source 的 Output 中设置 3 个音频对应输出的混合器组。最后播放游戏，即可在混合器面板中看到这 3 个声音混合后的音频波形了。

图 10-7　Audio Mixer

10.1.7 录音

Unity 提供了录音的接口。如图 10-8 所示，只要硬件支持，单击"开始录音"按钮，就可以说话了，单击"结束录音"按钮，将声音保存到 AudioClip 上，最终播放它即可。本示例的完整代码详见代码清单 10-2。

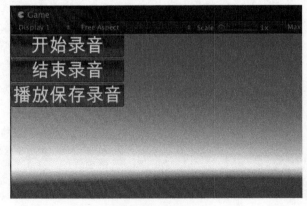

图 10-8　录音

代码清单 10-2　Script_10_02.cs 文件

```
using UnityEngine;
using System.Collections;
using UnityEngine.Events;
using System.IO;

public class Script_10_02 : MonoBehaviour
{
    public AudioSource source;

    public AudioClip m_Clip;

    void Start() {
        source.clip = Microphone.Start("Built-in Microphone", true, 10, 44100);
        source.Play();

    }

    void OnGUI()
    {
        if(GUILayout.Button("<size=50>开始录音</size>")) {
            m_Clip = Microphone.Start("Built-in Microphone", true, 10, 44100);

        }
        if(GUILayout.Button("<size=50>结束录音</size>")) {
            if(m_Clip) {
                Microphone.End("Built-in Microphone");
            }
        }

        if(GUILayout.Button("<size=50>播放保存录音</size>")) {
            if(m_Clip) {
                source.clip = m_Clip;
                source.Play();
            }
        }
    }

}
```

在上述代码中，我们使用 `Microphone.Start()` 和 `Microphone.End()` 来确定录音的内容，最终使用 `AudioSource` 播放它。

10.1.8　声音进度

如图 10-9 所示，拖动 Slider 即可调节音乐的进度。在代码中监听 Slider 拖动的值，可以动态更新音乐开始播放的位置。本示例的完整代码详见代码清单 10-3。

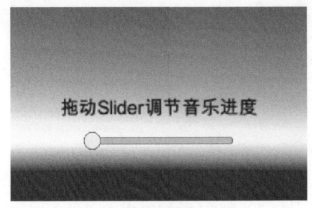

图 10-9 进度

代码清单 10-3　Script_10_03.cs 文件

```
using UnityEngine;
using System.Collections;
using UnityEngine.Events;
using System.IO;
using UnityEngine.UI;
public class Script_10_03 : MonoBehaviour
{

    public AudioSource source;

    public Slider slider;

    void Start()
    {
        slider.minValue = 0;
        slider.maxValue = source.clip.length;


        slider.onValueChanged.AddListener(delegate(float value) {
            source.Stop();
            source.time = value;
            source.Play();
        });

        source.Play();

    }

}
```

在上述代码中，我们通过监听 Slider 的拖动来重新定位音频播放的起始位置。

10.2 视频

视频在游戏中应用广泛，例如开场动画。Unity 提供了 Video Player 组件来专门处理视频，使用者只需要将视频文件直接拖入 Hierarchy 视图中就完成了。它就是这么简单！视频还提供了多种渲染模式以及分辨率自适应相关的参数，它们可以非常灵活地控制它。

10.2.1 视频文件

Unity 支持的视频文件包括 .mp4、.mov、.webm 和 .wmv.。另外，视频文件还支持从一个网址播放。如图 10-10 所示，Video Clip 表示播放工程内的一个视频文件，而 URL 表示播放工程外的视频，或者从网上边下载边播放。

图 10-10　视频类型

10.2.2 视频渲染模式

首先，需要给视频指定一个摄像机组件。如图 10-11 所示，我们可以在 Render Mode 中选择一种渲染模式，其中各个选项的含义如下。

- Camera Far Plane：表示视频将渲染在摄像机最远处，这样前景的 3D 对象就会挡住视频；
- Camera Near Plane：表示视频渲染在摄像机最近的位置，这样视频就会挡住所有的 3D 对象；
- Render Texture：表示可以将视频渲染在一张纹理上；
- Material Override：表示将视频覆盖渲染在一张指定材质上；
- API Only：表示需要使用脚本来动态设置视频渲染的目标。

图 10-11　视频渲染模式

这里我们设置的渲染模式是 Camera Far Plane，所以相机内前景的 3D 元素就会挡住视频了，如图 10-12 所示。

图 10-12　Camera Far Plane 渲染模式

10.2.3　视频自适应

Aspect Ratio 用于设置视频缩放的参数，如图 10-13 所示，其中各个选项的含义如下。

- Stretch：表示视频自适应，不过可能会被拉伸变形。
- Fit Vertically：表示锁定纵向，横向会无法自适应。
- Fit Horizontally：表示锁定横向，纵向会无法自适应。
- Fit Inside：表示整体锁定在最小区域。
- Fit Outside：表示整体锁定在最大区域。
- No Scaling：表示视频为原始大小，不会被拉伸。

图 10-13　自适应区域

在游戏中，我们还是尽可能使用 Stretch。如图 10-14 所示，无论如何修改屏幕分辨率，视频永远都会完整显示在屏幕中。

图 10-14 视频自适应

10.2.4 UI 盖在视频之上

在游戏中，在播放视频的时候一般都需要有个按钮，例如跳过当前视频一类的。由于 3D 摄像机和 UI 摄像机是分开的，所以需要在 Video Player 中指定渲染在 UI 摄像机上。接着在渲染模式中选择 Camera Far Plane 模式，这样视频就会显示在 UI 与 3D 之间了，UI 元素就默认盖在视频之上了，如图 10-15 所示。

图 10-15 UI 盖在视频之上

如代码清单 10-4 所示，点击"跳过视频"按钮后，会调用 `videoPlayer.Stop()` 方法来结束视频的播放。

代码清单 10-4　Script_10_04.cs 文件

```
using UnityEngine;
using System.Collections;
using UnityEngine.Events;
using System.IO;
```

```
using UnityEngine.UI;
using UnityEngine.Video;


public class Script_10_04 : MonoBehaviour
{
    public VideoPlayer videoPlayer;
    public Button button;

    void Start()
    {
        button.onClick.AddListener(delegate() {
            videoPlayer.Stop();
        });
    }

}
```

10.2.5 视频渲染在材质上

如图 10-16 所示,在场景中创一个 Plane 面板,接着给它绑定一个新材质,最后将 Mesh Renderer 组件拖入 Renderer 中,此时视频就被渲染在这个 Plane 面板上了。整个工程见 CodeList_10_05。

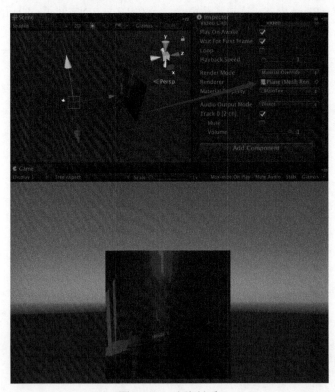

图 10-16　渲染材质

10.2.6 视频 Render Texture

首先，需要创建一个 Render Texture 文件，接着将 Render Texture 文件绑定在 Target Texture 上即可，如图 10-17 所示。然后，在场景中创建 UI 的 Raw Image 组件，并将其关联上 Render Texture，这样视频就可以直接显示在 UI 上了。整个工程见 CodeList_10_06。

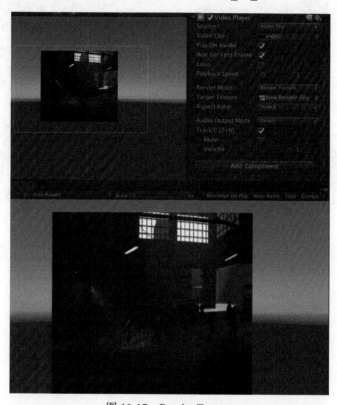

图 10-17 Render Texture

10.2.7 播放工程外视频

前面讲过，可以通过 URL 来播放网络视频，其实也可以播放工程外部的本地视频。首先，需要构建一个 macOS 的游戏包，运行游戏后，只需要输入一个本地视频的路径，单击"播放视频"按钮即可，如图 10-18 所示。本示例的完整代码详见代码清单 10-5。

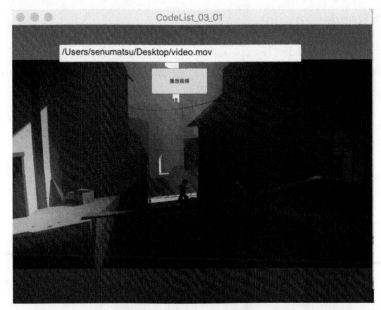

图 10-18 输入本地视频路径

代码清单 10-5　Script_10_07.cs 文件

```
using UnityEngine;
using System.Collections;
using UnityEngine.Events;
using System.IO;
using UnityEngine.UI;
using UnityEngine.Video;

public class Script_10_07 : MonoBehaviour
{
    public VideoPlayer videoPlayer;
    public InputField inputField;
    public Button button;

    void Start()
    {
        button.onClick.AddListener(delegate() {
            videoPlayer.source = VideoSource.Url;
            videoPlayer.url = string.Format("file://{0}",inputField.text);
            videoPlayer.Play();
        });
    }
}
```

在上述代码中,我们播放的是一个本地视频,其中 URL 以 file:// 开头。如果需要播放网络视频,则以 http:// 开头。

10.2.8 自定义视频显示

前面讲过,视频的渲染模式可以选择 API Only,它的意思就是,不依赖 Unity,自动将视频渲染出来,然后拿到视频底层输出的 Texture 贴图,接着自行将它渲染出来。例如,有多个视频文件,如果需要融合播放,就可以自己做个 Shader,将多个 videoPlayer.texture 传进去,最后将它们融合并渲染出来。

10.2.9 视频进度

在视频播放器中,拖动 Slider 即可调节视频的进度,如图 10-19 所示。在代码中,可以监听 Slider 拖动的值来动态更新视频开始播放的位置。本示例的完整代码详见代码清单 10-6。

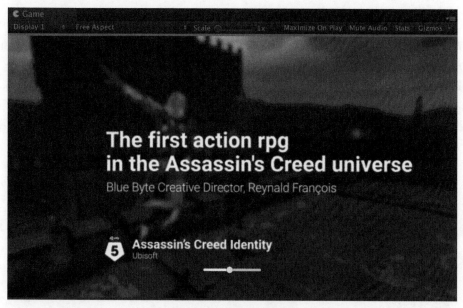

图 10-19 播放本地视频

代码清单 10-6 Script_10_08.cs 文件

```
using UnityEngine;
using System.Collections;
using UnityEngine.Events;
using System.IO;
using UnityEngine.UI;
using UnityEngine.Video;

public class Script_10_08 : MonoBehaviour
{

    public VideoPlayer videoPlayer;
```

```
public Slider slider;

void Start()
{
    slider.minValue = 0;
    slider.maxValue = (float)videoPlayer.clip.length;

    slider.onValueChanged.AddListener(delegate(float value) {
        videoPlayer.time = value;
    });
    videoPlayer.Play();
}
```
}

10.3 小结

本章介绍了如何使用音频和视频。如果音频文件比较多，有可能要同时播放多个音频，然而同时播放音频的数量是有限制的，所以要尽可能地将背景音乐的优先级设置成0，保证它永远都不会被剔除掉。另外，Unity还提供了音频混合功能，可以模拟山洞、闹市和房间等特殊场景带来的声音混合。为了达到逼真的效果，我们也可以自定义声音混合的参数。后面还介绍了视频播放，它提供了好几种渲染模式，可以很好地将视频夹在两个摄像机之间。另外，还提供了多种自适应的方式在屏幕中填充视频。

第 11 章

资源加载与优化

在游戏开发中，资源管理是非常重要的，管理不好，就容易导致内存溢出，引起闪退或者整体游戏卡顿的现象。Unity 提供了强大的 Profiler 工具，它专门查看每一帧内存的占用以及加载的耗时。此外，它还提供了自定义接口，开发者也可以自行定义一个查看区间。另外，Unity 还提供了 Frame Debugger，它可以查看每一帧渲染的详细过程。如果 DrawCall 很多的话，可以参照它来想办法合并。另外，Unity 也提供了丰富的资源加载接口，我们可以在编辑模式下加载资源，在运行模式下加载本地资源和下载资源。总之，管理好资源的加载以及优化，才能让游戏更加流畅。

11.1 编辑模式

编辑模式并非打包后的运行模式，而仅仅是在编辑器下运行。它可以访问加载到硬盘上的任意资源，这对拓展引擎内置的编辑器是非常好的。引擎可以读取任意资源来丰富编辑器。但是如果游戏发布后，就会有很多限制，比如编辑模式下可使用的大量代码在运行时无法使用。

11.1.1 加载资源

编辑模式下的资源可分为两类：一类是引擎可识别的资源，例如 Prefab、声音、视频、动画和 UI 等；另一类是引擎无法识别的资源，例如外部导入的资源，这类资源需要通过第三方工具将它的信息解析出来，最终组织成引擎内可识别资源才可以使用。

在编辑模式下，Unity 提供了一个标志性的类 `AssetDatabase`，它专门负责读取工程内的资源。如图 11-1 所示，需要保证所有资源必须放在项目的 Assets 目录下，不然 `AssetDatabase` 是无法读取的，只能使用 `File` 类或者其他第三方辅助类来读取。

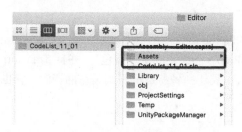

图 11-1 Assets 目录

接着，在编辑模式下实例化一个 Prefab 到 Scene 视图中。如图 11-2 所示，Prefab 被实例化到了 Main Camera 节点下，但是为什么它不是"蓝色"的呢？并且坐标为什么没有回归原点呢？因为 GameObject.Instantiate()只能创建新对象，这样将丢失 Prefab 的引用。坐标没有回归原点是因为 transform.SetParent 会继承世界坐标。通过这段代码可以看出，只要提供一个 Assets 内的相对路径，使用 AssetDatabase.LoadAssetAtPath()方法即可读取任意对象。本示例的完整代码详见代码清单 11-1。

图 11-2　实例化游戏对象（另见彩插）

代码清单 11-1　Script_11_01.cs 文件

```
using UnityEngine;
using UnityEditor;

public class Script_11_01
{
    [MenuItem("Assets/My Tools/Load",false,2)]
    static void MyLoad()
    {
        if(Selection.activeTransform){
            //读取 Prefab
            GameObject prefab = AssetDatabase.LoadAssetAtPath<GameObject>
                ("Assets/Cube.prefab");
            //实例化到 Scene 中
            GameObject go = GameObject.Instantiate<GameObject>(prefab);
            //设置它的父节点
            go.transform.SetParent(Selection.activeTransform);
        }
    }
}
```

11.1.2　实例化 Prefab

在编辑模式下，实例化 Prefab 需要使用 PrefabUtility.InstantiatePrefab()方法。如图 11-3 所示，实例化后的 Prefab 已经变成了"蓝色"，并且 Transform 的坐标已经回归原点。另外，除了 Prefab 以外，FBX 也属于一种特殊的引用关系的资源，它也可以使用这种方法实例化，这样会保持它的一些引用关系。

图 11-3 实例化 Prefab（另见彩插）

如代码清单 11-2 所示，将 Prefab 实例化到 Hierarchy 视图中。其中，`PrefabUtility.InstantiatePrefab()`方法可以保持原有的引用关系，并非克隆一份新的。另外，给`transform.SetParent()`的第二个参数传入 `false`，表示它不被父坐标影响，所以就会回归原点。

代码清单 11-2　Script_11_02.cs 文件

```
using UnityEngine;
using UnityEditor;

public class Script_11_02
{
    [MenuItem("Assets/My Tools/LoadPrefab",false,2)]
    static void LoadPrefab()
    {
        if(Selection.activeTransform){
            //读取 Prefab
            GameObject prefab = AssetDatabase.LoadAssetAtPath<GameObject>
                ("Assets/Cube.prefab");
            //实例化到 Scene 中
            //GameObject go = GameObject.Instantiate<GameObject>(prefab);
            GameObject go = PrefabUtility.InstantiatePrefab(prefab) as GameObject;
            //设置它的父节点
            //go.transform.SetParent(Selection.activeTransform);
            go.transform.SetParent(Selection.activeTransform,false);
        }
    }
}
```

11.1.3　创建 Prefab

使用`PrefabUtility.CreatePrefab()`方法可以创建 Prefab，此时需要提供保存的目录以及当前的游戏对象。其中，`ReplacePrefabOptions.ConnectToPrefab` 表示创建 Prefab 的同时自动关联到创建它的这个游戏对象。

如代码清单 11-3 所示，在 Hierarchy 视图中选择任意游戏对象，并且选择 Create Prefab 菜单项，在资源目录下创建 Prefab，并将它关联到创建的这个游戏对象本身。

代码清单 11-3　Script_11_03.cs 文件

```
using UnityEngine;
using UnityEditor;

public class Script_11_03
{
    [MenuItem("Assets/My Tools/Create Prefab",false,3)]
    static void CreatePrefab()
    {
        if(Selection.activeTransform){
            string path = "Assets/Prefab.prefab";
            //如果文件已经存在，删除它
            if(AssetDatabase.LoadAssetAtPath<GameObject>(path)) {
                AssetDatabase.DeleteAsset(path);
            }

            //创建新的 Prefab
            PrefabUtility.CreatePrefab("Assets/Prefab.prefab", Selection.
                activeGameObject,
                ReplacePrefabOptions.ConnectToPrefab);

            //刷新 Project 视图目录
            AssetDatabase.Refresh();
        }
    }
}
```

11.1.4　更新 Prefab

使用 `PrefabUtility.ReplacePrefab()` 方法可以更新 Prefab。不过在更新前，需要判断 Hierarchy 选择的游戏对象是否为 Prefab，如果不是 Prefab，也可以考虑创建新的。相关代码如代码清单 11-4 所示。

代码清单 11-4　Script_11_04.cs 文件

```
using UnityEngine;
using UnityEditor;

public class Script_11_04
{
    [MenuItem("Assets/My Tools/Applay Prefab",false,3)]
    static void ApplayPrefab()
    {
        if(Selection.activeTransform){
            //确保 Hierarchy 视图中当前选择的是 Prefab
            Object prefab = PrefabUtility.GetPrefabParent(Selection.activeGameObject);
            if(prefab) {
                //替换它
                PrefabUtility.ReplacePrefab(Selection.activeGameObject, prefab,
                    ReplacePrefabOptions.ConnectToPrefab);
            }
```

```
            //刷新 Project 视图目录
            AssetDatabase.Refresh();
        }
    }
}
```

在上述代码中，我们通过 `PrefabUtility.GetPrefabParent()` 判断当前选择的是否为 Prefab：如果是 Prefab，则更新它；如果不是 Prefab，则创建一个新的。

11.1.5 卸载资源

在编辑模式下，只能使用 `GameObject.DestroyImmediate()` 方法来卸载游戏对象。如果需要卸载游戏对象引用的资源，则第二个参数填 `true`，其默认值是 `false`。

如代码清单 11-5 所示，选择 Hierarchy 视图中的任意游戏对象，点击 Delete 菜单项即可删除它。

代码清单 11-5　Script_11_05.cs 文件
```
using UnityEngine;
using UnityEditor;

public class Script_11_05
{
    [MenuItem("Assets/My Tools/Delete",false,3)]
    static void Delete()
    {
        if(Selection.activeTransform){
            //第二个参数表示是否卸载游戏对象引用的资源
            GameObject.DestroyImmediate(Selection.activeGameObject,true);
        }
    }
}
```

11.1.6 游戏对象与资源的关系

游戏对象与资源是一种引用关系。例如一个模型是由贴图和 Mesh 组成，将它拖入场景中时，生成的游戏对象就会引用这两种资源。当程序调用 `GameObject.Destroy()` 或者 `GameObject.DestroyImmediate()` 方法时，只会卸载掉它的对象，它身上引用的贴图和 Mesh 还在内存中。

Unity 这么做是有原因的：很多游戏对象的加载与卸载是很频繁的，如果每次卸载都将引用的资源清理掉，无疑会造成 IO 的阻塞。但是如果长时间不卸载这些资源，那么内存必然涨上去，所以 Unity 又提供了一个方法来自动卸载无用资源。其中，无用资源表示没有被别的对象或者代码引用的资源。如代码清单 11-6 所示，调用 `EditorUtility.UnloadUnusedAssetsImmediate()` 方法即可卸载编辑器下无用的资源了。在运行时也存在类似的资源，后面会讲到。

代码清单 11-6　Script_11_06.cs 文件

```
using UnityEngine;
using UnityEditor;

public class Script_11_06
{
    [MenuItem("Assets/My Tools/UnloadUnusedAssetsImmediate",false,3)]
    static void UnloadUnusedAssetsImmediate()
    {
        EditorUtility.UnloadUnusedAssetsImmediate();
    }
}
```

11.2　版本管理

多人同时开发游戏时，需要对项目的版本进行管理，通常会将整个工程上传 SVN 或者 Git。然而资源在导入 Unity 的时候，会自动生成很多中间资源，这些资源是不需要上传的。如图 11-4 所示，只需要将 Assets、ProjectSettings 文件夹下的所有文件以及 .meta 文件上传即可。

图 11-4　版本管理

11.2.1　.meta 文件

.meta 文件是 Unity 自动生成的。每个游戏资源都会有一个对应的 .meta 文件，它会标记资源在引擎中的一些设置信息，我们可以在资源视图面板中重新设置这些资源的参数。此时 .meta 文件中将会保留这些参数。将资源拖入工程时，就会利用这些参数重新压缩资源。换句话说，资源在用户无感知的情况下被 Unity 优化了。如图 11-5 所示，首先需要在 Editor Settings 面板中设置显示 .meta 文件。

图 11-5　显示 .meta 文件

接着，随便打开一张贴图对应的 .meta 文件，如图 11-6 所示。每个 .meta 文件都会记录 guid 这个重要信息。guid 就是用来关联资源与游戏对象的引用的。比如场景中有个模型，引用了这张贴图，那么模型对应 Prefab 的 .meta 文件中就会引用这个 guid；如果模型不是 Prefab，只是保持在场景中，那么场景的 .meta 文件就会引用这个 guid。所以说，Unity 中所有资源的引用关系都是这样来计算的。

再回到这个贴图资源。引擎内可以设置每张贴图的压缩格式、大小和 mipmap 等，这些信息都会保存在 .meta 文件中，Unity 会根据这些参数重新压缩这个贴图。最终呈现给玩家的贴图，已经不是当初我们放入 Unity 中的了。所以说，开发者只需要设置一些参数，Unity 就实现了无感知的优化。

另外，.meta 文件也一定要上传到 SVN 中，不然别人在更新到这个资源的时候，无法用它本地的 Unity 对资源做正确的设置了。

图 11-6　.meta 文件

11.2.2　多工程

游戏项目会多角色进行参与，如程序员、策划人员和美术人员等，如果不希望一部分代码让美术人员和策划人员看到，那么项目就需要分成两个工程了。分工程很容易，但是合并两个工程就比较麻烦了。如图 11-7 所示，在 Project 视图中选择需要导出的资源，单击鼠标右键，从弹出的快捷菜单中选择 Export Package 命令，可导出一个包。打开新的工程，把刚刚导出的包导入，即可实现两个工程的同步。

图 11-7 导出包

11.2.3 同步文件

首先，需要保证美术资源都放在一个最顶层的文件夹下。资源可以按子文件夹归类，但是引用关系只能在最顶层的文件夹中。同步文件其实就是同步目录，此时使用 Export Package 命令就不太方便了，可以使用代码直接复制文件夹。这里可能有人有疑问：直接复制会不会引起资源依赖的改变？其实是不会的。因为美术资源都放在同一个顶层文件夹下，这样依赖只维持在这个内部文件夹下，即使复制出来以后，还会保留原有的依赖。如代码清单 11-7 所示，我们来实现文件夹的复制同步功能。

代码清单 11-7 Script_11_07.cs 文件

```
using UnityEngine;
using UnityEditor;
using System.IO;
public class Script_11_07
{
    ///<summary>
    ///复制目录
    ///</summary>
    ///<param name="raw">Raw.</param>
    ///<param name="copy">Copy.</param>
    static public void CopyFolder(string strSource,string strDestination)
    {
        ClearFolder(strDestination);
        foreach(string from in Directory.GetFiles(strSource, "*.*",
            SearchOption.AllDirectories)) {
            if(!from.Contains(".svn")) {
                CopyFile(from, from.Replace(strSource, strDestination));
```

```csharp
        }
    }
}

///<summary>
///复制文件
///</summary>
///<param name="raw">Raw.</param>
///<param name="copy">Copy.</param>
static private void CopyFile(string raw,string copy)
{
    string extension = Path.GetExtension(raw);
    if(extenion != ".DS_Store") {
        if(File.Exists(copy)) {
            File.Delete(copy);
        }
        if(File.Exists(raw)) {
            string path = Path.GetDirectoryName(copy);
            if(!Directory.Exists(path)) {
                Directory.CreateDirectory(path);
            }
            File.Copy(raw, copy);
        }
    }
}

///<summary>
///清空文件夹下的所有资源
///</summary>
///<param name="path">Path.</param>
static private void ClearFolder(string path)
{
    if(Directory.Exists(path)) {
        Directory.Delete(path, true);
        AssetDatabase.Refresh();
    }
}
```

11.2.4　SVN 外链

合并资源还有一种更方便的方式，那就是添加 SVN 的外链。客户端的 SVN 仓库在 Assets 文件夹下外链一个美术人员的 SVN 仓库文件夹，那么美术人员在自己的 SVN 仓库中提交资源时，客户端只需要更新就行了，不需要考虑资源的合并。这样是最方便的，但是也有个问题，那就是未来如果打 SVN 版本分支的时候，外链也必须打分支，不然未来美术人员更新了新资源，老的客户端 SVN 仓库就会把新的资源外链下来。游戏就是使用这种方法来管理客户端和美术资源的，这样确实很方便。

11.3 运行模式

运行模式和编辑模式是完全不同的，编辑模式下可以放成千上万的资源，那么这些资源是否需要都打包在发布的游戏包中呢？显然是不能的！打包时，Unity 会自动删除掉没有引用的资源，只会保留 Resources 目录以及 StreamingAssets 目录下的资源。

11.3.1 引用资源

只有被引用到的资源 Unity 才会打包，那么如何分辨资源是否被引用呢？在图 11-8 中：

❑ B 贴图被 New Material 材质引用；
❑ New Material 材质被 Cube 引用；
❑ Cube 被 Scene 引用。

图 11-8 引用资源

如图 11-9 所示，如果这个 Scene 被添加到了 Scenes In Build 中，那么以上这几种资源都会被打入游戏包中。但是这个 Cube 在代码中是无法直接操作的，因为没有办法去加载它，只能在打开场景时自动实例化它。

图 11-9 引用资源

11.3.2 Resources

  Resources 文件夹是 Unity 中标志性的目录，这个目录下的资源无论是否有引用关系，都会被强制打在游戏包中。如图 11-10 所示，Resources 文件夹可以是顶层目录，也可以是某个文件夹的子目录。打包后，Unity 会自动将它们合并在一起，接着在代码中动态读取这些资源，并且加载它。

图 11-10　Resources 文件夹

  如代码清单 11-8 所示，我们可以通过 Resources.Load<T> 来加载各类游戏资源。

代码清单 11-8　Script_11_08.cs 文件

```
using UnityEngine;
using UnityEditor;
using System.IO;
public class Script_11_08 : MonoBehaviour
{
    void Start()
    {
        //读取材质
        Material material = Resources.Load<Material>("New Material");
        //读取贴图
        Texture texture = Resources.Load<Texture>("B");
        //读取 Prefab
        GameObject prefab = Resources.Load<GameObject>("Cube");
        //实例化游戏对象
        GameObject go = GameObject.Instantiate<GameObject>(prefab);
        //挂在主摄像机节点下
        go.transform.SetParent(Camera.main.transform,false);
    }
}
```

注意：由于 Resources 目录下的资源都会被打包，所以尽可能不要把不需要运行时加载的资源，或者已经废弃掉的资源放进去，因为这无疑会增大包体。此外，Resources 目录下的资源尽量不要直接引用在场景中，不然这个资源会被场景和 Resources 打成两份。

11.3.3 删除对象

运行时,需要使用 GameObject.Destroy() 和 GameObject.DestroyImmediate() 方法删除游戏对象。其中,GameObject.Destroy() 会等一帧再彻底删除。因为有可能在这一帧的后面还有地方在操作这个对象,所以一般建议使用它来删除对象。GameObject.DestroyImmediate() 表示立即删除。如果这句代码后面有地方操作删除的对象,立刻就会报错,具体如代码清单 11-9 所示。

代码清单 11-9　Script_11_09.cs 文件

```
using UnityEngine;
using System.IO;
public class Script_11_09 : MonoBehaviour
{
    public GameObject g1;
    public GameObject g2;
    public GameObject g3;

    void Start()
    {
        //删除游戏对象
        GameObject.Destroy(g1);
        //一秒后删除
        GameObject.Destroy(g2, 1f);
        //立即删除
        GameObject.DestroyImmediate(g3);
    }
}
```

11.3.4 删除资源

前面我们也讲过游戏对象和游戏资源的关系,游戏对象删除了,它引用的资源其实并没有删除。如图 11-11 所示,在 Profiler 中依然能看到这个贴图在内存中。不过我们可以使用 Resources.UnloadAsset() 以及 Resources.UnloadUnusedAssets() 方法强制卸载资源。由于卸载资源是异步操作,所以可以使用 isDone 来判断是否完成。

图 11-11　Profiler

如代码清单 11-10 所示，释放游戏对象后，`Resources.UnloadUnusedAssets()` 会释放无用资源，然后在 `Update()` 方法中判断 `isDone` 是否释放结束。

代码清单 11-10 Script_11_10.cs 文件

```csharp
using UnityEngine;
using System.IO;
public class Script_11_10 : MonoBehaviour
{
    public GameObject g1;
    private AsyncOperation m_Operation;
    void Start()
    {
        GameObject.Destroy(g1);
        m_Operation = Resources.UnloadUnusedAssets();

        //也可以强制卸载对象引用的资源
        //Resources.UnloadAsset(g1);
    }

    void Update()
    {
        if(m_Operation !=null) {
            if(m_Operation.isDone) {
                m_Operation = null;
                Debug.Log("资源卸载完成");
            }
        }
    }
}
```

11.3.5 GC

在 C# 中，可能还会有很多临时对象引用这个游戏资源，这很可能会导致 `Resources.UnloadUnusedAssets()` 无法释放掉。因此，在卸载无用资源前，需要保证 C# 完成垃圾收集工作，而且有时候进行一遍垃圾回收工作是没用的，最好调用两遍 `GC()` 和 `Resources.UnloadUnusedAssets()`。

如代码清单 11-11 所示，这里封装了一个内部的 GC 类，调用完 `UnloadUnusedAssets()` 后，再调用一次 `UnloadUnusedAssets()` 进行充分的垃圾回收。

代码清单 11-11 Script_11_11.cs 文件

```csharp
using UnityEngine;
using System.IO;
using UnityEngine.Events;


public class Script_11_11 : MonoBehaviour
{
    public GameObject g1;
```

```csharp
private AsyncOperation m_Operation;
void Start()
{
    GameObject.Destroy(g1);
    GC gc = GetComponent<GC>() ?? gameObject.AddComponent<GC>();
    gc.UnloadUnusedAssets(delegate() {
        gc.UnloadUnusedAssets(delegate() {
            Debug.Log("彻底卸载掉资源!!");
        });
    });
}

public class GC : MonoBehaviour
{
    public  AsyncOperation m_Operation;
    public UnityAction m_Callback;

    public void UnloadUnusedAssets(UnityAction callback){
        m_Callback = callback;
        System.GC.Collect();
        m_Operation = Resources.UnloadUnusedAssets();
    }

    void Update()
    {
        if(m_Operation !=null) {
            if(m_Operation.isDone) {
                m_Operation = null;
                m_Callback();
                //删除自身
                DestroyImmediate(this);
            }
        }
    }
}
```

11.4 AssetBundle

在网络游戏中，可能需要在运行时下载并更新资源，而 Unity 提供了 AssetBundle 组件，可以将指定的一部分资源构建成 AssetBundle 文件。如果需要下载，那么需要将这些 AssetBundle 文件上传到 CDN 上。

11.4.1 设置 AssetBundle

如图 11-12 所示，选择需要构建 AssetBundle 的资源，其中右下角需要写入它的资源名以及资源的后缀名。代码中调用 `BuildPipeline.BuildAssetBundles()`方法，只需提供一个输出的目录就可以构建 AssetBundle 了，最终将输出在 StreamingAssets 目录下，具体如代码清单 11-12 所示。

11.4 AssetBundle

图 11-12 设置 AssetBundle

代码清单 11-12　Script_11_12.cs 文件

```
using UnityEngine;
using System.IO;
using UnityEngine.Events;
using UnityEditor;

public class Script_11_12
{
    [MenuItem("Tools/BuildAssetbundle")]
    static void BuildAssetbundle()
    {
        string outPath = Path.Combine(Application.dataPath, "StreamingAssets");

        //如果目录已经存在，则删除它
        if(Directory.Exists(outPath)) {
            Directory.Delete(outPath,true);
        }
        Directory.CreateDirectory(outPath);

        //构建 Assetbundle
        BuildPipeline.BuildAssetBundles(outPath, BuildAssetBundleOptions.
            ChunkBasedCompression, BuildTarget.StandaloneOSX);
        //刷新
        AssetDatabase.Refresh();
    }
}
```

注意：每个平台下的 Bundle 文件是不一样的，因此需要指定 BuildTarget 的构建平台。例如，在编辑模式下运行游戏，加载 iOS 或者 Android 的 AssetBundle，尤其是 Shader，都会显示错误，只能在真机上才能看到正确的效果。

11.4.2 依赖关系

如果有两个 Prefab 都依赖了同一份材质和贴图文件,那么按照上面的打包方式,材质和贴图会生成两份了。如图 11-13 所示,可以在有可能出现冗余的资源上输入它的 AssetBundle 名称,然后再构建 AssetBundle,这样这两个 Prefab 就会自动依赖材质和贴图了。构建 AssetBundle 后,会生成资源依赖关系文本,例如 cube.unity3d.manifest。

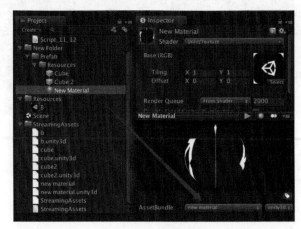

图 11-13 依赖关系

分别打开这两个 Prefab 的关系文件,如图 11-14 所示,我们能清楚地看到它们共同依赖同一个材质。

图 11-14 依赖关系

11.4.3 通过脚本设置依赖关系

游戏中的资源非常多，总不能每一个都手动设置 AssetBundle 的名称和依赖吧，所以需要在脚本中指定设置它们的依赖关系。

如代码清单 11-13 所示，将需要构建的 AssetBundle 添加至 List<AssetBundleBuild> 列表中，最后使用 `BuildPipeline.BuildAssetBundles()`方法构建它们即可。

代码清单 11-13　Script_11_13.cs 文件

```
using UnityEngine;
using System.IO;
using UnityEngine.Events;
using UnityEditor;
using System.Collections.Generic;

public class Script_11_13
{
    [MenuItem("Tools/BuildAssetbundle")]
    static void BuildAssetbundle()
    {
        string outPath = Path.Combine(Application.dataPath, "StreamingAssets");

        //如果目录已经存在，则删除它
        if(Directory.Exists(outPath)) {
            Directory.Delete(outPath,true);
        }
        Directory.CreateDirectory(outPath);

        List<AssetBundleBuild> builds = new List<AssetBundleBuild>();
        //设置AssetBundle名，将多个资源构建在同一个AssetBundle内
        builds.Add(new AssetBundleBuild(){assetBundleName="Cube.unity3d",assetNames
            =new string[]{"Assets/Cube.prefab","Assets/Cube 2.prefab"}});
        builds.Add(new AssetBundleBuild(){assetBundleName="New Material.unity3d",
            assetNames =new string[]{"Assets/New Material.mat"}});

        //构建AssetBundle
        BuildPipeline.BuildAssetBundles(outPath,builds.ToArray(),BuildAssetBundle
            Options.ChunkBasedCompression, BuildTarget.StandaloneOSX);
        //刷新
        AssetDatabase.Refresh();
    }
}
```

11.4.4 压缩格式

AssetBundle 提供了如下 3 种可选的压缩格式。

- ❑ **LZMA 压缩**：如果不做特殊指定，AssetBundle 默认会以这种方式压缩。它的优点是 Bundle 会被压缩得非常小，缺点是每次使用都需解压，可能会带来卡顿。因此，不建议在项目中使用。

- **BuildAssetBundleOptions.UncompressedAssetBundle** 不压缩：它的缺点是构建出来的 AssetBundle 比较大，优点是加载得非常快。可以将不压缩的 AssetBundle 构建出来，用第三方压缩算法压缩它，再将它上传到 CDN 上。这样下载的时候还是压缩过的 AssetBundle，所以不影响下载流量。接着，使用第三方解压算法将它写在硬盘上，这样用户在读取的时候就会非常快了。
- **BuildAssetBundleOptions.ChunkBasedCompression** LZ4 压缩方式：它是 LZMA 与不压缩之间的折中方案，构建出来的 Bundle 会比不压缩的小一点，加载速度会比 LZMA 压缩的快一点，建议在项目中使用它。

11.4.5 加载包体内的 AssetBundle

包体内的 AssetBundle 只能放在 StreamingAssets 目录下，别的目录是无法读取的。可以使用 `AssetBundle.LoadFromFile()` 或者 `AssetBundle.LoadFromFileAsync()` 方法同步或者异步加载。这里需要注意的是，加载 AssetBundle 之前，需要使用 AssetBundleManifest 提取每个 AssetBundle 的相互依赖关系。运行游戏后，立方体已经从 Bundle 中加载到游戏中了，如图 11-15 所示。

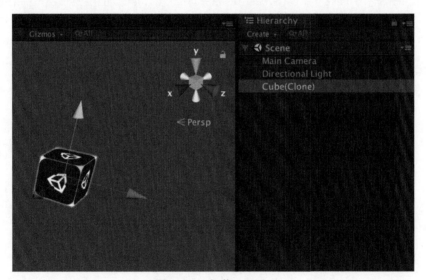

图 11-15 加载 Bundle

如代码清单 11-14 所示，首先 `manifest.GetAllDependencies()` 获取 AssetBundle 的依赖部分，然后使用 `AssetBundle.LoadFromFile()` 加载需要的资源。

代码清单 11-14 Script_11_14.cs 文件

```
using System.Collections;
using System.Collections.Generic;
using UnityEngine;
```

```csharp
using System.IO;

public class Script_11_14 : MonoBehaviour {

    void Start() {
        AssetBundle assetbundle = AssetBundle.LoadFromFile(Path.Combine(Application.
            streamingAssetsPath, "StreamingAssets"));
        AssetBundleManifest manifest = assetbundle.LoadAsset<AssetBundleManifest>
            ("AssetBundleManifest");

        //加载 AssetBundle 前，需要加载依赖的 Bundle
        foreach(var item in manifest.GetAllDependencies("cube.unity3d")) {
            AssetBundle.LoadFromFile(Path.Combine(Application.streamingAssetsPath,
                item));
        }
        //读取 Bundle
        assetbundle = AssetBundle.LoadFromFile(Path.Combine(Application.
            streamingAssetsPath, "cube.unity3d"));
        //从 Bundle 中读取资源
        GameObject prefab = assetbundle.LoadAsset<GameObject>("Cube");
        //实例化资源
        GameObject.Instantiate<GameObject>(prefab);
    }

}
```

11.4.6 下载 AssetBundle

Unity 提供了 WWW 类来进行下载。它可以下载包括 AssetBundle 在内的任意资源，此时只需要提供一个 URL 下载地址即可。如果下载本地任意资源，则在前面加上 `file://`即可。如图 11-16 所示，首先将 Bundle 下载至 Application.persistentDataPath 可读写目录下，接着再读取它并将其实例化在游戏中。

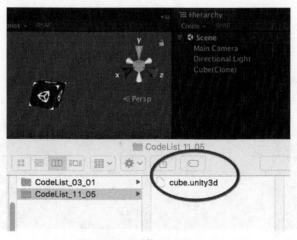

图 11-16 下载 AssetBundle

如代码清单 11-15 所示，使用 www 来下载 AssetBundle，接着通过 File.WriteAllBytes() 将 AssetBundle 写在本地中，最后使用 AssetBundle.LoadFromFile() 从硬盘中加载它。

代码清单 11-15　Script_11_15.cs 文件

```csharp
using System.Collections;
using System.Collections.Generic;
using UnityEngine;
using System.IO;

public class Script_11_15 : MonoBehaviour {

    IEnumerator Start() {

        string path = Path.Combine(Application.streamingAssetsPath, "cube.unity3d");

        //下载 Bundle
        WWW www = new WWW("file://"+path);
        yield return null;
        string name = Path.GetFileName(path);
        string filePath = Path.Combine(Application.persistentDataPath, name);
        if(File.Exists(filePath)) {
            File.Delete(filePath);
        }
        //下载结束后写入本地
        File.WriteAllBytes(filePath, www.bytes);
        Debug.Log(filePath);

        //从硬盘中读取 Bundle
        AssetBundle assetbundle = AssetBundle.LoadFromFile(filePath);
        //从 Bundle 中读取资源
        GameObject prefab = assetbundle.LoadAsset<GameObject>("Cube");
        //实例化资源
        GameObject.Instantiate<GameObject>(prefab);
    }
}
```

11.4.7　加载场景

场景构建 AssetBundle 的方法和资源类似。如图 11-17 所示，从 Bundle 中加载场景是不需要在 Scenes In Build 中添加这个场景的。

图 11-17　场景

如代码清单 11-16 所示，首先使用 `AssetBundle.LoadFromFile()` 加载场景，接着读取场景即可。

代码清单 11-16　Script_11_16.cs 文件

```csharp
using System.Collections;
using System.Collections.Generic;
using UnityEngine;
using System.IO;
using UnityEngine.SceneManagement;

public class Script_11_16 : MonoBehaviour {
    void Start() {
        string path = Path.Combine(Application.streamingAssetsPath, "scene-ab.unity3d");
        //加载场景 Bundle
        AssetBundle.LoadFromFile(path);
        SceneManager.LoadScene("scene-ab");
    }
}
```

11.4.8　卸载 AssetBundle

在同一个 `AssetBundle` 文件中，可以同时构建多个资源文件。正确地从 `AssetBundle` 中实例化一个 Prefab 的步骤如下。

(1) 从硬盘中加载 `AssetBundle` 对象。
(2) 从 `AssetBundle` 对象中加载需要的资源对象。
(3) 从资源读取对象并且将其实例化到 Hierarchy 视图中，变成真正的游戏对象。

由此可见，一次加载需要产生 3 种对象：`AssetBundle` 对象、资源对象和游戏对象。接着，开始卸载对象。

- `GameObject.Destroy()` 方法只能卸载游戏对象，资源对象还静静地在内存中。
- `Resources.UnloadUnusedAssets()` 方法也无法卸载资源对象，因为它被 `AssetBundle` 对象引用着。
- `AssetBundle.UnloadAllAssetBundles(true);` 方法终于可以全部卸载掉了，其中参数 `true` 表示同时卸载 `AssetBundle` 对象以及资源对象。资源对象一旦卸载掉，下次再加载时又需要耗时处理，所以有时候只希望卸载 `AssetBundle` 对象而不卸载资源对象，这时参数就可以填 `false` 了。

11.5　游戏对象

通过上面的学习，我们应该明白，游戏对象就是个空壳子，它关联着所有资源以及组件的引用。游戏对象也有一套自己的管理方式，Unity 也提供了丰富的接口来操作它，比如创建、删除、修改和查询等。

11.5.1 创建游戏对象

创建游戏对象的方式有两种。第一种是从资源中创建对象，例如前面讲过的 Resources 读取一个 Prefab，接着通过 GameObject.Instantiate() 实例化创建。另一种就是通过代码创建，即使用 new GameObject()，接着在后面挂脚本或者设置参数等。如代码清单 11-17 所示，我们可以在运行时创建一个空的游戏对象，并将其绑定在一个摄像机脚本中。

代码清单 11-17　Script_11_17.cs 文件

```csharp
using System.Collections;
using UnityEngine;

public class Script_11_17 : MonoBehaviour {

    void Start() {
        Camera camera = new GameObject("MyCamera").AddComponent<Camera>();
        camera.transform.position = Vector3.one * 10f;
    }
}
```

11.5.2 Transform 设置排序

Transform 使用 SetSiblingIndex() 方法即可设置子节点下所有对象的排序。如代码清单 11-18 所示，SetAsFirstSibling() 和 SetAsLastSibling() 设置在相同节点下的首位和末尾。

代码清单 11-18　Script_11_18.cs 文件

```csharp
using System.Collections;
using UnityEngine;

public class Script_11_18 : MonoBehaviour {

    public GameObject go1;
    public GameObject go2;
    void Start()
    {
        //设置在同节点下的首位
        go1.transform.SetAsFirstSibling();

        //设置在同节点下的末尾
        go1.transform.SetAsLastSibling();

        //设置在 go2 节点的上面
        go1.transform.SetSiblingIndex(go2.transform.GetSiblingIndex());

        //设置在 go2 节点的下面
        go1.transform.SetSiblingIndex(go2.transform.GetSiblingIndex() + 1);
    }
}
```

11.5.3 删除节点

使用 GameObject.Destroy() 方法，可以将某个节点（包括自身下的所有元素）删除。但是有时候需要保留父节点，只删除子节点。如代码清单 11-19 所示，通过一个 while 循环来立即删除每个子节点。

代码清单 11-19　Script_11_19.cs 文件

```
using System.Collections;
using UnityEngine;

public class Script_11_19 : MonoBehaviour {

    void Start()
    {
        while(transform.childCount > 0) {
            GameObject.DestroyImmediate(transform.GetChild(0).gameObject);
        }
    }
}
```

在上述代码中，我们使用 GameObject.DestroyImmediate() 来立即删除游戏对象。如果需要安全删除游戏对象，就需要使用 GameObject.Destroy()。为了保证删除的顺序不会错乱，需要从后向前删除，代码如下所示：

```
//从后向前删除
for(int i = transform.childCount - 1; i >= 0; i--)
{
    GameObject.Destroy(transform.GetChild(i).gameObject);
}
```

11.5.4 获取游戏对象

我们既可以通过完整的路径获取游戏对象，也可以根据某个游戏对象获取它的子对象。此外，还可以通过 tag 的方式来获取。如图 11-18 所示，我们可以自行添加新的 tag。

图 11-18　添加 tag

如代码清单 11-20 所示，我们可以通过游戏对象名称、路径和 Tag 来获取它。

代码清单11-20　Script_11_20.cs 文件

```csharp
using System.Collections;
using UnityEngine;

public class Script_11_20 : MonoBehaviour {

    void Start()
    {
        //根据Scene完整路径获取游戏对象
        GameObject g1 = GameObject.Find("Main Camera/GameObject");

        //根据相对路径获取某游戏对象下的子游戏对象
        GameObject g2 = g1.transform.Find("GameObject/child").gameObject;

        //根据相对索引获取某游戏对象下的子游戏对象
        GameObject g3 = g1.transform.GetChild(0).gameObject;

        //根据tag获取单个游戏对象
        GameObject g4 =  GameObject.FindGameObjectWithTag("momo");
        //根据tag获取多个游戏对象
        foreach(var item in GameObject.FindGameObjectsWithTag("momo")) {

        }
    }
}
```

11.5.5　管理游戏组件

任意游戏对象都可以绑定多个游戏组件。Unity 也提供了管理组件的方法，比如添加、获取、删除和修改等。如代码清单 11-21 所示，我们在代码中添加组件、判断组件是否存在、获取组件、删除组件等。

代码清单11-21　Script_11_21.cs 文件

```csharp
using System.Collections;
using UnityEngine;

public class Script_11_21 : MonoBehaviour {

    void Start()
    {
        GameObject go = new GameObject();

        //添加游戏组件
        go.AddComponent<Camera>();

        //没有组件时添加组件
        if(go.GetComponent<Camera>()) {
            go.AddComponent<Camera>();
        }
        //获取子对象中的组件（包括自身）
        go.GetComponentInChildren<Camera>();
```

```
//获取所有子对象中的组件（包括自身）
foreach(var item in go.GetComponentsInChildren<Camera>()) {
}
//获取父对象中的组件（包括自身）
go.GetComponentInParent<Camera>();

//获取所有父对象中的组件（包括自身）
foreach(var item in go.GetComponentsInParent<Camera>()) {
}
//获取游戏场景内单个组件
GameObject.FindObjectOfType<Camera>();
//获取游戏场景内所有组件
foreach(var item in GameObject.FindObjectsOfType<Camera>()) {
}
//删除组件
GameObject.Destroy(GetComponent<Camera>());
//立即删除组件
GameObject.DestroyImmediate(GetComponent<Camera>());
    }
}
```

11.6 优化工具

Unity 提供了很多强有力的工具来辅助开发者进行优化。其中，Profiler 工具可以查看每一帧游戏的渲染、加载和内存，并且它可以精确到耗时资源本身。Frame Debugger 工具用来查看渲染的 DrawCall 顺序，进一步查出为什么 DrawCall 没有合并。

11.6.1 重复无用资源

在游戏开发的过程中，经常会换 UI、模型和资源等，时间长了就会积累出很多无用的资源，其中有些资源已经废弃了但还被用着，或者资源发生了重复并且重复的资源还分别被有些对象所依赖，所以需要做一个工具来自动找出这些不合理的资源，统一处理掉。如图 11-19 所示，Resources 下有两个 Prefab 并且它们分别引用了两个相同的贴图资源，我们需要将它们自动找出来。其原理就是使用 `AssetDatabase.GetDependencies()` 自动查找资源依赖，最后比对文件的 MD5，找出重复的资源，具体如代码清单 11-22 所示。

图 11-19 重复

代码清单 11-22　Script_11_22.cs 文件

```csharp
using System.Collections;
using UnityEngine;
using UnityEditor;
using System.Security.Cryptography;
using System;
using System.IO;
using System.Collections.Generic;

public class Script_11_22 {

    [MenuItem("Tools/Report/查找重复贴图")]
    static void ReportTexture()
    {
        Dictionary<string,string> md5dic = new Dictionary<string, string>();
        string[] paths = AssetDatabase.FindAssets("t:prefab",new string[]
            {"Assets/Resources"});

        foreach(var prefabGuid in paths) {
            string prefabAssetPath = AssetDatabase.GUIDToAssetPath(prefabGuid);
            string[] depend = AssetDatabase.GetDependencies(prefabAssetPath,true);
            for(int i = 0; i < depend.Length; i++) {
                string assetPath = depend [i];
                AssetImporter importer = AssetImporter.GetAtPath(assetPath);
                //满足贴图和模型资源
                if(importer is TextureImporter || importer is ModelImporter) {
                    string md5 = GetMD5Hash(Path.Combine(Directory.
                        GetCurrentDirectory(),assetPath));
                    string path;

                    if(!md5dic.TryGetValue(md5, out path)) {
                        md5dic [md5] = assetPath;
                    }else {
                        if(path != assetPath) {
                            Debug.LogFormat("{0} {1} 资源发生重复！ ", path, assetPath);
                        }
                    }
                }
            }
        }
    }
    ///<summary>
    ///获取文件 MD5
    ///</summary>
    ///<returns>The Md5 hash.</returns>
    ///<param name="filePath">File path.</param>
    static string GetMD5Hash(string filePath)
    {
        MD5 md5 = new MD5CryptoServiceProvider();
        return BitConverter.ToString(md5.ComputeHash(File.ReadAllBytes(filePath))).
            Replace("-", "").ToLower();
    }
}
```

11.6.2 查看内存

要优化内存，可以使用 Unity 自带的 Profiler 工具，它可以查看每一帧中内存的各项占用，并且可以精确到每一个资源。虽然编辑器中也可以查看 Profiler，但它非常不准。例如，在 Project 视图中选择一个资源后，编辑器就会缓存它，Profiler 中也就会出现它，但实际上这个资源并不是游戏内产生的内存。

如果需要查看得更准确，必须打包来测试。打包前，只需选中 Development Build 复选框即可。一般情况下，打手机包测试比较麻烦，可以打一个 PC 包，这样连上 Unity 的 Profiler 就可以准确地定位内存占用了。如图 11-20 所示，连上 Profiler 以后，可以看到游戏内存占用最多的一般都是贴图文件。展开 Texture2D，可以查看每一张图具体的内存占用了。如果图片压缩格式不合理，就及时修改。

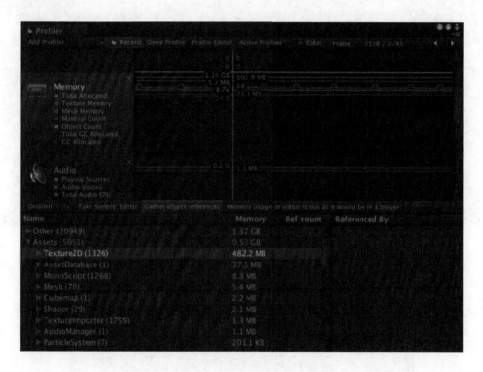

图 11-20　Profiler

11.6.3 查看 CPU 效率

使用 Profiler 工具，还可以查看 CPU。一般情况下，游戏中都是 CPU 比较耗时，尤其是同步读取大量资源的时候，此时 Profiler 中就能看到明显的尖刺。如图 11-21 所示，将帧移动到尖刺的地方，即可看到这一帧中最为耗时的地方。Total 表示占总效率的百分比，Self 表示它自身占效率的百分比（不包括子元素的耗时），Calls 表示调用次数，GC Alloc 表示申请的堆内存大小，Time 表示总共耗费的毫秒数，Self 表示自身耗时的毫秒数（不包括子元素的耗时）。通过面板中的信息，就可以更快定位耗时的位置了。

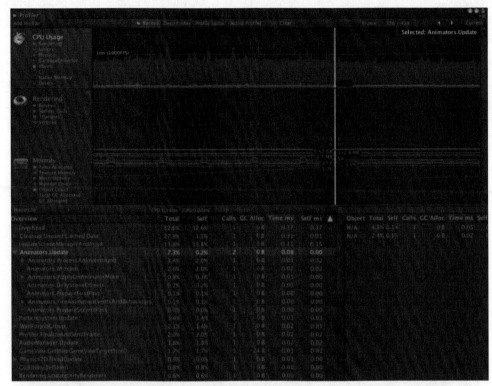

图 11-21　CPU 效率

11.6.4　自定义观察区间

如果对某一段代码的效率不太放心，可以设置 Profiler 的自定义观察区间来观察它的代码效率。如图 11-22 所示，给某段代码添加一个观察区间，这样就能清楚地看到执行的耗时时间了。

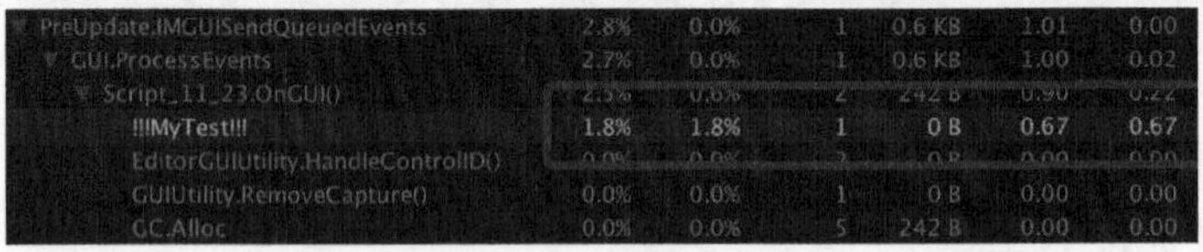

图 11-22　观察区间

如代码清单 11-23 所示，将 `Profiler.BeginSample()` 和 `Profiler.EndSample()` 添加在需要查看效率的代码中，即可在 Profiler 中查看它整体的消耗情况。

代码清单 11-23   Script_11_23.cs 文件

```
using UnityEngine;
using UnityEditor;
using UnityEngine.Profiling;

public class Script_11_23: MonoBehaviour
{
    void Test()
    {
        Profiler.BeginSample("!!!MyTest!!!");
        for(int i = 0; i < 100000; i++) {
        }
        Profiler.EndSample();
    }

    void OnGUI()
    {
        if(GUILayout.Button("test")) {
            Test();
        }
    }
}
```

11.6.5　Profiler 信息的导出与导入

Profiler 的信息可以在真实设备上导出。比如，手机上调试不太方便时，可以通过代码把信息导出来在计算机上查看。如图 11-23 所示，在 Profiler 面板中还可以单击 Load 按钮，选择之前导出的信息重新导入。

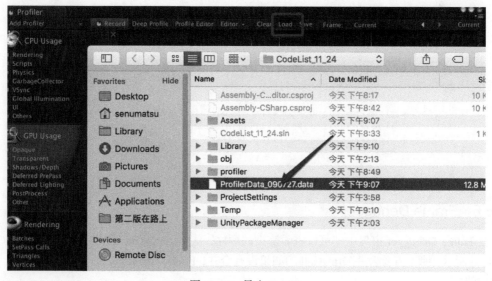

图 11-23　导入 Profiler

如代码清单 11-24 所示，调用 `StartRecord()` 方法开始记录，调用 `StopRecord()` 方法结束记录。`Profiler.logFile` 就是 Profiler 记录的数据。我们可以在手机上可以导出这些数据，在 PC 上查看它们。

代码清单 11-24　Script_11_24.cs 文件

```csharp
using UnityEngine;
using UnityEngine.Profiling;
using System.IO;
using System;

public class Script_11_24: MonoBehaviour
{
    void OnGUI()
    {
        if(GUILayout.Button("开始记录 Profiler")) {
            StartRecord();
        }
        if(GUILayout.Button("结束记录 Profiler")) {
            StopRecord();
        }
    }
    //开始记录 Profiler
    public static void StartRecord()
    {
        string fileName = string.Format("ProfilerData_{0}.data",DateTime.Now.
            ToString("hhmmss"));
        if(Application.isEditor)
            Profiler.logFile = Path.Combine(Application.dataPath+"/..", fileName);
        else
            Profiler.logFile = Path.Combine(Application.persistentDataPath, fileName);
        Profiler.enabled = true;
        Profiler.enableBinaryLog = true;
    }
    //结束记录 Profiler
    public static void StopRecord()
    {
        UnityEngine.Profiling.Profiler.enabled = false;
        UnityEngine.Profiling.Profiler.enableBinaryLog = false;
        UnityEngine.Profiling.Profiler.logFile = "";
    }
}
```

11.6.6　Frame Debugger

在 Frame Debugger 窗口中，可以查看渲染的顺序，了解 DrawCall 的数量是如何产生的。如图 11-24 所示，原本需要合并 DrawCall 的如果没有合并，Frame Debugger 还会告诉我们是什么原因导致的。

图 11-24　Frame Debugger

11.7　资源管理实例

游戏资源管理是一门很大的学问。前面已经介绍了资源加载以及优化的相关知识,这里会结合相关实例进一步加强学习。

11.7.1　特殊内置的文件夹

在 Unity 项目中,有些文件夹的名字具有内置的含义,比如以下几种。

- **Assets 文件夹**：游戏资源的顶层文件夹。AssetDatabase 方法可以访问里面的任意资源。
- **Editor 文件夹**：编辑模式下的代码需要放在这里。打包以后,会自动剥离它。
- **Editor Default Resources 文件夹**：资源文件夹。编辑模式下的资源尽可能放在这里。
- **Gizmos 文件夹**：一般放一些工具类图标。使用 Gizmos.DrawIcon() 方法来显示它们。
- **Plugins 文件夹**：这个之前我们讲过,此文件夹下的代码会优先编译成 DLL 文件。
- **Resources 文件夹**：该目录下的所有资源都会自动构建在包体内,可使用 Resources.load() 方法加载资源。
- **Standard Assets 文件夹**：导入 Unity 标准资源的 package 文件夹。这里脚本的执行顺序会设置得靠前。
- **StreamingAssets 文件夹**：AssetBundle 文件适合放在这里,因为这个文件下的资源不会被自动压缩或者改变。

11.7.2 隐藏文件夹

在工程中,可以设置隐藏文件夹或文件,这有什么用呢?游戏包很有可能需要支持小包或者大包模式,小包就是需要额外下载游戏资源,而大包就是所有游戏资源都放在包内。大包由于很多资源都放在 Resources 目录下,如果此时需要打小包,就必须把 Resources 改个名字或者挪出工程。如图 11-25 所示,可以在 Resources 名字后面加个波浪号(~),表示被 Unity 工程所忽略,这样在打包的时候这下面的资源就不会被打包进包体内了。另外,还有一些隐藏规则。隐藏开头是点(.)符号的文件夹以及文件,隐藏结尾是波浪号(~)的文件夹以及文件,隐藏名字是 cvs 的文件夹以及文件,隐藏扩展名是 .tmp 的文件。

图 11-25 隐藏文件夹

11.7.3 Resources 与 AssetBundle 无缝切换

平时开发中,如果一直都在用 Resources.Load() 方法,那么项目到后期想切换成 AssetBundle,就非常麻烦了,所以需要一套 Resources 和 AssetBundle 无缝切换的方案。思路就是使用一个自定义的类来封装 Resources 和 AssetBundle 的读取方法,构建 AssetBundle 的时候需要记录资源和 Bundle 之间的引用关系,通过 Resources 加载的目录立即取到对应 AssetBundle 的加载目录进行无缝切换。如代码清单 11-25 所示,构建资源并且生成资源描述文件。

代码清单 11-25 BuildAB.cs 文件

```
using UnityEngine;
using System.IO;
using UnityEngine.Events;
using UnityEditor;
using System.Collections.Generic;

public class BuildAB
{
    [MenuItem("Tools/BuildAssetbundle")]
    static void BuildAssetbundle()
    {
```

```csharp
        string outPath = Path.Combine(Application.dataPath, "StreamingAssets");

        //如果目录已经存在, 则删除它
        if(Directory.Exists(outPath)) {
            Directory.Delete(outPath,true);
        }
        Directory.CreateDirectory(outPath);

        List<AssetBundleBuild> builds = new List<AssetBundleBuild>();
        //设置Bundle名, 确定多少资源构建在同一个Bundle中
        builds.Add(new AssetBundleBuild(){assetBundleName="Cube.unity3d",
            assetNames =new string[]{"Assets/Resources/Cube.prefab",
                "Assets/Resources/Cube 1.prefab"}});

        //构建AssetBundle
        BuildPipeline.BuildAssetBundles(outPath,builds.ToArray(),
            BuildAssetBundleOptions.ChunkBasedCompression | BuildAssetBundleOptions.
            DeterministicAssetBundle, BuildTarget.StandaloneOSX);

        //生成描述文件
        BundleList bundleList = ScriptableObject.CreateInstance<BundleList>();
        foreach(var item in builds) {
            foreach(var res in item.assetNames) {
                bundleList.bundleDatas.Add(new BundleList.BundleData(){ resPath =
                    res, bundlePath = item.assetBundleName });
            }
        }
        AssetDatabase.CreateAsset(bundleList,"Assets/Resources/bundleList.asset");

        //刷新
        AssetDatabase.Refresh();
    }
}
```

如代码清单 11-26 所示, 构建 AssetBundle 时, 序列化资源的路径和 AssetBundle 加载路径。

代码清单 11-26 BundleList.cs 文件

```csharp
using System.Collections;
using System.Collections.Generic;
using UnityEngine;

[System.Serializable]
public class BundleList : ScriptableObject {

    public List<BundleData> bundleDatas = new List<BundleData>();

    //保存每个res路径对应的Bundle路径
    [System.Serializable]
    public class BundleData
    {
        public string resPath = string.Empty;
        public string bundlePath = string.Empty;
    }
}
```

如代码清单 11-27 所示，这里提供了一个 Assets 类来代替 Resources 类，这样代码中就不要再使用 Resources 而要采取 Assets.Load<T>()。该方法需要智能地判断是否有 AssetBundle，如果有 AssetBundle，就采用 AssetBundle 加载；如果没有 AssetBundle，就采用 Resources 加载。

代码清单 11-27　Assets.cs 文件

```csharp
using System.Collections;
using System.Collections.Generic;
using UnityEngine;
using System.IO;

public class Assets
{
    static Dictionary<string,string> m_ResAbDic = new Dictionary<string, string>();
    static Dictionary<string,AssetBundle> m_BundleCache = new Dictionary<string, AssetBundle>();
    static Assets()
    {
        //读取依赖关系
        BundleList list = Resources.Load<BundleList>("bundleList");
        foreach(var bundleData in list.bundleDatas) {
            m_ResAbDic [bundleData.resPath] = bundleData.bundlePath;
        }
    }

    static public T LoadAsset<T>(string path) where T : Object
    {
        //从 AssetBundle 中加载资源，最好提供后缀名，不然无法区分同名文件
        string bundlePath;
        string resPath = Path.Combine("Assets/Resources", path);
        if(typeof(T) ==  typeof(GameObject)) {
            resPath = Path.ChangeExtension(resPath, "prefab");
        }
        //如果 Bundle 有这个资源，则从 Bundle 中加载
        if(m_ResAbDic.TryGetValue(resPath, out bundlePath)) {
            AssetBundle assetbundle;
            if(!m_BundleCache.TryGetValue(bundlePath,out assetbundle)){
                assetbundle = m_BundleCache [bundlePath] = AssetBundle.LoadFromFile
                    (Path.Combine(Application.streamingAssetsPath, bundlePath));
            }
            return assetbundle.LoadAsset<T>(resPath);
        }
        //如果 Bundle 中没有这个资源，则从 Resources 目录中加载
        return Resources.Load<T>(path);
    }
}
```

如代码清单 11-28 所示，使用 Assets.LoadAsset<T> 智能读取资源，最终将它实例化进 Hierarchy 视图中。

代码清单 11-28　Script_11_25.cs 文件

```csharp
using UnityEngine;
using UnityEngine.Profiling;
using System.IO;
```

```
using System;

public class Script_11_25: MonoBehaviour
{
    void Start(){

        GameObject.Instantiate<GameObject>(Assets.LoadAsset<GameObject>("Cube"));
        GameObject.Instantiate<GameObject>(Assets.LoadAsset<GameObject>("Cube 1"));
    }
}
```

11.7.4 资源加载策略

上面的代码中，我们实现了本地的 AssetBundle 和 Resources 的无缝切换，其实可能还有下载 AssetBundle 这一步骤。整个 AssetBundle 的加载策略如下。

(1) 从 Application.persistentDataPath 目录中查找读写目录下是否有需要加载的 AssetBundle。

(2) 如果第(1)步没加载到资源，接着在 Application.streamingAssetsPath 目录中查找本地是否有需要加载的 AssetBundle。

(3) 如果第(2)步没加载到资源，接着在 Resources 目录中加载文件。

按照这个加载策略，就可以保证用户加载的永远是最新的资源。另外，整个加载策略还需要考虑 AssetBundle 卸载的时机。不然，由于 AssetBundle 引用资源，导致资源无法被卸掉了。

11.7.5 资源更新

资源更新就是将 CDN 上的资源下载并保存在 Application.persistentDataPath 目录中，然后按照上一节的加载顺序加载资源。首先，要维护一个下载资源列表，里面记录着每个资源的散列值，这样应用程序启动的时候，用包体内资源的散列值和远端 CDN 中保存的散列值做比较，决定需不需要下载。

为了避免重复下载，必须要保证每次相同资源构建出来的 AssetBundle 是一致的。在构建方法中必须设置 `BuildAssetBundleOptions.DeterministicAssetBundle`，这样可以保证打包资源的一致性。另外，AssetBundle 文件是不能信任 MD5 的。Unity 提供了一个散列值来约束一致性。代码中需要调用 `AssetBundleManifest.GetAssetBundleHash()` 方法来提取散列值，从而来比较是否需要更新，这样就能避免重复下载的情况了。

11.7.6 资源引用关系

Unity 只提供了某个资源引用了什么资源，但是并没有提供某个资源被什么资源所引用。右击任意资源后，会弹出如图 11-26 所示的快捷菜单，部分菜单的含义如下。

❑ Find References In Scene：表示查找这个资源在当前场景中的什么地方引用。
❑ Select Dependencies：表示查找这个资源应用了哪些资源。

图 11-26 查找引用关系

为了更方便查找资源，我们可以自己写一种查找方法，专门查询这个资源被哪些资源所引用。如图 11-27 所示，我们拓展一个右键菜单项，选择任意资源后，点击 Find References 菜单后，即可查找该资源被哪里引用了。可以看出，这个资源被 New Material.mat 材质所引用，相关代码如代码清单 11-29 所示。

图 11-27 查找被引用关系

代码清单 11-29　FindReferences.cs 文件

```
using UnityEngine;
using UnityEditor;
using System.IO;
using System.Linq;
using System.Text.RegularExpressions;
using System.Collections.Generic;

public class FindReferences
{
    [MenuItem("Assets/Find References", false, 10)]
    static private void Find()
    {
        Dictionary<string,string>guidDics =new Dictionary<string, string>();
        foreach(Object o in Selection.objects){
            string path = AssetDatabase.GetAssetPath(o);
```

```csharp
            if(!string.IsNullOrEmpty(path)){
                string guid = AssetDatabase.AssetPathToGUID(path);
                if(!guidDics.ContainsKey(guid)){
                    guidDics[guid] = o.name;
                }
            }
        }

        if(guidDics.Count > 0)
        {
            List<string> withoutExtensions = new List<string>(){".prefab",
                ".unity",".mat",".asset"};
            string[] files = Directory.GetFiles(Application.dataPath, "*.*",
                SearchOption.AllDirectories)
                    .Where(s => withoutExtensions.Contains(Path.GetExtension(s).
                        ToLower())).ToArray();
            for(int i=0; i<files.Length;i++)
            {
                string file = files[i];
                if(i%20==0){
                    bool isCancel = EditorUtility.DisplayCancelableProgressBar
                        ("匹配资源中", file, (float)i / (float)files.Length);
                    if(isCancel){
                        break;
                    }
                }
                foreach(KeyValuePair<string,string> guidItem in  guidDics){
                    if(Regex.IsMatch(File.ReadAllText(file), guidItem.Key))
                    {
                        Debug.Log(string.Format("name: {0} file: {1}",guidItem.
                            Value,file), AssetDatabase.LoadAssetAtPath<Object>
                            (GetRelativeAssetsPath(file)));
                    }
                }
            }
            EditorUtility.ClearProgressBar();
            Debug.Log("匹配结束");
        }
    }

    [MenuItem("Assets/Find References", true)]
    static private bool VFind()
    {
        string path = AssetDatabase.GetAssetPath(Selection.activeObject);
        return(!string.IsNullOrEmpty(path));
    }

    static private string GetRelativeAssetsPath(string path)
    {
        return "Assets" + Path.GetFullPath(path).Replace(Path.GetFullPath
            (Application.dataPath), "").Replace('\\', '/');
    }
}
```

11.7.7 系统资源修改

Unity 有很多系统资源，其后缀名是 .asset。在各类设置面板中，Unity 会将很多配置数据序列化进去，但是可能并没有对外提供修改的接口，特定的时候我们需要脚本自动来修改它。如图 11-28 所示，ProjectSettings 中的这些资源都是系统资源，Assets 目录下创建的一些资源也是系统资源。

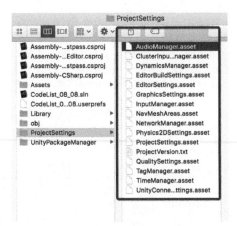

图 11-28　修改系统资源

用记事本随便打开一个资源，随便找个属性来修改，比如 Splash 背景颜色这一栏，如图 11-29 所示。

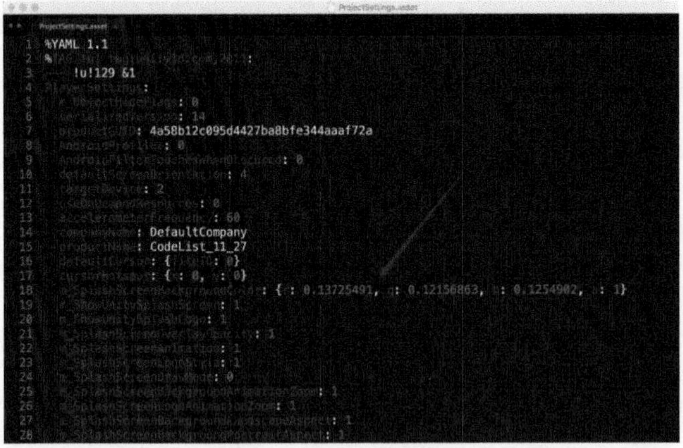

图 11-29　查看系统资源

如代码清单 11-30 所示，首先使用 `AssetDatabase.LoadAssetAtPath()` 读取一个资源对象，接着使用 `FindProperty()` 获取该资源对象序列化中的某个数据对象，最后就可以修改每个序列化的数据了。

代码清单 11-30　Script_11_27.cs 文件

```
using UnityEngine;
using UnityEditor;

public class Script_11_30
{
    [MenuItem("Tool/EditValue")]
    static void EditValue()
    {
        //读取资源
        const string projectSettingAssetPath = "ProjectSettings/ProjectSettings.asset";
        SerializedObject projectSetting = new SerializedObject(AssetDatabase.
            LoadAssetAtPath(projectSettingAssetPath,typeof(Object)));
        SerializedProperty m_SplashScreenBackgroundColor = projectSetting.FindProperty
            ("m_SplashScreenBackgroundColor");
        Debug.LogFormat("当前的颜色值:{0}", m_SplashScreenBackgroundColor.colorValue);
        //修改它的颜色
        m_SplashScreenBackgroundColor.colorValue = Color.black;
        //修改后保存
        projectSetting.ApplyModifiedProperties();
        AssetDatabase.SaveAssets();
    }
}
```

注意：只要是 Unity 自己的 .asset 资源，都可以使用上述方法来自动修改它。但是它只支持在编辑模式下修改，运行模式下是不行的。

11.7.8　AssetBundle 里的脚本

Unity 在构建 AssetBundle 的时候，可以指定 Prefab 文件。Prefab 自身是可以挂游戏脚本的，那是不是说脚本也可以热更新了？答案是否定的！前面讲过，资源之间的依赖关系是通过 guid 来关联的，其实脚本也一样。如图 11-30 所示，打开脚本对应的 .meta 文件后，可以看到每个脚本都会对应一个 guid 数值。

图 11-30　脚本的 guid

接着，将此脚本绑定在某个 Prefab 上。如图 11-31 所示，将脚本挂在 Main Camera 上，然后直接将它用文本打开，就可以发现这个脚本的 guid 了。所以说，即使 AssetBundle 包含了脚本，也是无法热更新的。因为实际运行时，Unity 只会找游戏包里对应这个 guid 的脚本。

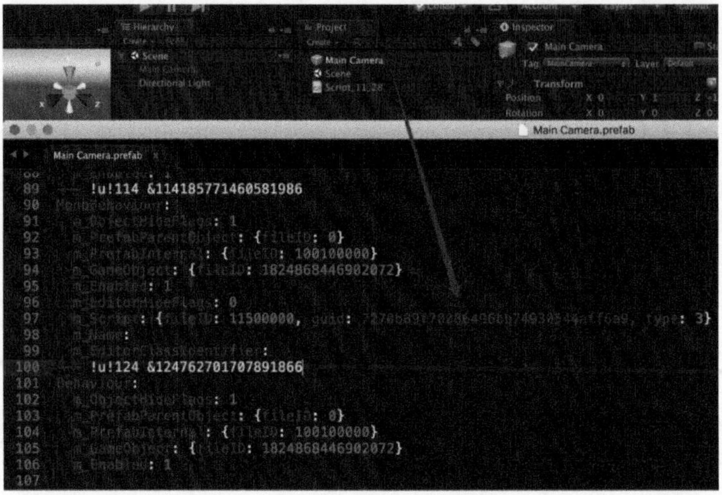

图 11-31　Prefab 关联脚本

11.7.9　热更新代码

Unity 的脚本是无法热更新的，但是 DLL 是可以热更新的。我们可以将多个 .cs 文件打入 DLL 中，通过下载的方式来更新游戏代码。如代码清单 11-31 所示，首先写一个静态类，封装一个静态方法。我们将这个类封装成 DLL 文件，临时放在 StreamingAssets/HotUpdate.dll 下。

代码清单 11-31　HotUpdate.cs 文件

```
public class HotUpdate {

    static public string GetName()
    {
        return "HotUpdate";
    }
}
```

如代码清单 11-32 所示，读取 DLL 文件后，通过反射的方法来执行它内部的方法。如图 11-32 所示，最终构建游戏包后，将返回值 HotUpdate 显示在屏幕中。

代码清单 11-32　Script_11_28.cs 文件

```
using System.Collections;
using System.Collections.Generic;
using UnityEngine;
using System.Reflection;
using System;
```

```
public class Script_11_28 : MonoBehaviour {

    private string m_HotUpdateInfo = string.Empty;
    void Start() {
        //加载 DLL 文件
        Assembly assembly = Assembly.LoadFrom(Application.streamingAssetsPath+
            "/HotUpdate.dll");
        //获取某个类
        Type type=assembly.GetType("HotUpdate");
        //获取类中的某个方法
        MethodInfo mi=type.GetMethod("GetName");
        //反射调用 GetName 方法, 并且获取返回值
        m_HotUpdateInfo = mi.Invoke(null,new object[]{}).ToString();
    }

    void OnGUI()
    {
        //显示读取的内容
        GUILayout.Label(string.Format("<size=80>{0}</size>",m_HotUpdateInfo));
    }
}
```

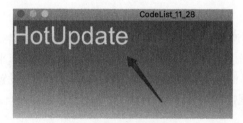

图 11-32　热更新 DLL

注意：由于苹果禁止了 JIT（即时编译器），所以热更新 DLL 文件在 iOS 平台下是不支持的，在 Android 平台以及 PC 平台下都是支持的。

11.8　小结

本章介绍了资源的加载以及优化。Unity 提供了 Profiler 工具，使用它，不仅可以查看每一帧 CPU、GPU、UI 和物理等性能耗时，还可以自定义代码区间来查看自己代码的性能指标。发布游戏后，可以连接 Profiler 工具查看性能，还可以将 Profiler 信息导出后，再在 PC 上查看。总之，Profiler 是分析性能最重要的工具。另外，引擎还提供了 Frame Debugger 工具，该工具可以查看渲染层次以及 DrawCall 没有合并的原因。此外，还介绍了编辑模式下和运行模式下资源的加载以及注意事项，包括 AssetBundle 的下载、导入和导出等。通过本章的学习，你应该对 Unity 的各项资源有了更清晰的认识。

第 12 章

自动化与打包

通过第 11 章的学习，我们已经对资源管理有了深刻的认识。游戏中的资源是海量的，这就带来新的问题：大量的资源格式需要设置，比如贴图格式、模型格式、音频格式和文件格式等。如果是人为手动设置，效率就太低了而且还容易出错，所以我们需要结合自动化来管理资源。另外，游戏打包这一步也需要自动化。例如，每次打包前，需要对包体做一些修改，如果全都手动操作，就非常麻烦了。本章将向读者介绍 Unity 的自动化接口。学完这一章，希望可以解放你的双手。

12.1 资源导入

在 Unity 中导入任何资源后，都可以通过设置参数来修改它的格式，而引擎会自动重新压缩资源。游戏中的大量资源都是美术人员或者策划人员导入的，这听起来好像就是设置一些参数而已，其实这个过程中很容易设置成错误的资源类型，或者由于没有设置被设置成默认的效率不高的资源类型。为了避免出现这种低级错误，我们要做的就是帮它们自动设置。

12.1.1 监听导入事件

资源导入后，Unity 会根据设置信息自动压缩。而导入事件又分两种：压缩前的事件和压缩后的事件。如果需要修改导入参数，那么就可以监听压缩前的事件来修改它，后面 Unity 就会自动执行导入操作了。如果想等导入彻底结束后生成点什么对象，可以监听导入后的事件。如代码清单 12-1 所示，继承 AssetPostprocessor 后，就可以监听它们的事件了。

代码清单 12-1　Script_12_01.cs 文件

```
using UnityEngine;
using UnityEditor;

public class Script_12_01 : AssetPostprocessor
{
    //导入声音前
    void OnPreprocessAudio()
    {
        AudioImporter audioImporter = (AudioImporter)assetImporter;
    }
```

```csharp
//导入动画前
void OnPreprocessAnimation()
{
    ModelImporter modelImporter = (ModelImporter)assetImporter;
}
//导入模型前
void OnPreprocessModel()
{
    ModelImporter modelImporter  = (ModelImporter)assetImporter;
}
//导入贴图前
void OnPreprocessTexture()
{
    TextureImporter textureImporter  = (TextureImporter)assetImporter;
}
//导入声音后
void OnPostprocessAudio(AudioClip clip)
{
    Debug.Log(AssetDatabase.GetAssetPath(clip));
}
//导入模型后
void OnPostprocessModel(GameObject g)
{
    Debug.Log(AssetDatabase.GetAssetPath(g));
}
//导入材质后
void OnPostprocessMaterial(Material material)
{
    Debug.Log(AssetDatabase.GetAssetPath(material));
}
//导入精灵后
void OnPostprocessSprites(Texture2D texture ,Sprite[] sprites ) {
    Debug.Log("Sprites: " + sprites.Length);
}
//导入贴图
void OnPostprocessTexture(Texture2D texture) {
    Debug.Log("Texture2D: (" + texture.width + "x" + texture.height + ")");
}
```

资源的种类有很多。在上述代码中，我们可以分别监听资源导入前以及导入后的事件，导入前事件用于设置资源的压缩参数，而导出后事件可以修改最终保存的资源文件。

12.1.2 自动设置贴图压缩格式

比如，项目有两个目录：一个是 UI，另一个是 Texture。当贴图拖入 UI 目录下时，自动设置 UI 贴图的类型；而当贴图拖入 Texture 目录下时，则自动设置普通模型使用的贴图。如图 12-1 所示，当贴图拖入 UI 目录下时，贴图将自动设置 RGBA32 的压缩格式。

第 12 章 自动化与打包

图 12-1 设置压缩格式

如代码清单 12-2 所示,首先监听贴图导入前事件,然后根据资源路径设置具体的压缩格式。其中,SetPlatformTextureSettings()方法同时设置多个平台下该资源的压缩格式。

代码清单 12-2　Script_12_02.cs 文件

```
using UnityEngine;
using UnityEditor;

public class Script_12_02 : AssetPostprocessor
{
    //导入贴图前
    void OnPreprocessTexture()
    {
        TextureImporter textureImporter  = (TextureImporter)assetImporter;

        if(textureImporter.assetPath.Contains("UI")) {
            textureImporter.textureType = TextureImporterType.Sprite;
            textureImporter.mipmapEnabled = false;
            //设置 UI 贴图在 3 个平台下的压缩格式以及大小
            textureImporter.SetPlatformTextureSettings("Standalone",2048,
                TextureImporterFormat.RGBA32);
            textureImporter.SetPlatformTextureSettings("iPhone", 2048,
                TextureImporterFormat.RGBA32 , 100, true);
            textureImporter.SetPlatformTextureSettings("Android",2048,
                TextureImporterFormat.RGBA32,true);

        } else if(textureImporter.assetPath.Contains("Texture")) {
```

```
            textureImporter.textureType = TextureImporterType.Default;
            textureImporter.mipmapEnabled = true;
            //设置模型贴图在3个平台下的压缩格式以及大小
            textureImporter.SetPlatformTextureSettings("Standalone",2048,
                TextureImporterFormat.DXT5);
            textureImporter.SetPlatformTextureSettings("iPhone", 2048,
                TextureImporterFormat.ASTC_RGBA_4x4 , 100, true);
            textureImporter.SetPlatformTextureSettings("Android",2048,
                TextureImporterFormat.ETC_RGB4,true);
        }
    }
}
```

另外,资源并不是从外部导入的,而是 Project 中别的目录移动到该目录下的资源。如图 12-2 所示,移动完毕后,单击鼠标右键,从弹出的快捷菜单中选择 Reimport 命令,即可重新执行导入流程。

图 12-2　重新导入

12.1.3　自动设置模型

模型是 FBX 文件,我们可以监听模型的导入事件,自动设置它的一些参数。当导入完成后,会自动生成 Prefab 文件。如图 12-3 所示,当模型拖入 Project 视图中,就会立刻在根目录下生成一个和它同名的 Prefab。本示例的完整代码详见代码清单 12-3。

图 12-3　导入模型

代码清单 12-3　Script_12_03.cs 文件
```
using UnityEngine;
using UnityEditor;

public class Script_12_03 : AssetPostprocessor
{
    //导入模型前
    void OnPreprocessModel()
    {
        ModelImporter modelImporter = (ModelImporter)assetImporter;
        //设置模型动画
        modelImporter.animationType = ModelImporterAnimationType.Generic;
    }


    //导入模型后
    void OnPostprocessModel(GameObject g)
    {
        //自动生成 Prefab
        PrefabUtility.CreatePrefab(string.Format("Assets/{0}.prefab", g.name), g);
    }
}
```

在上述代码中，我们监听 OnPostprocessModel() 导入 FBX 模型事件，执行 PrefabUtility.CreatePrefab() 来自动生成 Prefab 对象。

12.1.4　禁止模型生成材质文件

如果美术人员在 3ds Max 中指定了模型的材质，那么在导入模型的时候，会自动生成这个材质。如图 12-4 所示，如果后面生成模型 Prefab 时没有使用这个材质，会有严重的隐患。例如，有两个模型已经在 Unity 中重新指定不同的材质了，但是由于它们的 FBX 包含了相同的引用材质，这将导致最后生成的 AssetBundle 会重复打包这个资源，所以可以考虑关掉自动导入材质。

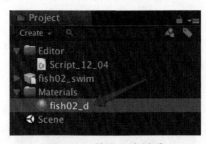

图 12-4　禁止生成材质

如代码清单 12-4 所示，监听模型导入前事件，设置 importMaterials 为 false，这样模型就不会每次导入时，都生成材质文件了。

代码清单 12-4　Script_12_04.cs 文件

```
using UnityEngine;
using UnityEditor;

public class Script_12_04 : AssetPostprocessor
{
    void OnPreprocessModel()
    {
        ModelImporter modelImporter = (ModelImporter)assetImporter;
        //禁止导入材质
        modelImporter.importMaterials = false;
    }
}
```

12.1.5　删除移动资源事件

导入资源不一定能满足所有需求。很多时候，资源很可能只是在内部移动一个位置，比如从一个目录移动到另一个目录下，此时我们可以监听资源删除和移动的事件。

如代码清单 12-5 所示，在 OnPostprocessAllAssets() 方法中，就可以监听新导入资源、删除的资源、移动后的资源、移动前的资源了。

代码清单 12-5　Script_12_05.cs 文件

```
using UnityEngine;
using UnityEditor;

public class Script_12_05 : AssetPostprocessor
{
    static void OnPostprocessAllAssets(string[] importedAssets, string[] deletedAssets,
        string[] movedAssets, string[] movedFromAssetPaths)
    {
        foreach(string str in importedAssets)
        {
            Debug.LogFormat("新导入的资源：{0}",str);
        }

        foreach(string str in deletedAssets)
        {
            Debug.LogFormat("删除的资源：{0}", str);
        }

        for(int i = 0; i < movedAssets.Length; i++)
        {
            Debug.LogFormat("移动资源位置：from: {0} to :{1}",movedFromAssetPaths[i],
                movedAssets[i]);
        }
    }
}
```

在上述代码中，我们继承了 AssetPostprocessor，因此可以监听资源的导入事件，可以判断资源路径来特殊处理工程中的某些资源。

12.1.6 选择性自动设置

通常情况下，我们会设置某一个文件夹下的资源格式保持一致。但是这也未必好，比如和美术人员约定了每个文件夹下的贴图必须是压缩格式，突然美术人员又设计了一张新的贴图，但它不能设置压缩格式，这时如果这个贴图不放在这个目录下，程序的加载代码就需要改了。所以我们不能太强制地锁定设置选项，可以设置一个导入的默认项，这样就能满足 90% 的情况，剩下的10% 可以由美术人员自己来把握。例如，以移动平台为例，Android 下贴图导入自动设置成 ETC2 压缩格式，如果某张图确实无法用 ETC2 压缩，美术人员就自行勾选 RGBA32，这样程序的逻辑完全不受影响了。

如代码清单 12-6 所示，在 `OnPreprocessTexture()` 方法中，首先判断当前资源的压缩格式是否满足 RGBA32 或者 ETC2，如果不满足，则强制设置为 ETC2 压缩格式。

代码清单 12-6　Script_12_06.cs 文件

```
using UnityEngine;
using UnityEditor;

public class Script_12_06 : AssetPostprocessor
{
    //导入贴图前
    void OnPreprocessTexture()
    {
        TextureImporter textureImporter  = (TextureImporter)assetImporter;

        TextureImporterPlatformSettings settings = textureImporter.
            GetPlatformTextureSettings("Android");
        //如果贴图当前的压缩格式不是 RGBA32 或者 ETC2_RGBA8，那么可强制转成 ETC2_RGBA8 格式
        if(settings.format != TextureImporterFormat.RGBA32 && settings.format != 
            TextureImporterFormat.ETC2_RGBA8) {
            textureImporter.SetPlatformTextureSettings("Android", 2048,
                TextureImporterFormat.ETC2_RGBA8, true);
        }
    }
}
```

12.1.7 主动设置

前面的代码都是依赖资源导入的回调来做的操作，有时并不需要导入资源而是主动去设置某个资源格式。如图 12-5 所示，右击一个贴图文件，从弹出的快捷菜单中选择 `SetTextureFormat` 菜单项来主动设置贴图的格式为 RGBA32。本示例的完整代码详见代码清单 12-7，其中需要先获取资源的 `AssetImporter` 对象，接着就可以设置参数了，最后调用 `SaveAndReimport()` 方法就可以保存并且重新导入资源了。

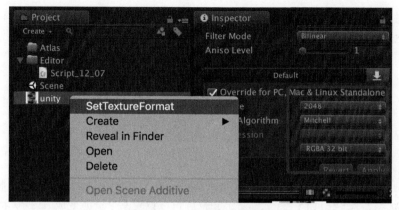

图 12-5 主动设置压缩格式

代码清单 12-7 Script_12_07.cs 文件

```
using UnityEngine;
using UnityEditor;

public class Script_12_07 : AssetPostprocessor
{
    [MenuItem("Assets/SetTextureFormat",false,-1)]
    static void SetTextureFormat()
    {
        //确保在 Project 视图中选择一个文件
        if(Selection.assetGUIDs.Length > 0){

            AssetImporter import = AssetImporter.GetAtPath(AssetDatabase.
                GetAssetPath(Selection.activeObject));
            //确保选择的是一个贴图文件
            if(import is TextureImporter) {
                (import as TextureImporter).SetPlatformTextureSettings("Standalone",
                    2048, TextureImporterFormat.RGBA32, true);
                //保存并且重新导入
                import.SaveAndReimport();
            }
        }
    }
}
```

在上述代码中，我们主动设置了某个资源的压缩格式。利用 `AssetImporter` 对象，还可以设置模型、声音等常用资源。参数设置完毕后，最终调用 `SaveAndReimport()` 方法保存即可。

12.1.8 待保存状态

只有对象变成 `dirty` 后，才可以进行保存。如图 12-6 所示，如果主动修改场景中的任意对象，它就会变成 `dirty` 状态，此时场景旁边就会出现一个*符号，按 Command+S 快捷键即可保存。但是有时通过代码去设置游戏对象或者对象身上的序列化信息时，很可能就不会造成 `dirty`

状态,这样数据是无法保存的,重新打开 Unity 数据依然是旧的,所以必要的时候可以再通过代码强制设置 dirty 状态。

图 12-6　dirty 状态

如代码清单 12-8 所示,`EditorSceneManager.MarkSceneDirty()`用于强制设置某个场景为 dirty 状态,`EditorUtility.SetDirty()`可以设置某个资源变成 dirty 状态。

代码清单 12-8　Script_12_08.cs 文件

```
using UnityEngine;
using UnityEditor;
using UnityEditor.SceneManagement;

public class Script_12_08 : AssetPostprocessor
{
    [MenuItem("Assets/SetSceneDirty",false,-1)]
    static void SetSceneDirty()
    {
        //设置场景的 dirty 状态
        EditorSceneManager.MarkSceneDirty(EditorSceneManager.GetActiveScene());
        //设置 Prefab dirty
        EditorUtility.SetDirty(AssetDatabase.LoadAssetAtPath<GameObject>
            ("Assets/Cube.prefab"));
    }
}
```

12.1.9　自动执行 MenuItem

Unity 内部有很多 MenuItem,通常我们只能在编辑器中手动选择后使用。但是有些功能希望通过脚本来自动完成,例如在 Project 视图中实现按 Command+D 快捷键复制场景以及 Prefab 的操作,如图 12-7 所示,相关代码如代码清单 12-9 所示。

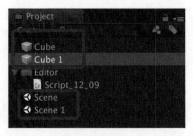

图 12-7　复制

代码清单 12-9 Script_12_09.cs 文件

```csharp
using UnityEngine;
using UnityEditor;

public class Script_12_09 : AssetPostprocessor
{
    [MenuItem("Assets/AutoMenuItem",false,-1)]
    static void AutoMenuItem()
    {
        //自动选择场景
        Selection.activeObject = AssetDatabase.LoadAssetAtPath("Assets/Scene.unity",
            typeof(Object));
        //按Command+D 快捷键复制场景
        EditorApplication.ExecuteMenuItem("Edit/Duplicate");

        //自动选择 Prefab
        Selection.activeObject = AssetDatabase.LoadAssetAtPath("Assets/Cube.prefab",
            typeof(Object));
        //按Command+D 快捷键复制场景
        EditorApplication.ExecuteMenuItem("Edit/Duplicate");
    }
}
```

12.2 配置错误

在游戏开发中，常见的经典资源错误如下。

- 策划人员把表配错了，程序运行报错。
- 美术人员制作的模型缺少规定的挂载点，或者起名不规范，程序运行后，找不到就报错了。

围绕着这两种错误衍生出来的错误简直太多了。通常，这个时候程序员就要去查了，查了半天还是资源配置错了！要想解决这个问题，我也思考了很久，后来得出的结论是，只要是人为手动去管理资源，那么必然会犯错。这是一个无法改变的事实，但是程序员不能为这种错误去买单，要想办法避免它们犯错，就是为自己节省时间。

程序员有代码编辑工具，代码写错了，立即就会报错或者运行不通过，所以程序员很难上传错误的代码。但是策划人员和美术人员没有这样的工具，他们只是按照和程序员规定的一套规范在配置资源，所以要想彻底解决这个问题，也要给他们提供资源检查工具。

12.2.1 主动检查工具

检查工具很好理解，就是程序人员开发一套工具来检查资源的规范性，策划人员或者美术人员在需要提交资源之前，运行一下检查工具，如果资源不符合规范，就会输出错误日志，接着修改好后再重复检查，直到没问题后再提交资源。这听起来好像很完美，但是实施起来却比较困难。如果在提交资源前忘记了执行检查工具，那么依然会把错误的资源上传上来，所以这并不是最好的一套方案。

12.2.2 被动检查工具

被动检查工具就是依赖前面我们介绍过的监听资源导入事件，程序来动态分析资源的合理性。这种检查方法比主动检查工具更安全了，因为资源只要拖入就会自动检查，不需要使用者主动操作，这样就不会有遗漏检查的情况。这听起来好像更完美了，但是实施起来也很困难。问题确实是检查出来了，但是如果美术人员或者策划人员没有看到输出检查的错误，那么依然会把错误的资源上传上来，所以这只是比上一个好那么一点点的方案。

12.2.3 导出类检查工具

无论是主动检查还是被动检查，其实最核心的问题是，策划人员和美术人员可以跳过检查直接提交 SVN。理论上，他们可以上传各种无法想象的资源。所以要想彻底解决这个问题，程序就不能直接使用他们提交的资源，而采取间接使用的方式。

程序提供了一个导出工具，即根据美术资源来自动生成一个新资源（程序用的资源），导出的同时做资源检查，如果资源不符合规范，那么提示错误并且不生成新资源。因为程序使用的是导出后的资源，所以错误的资源并不能被导出成新的，也就不存在上传错误资源的情况，此时这个问题就可以彻底解决了。数据表、UI 界面 Prefab、模型 Prefab、场景 scene 都可以导出，导出的时候不仅检查错误，还可以智能地修改一些资源，比如自动生成 UI 代码，自动删除场景中没有的模型等。

12.2.4 导出 UI

在 Hierarchy 视图中做好 UI 以后，可以给 UI 的顶层节点添加一个自定义的 Tag，如图 12-8 所示，导出界面会以它的父节点自动生成 Prefab 以及部分代码。

图 12-8 导出 UI

如图 12-9 所示，UIMain.prefab 将自动生成在 UIPrefab 文件夹下，而 `BaseUIMain` 就是导出界面的代码类了。

图 12-9 生成界面

12.2.5 生成 UI 代码

界面中操作得最多的就是 Button、Image、Text 和 RectTransform 这几个对象了。我们可以在导出界面 Prefab 的时候，生成基类代码。如图 12-10 所示，生成对应的代码并且在 Awake() 方法中通过代码来获取它们的对象。

图 12-10 生成代码

接着，可以创建一个 UIMain 继承导出的界面类 BaseUIMain，在子类中可以很方便地使用 this 访问基类导出成员变量了，具体如代码清单 12-10 所示。

代码清单 12-10　Main.cs 文件

```
using System.Collections;
using System.Collections.Generic;
using UnityEngine;

public class UIMain : BaseUIMain
{
    void Start() {

        this.m_image.sprite = null;
        this.m_text.text = "雨松MOMO";
        this.m_button.onClick.AddListener(delegate () {
            Debug.Log("点击事件");
        });
        this.m_myroot.anchoredPosition = Vector3.zero;
    }
}
```

在 UI 中，并不是所有的 Image 和 Text 都需要导出，所以可以再设置一个 Tag 来表示是否需要导出，目前定义的 Tag 名称是 UIProperty。最后，在导出界面时遍历所有 UI 对象。只有对象标记了 UIProperty，才可以导出代码。如图 12-11 所示，导出前需要一个模板，这样可以将导出的代码填充在 #Name#、#Property# 和 #Get# 中。

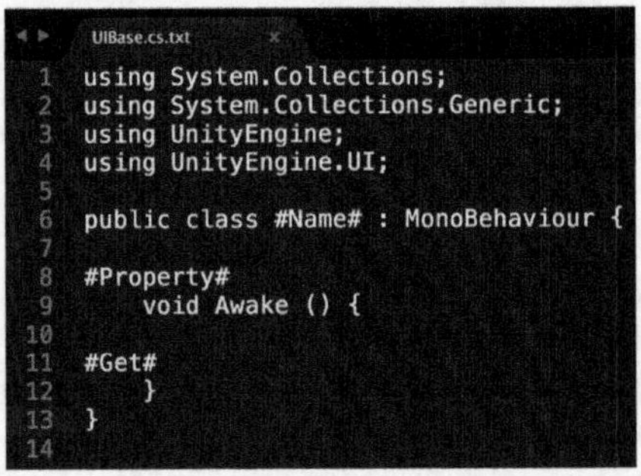

```
UIBase.cs.txt
1  using System.Collections;
2  using System.Collections.Generic;
3  using UnityEngine;
4  using UnityEngine.UI;
5
6  public class #Name# : MonoBehaviour {
7
8  #Property#
9      void Awake () {
10
11 #Get#
12     }
13 }
14
```

图 12-11　模板

如代码清单 12-11 所示，导出代码就是遍历 UI 的 Prefab。将控件代码生成在代码中供子类使用即可。

代码清单 12-11　Script_12_10.cs 文件

```
using UnityEngine;
using UnityEditor;
using System.Text;
using UnityEngine.UI;
using System.IO;
using System.Collections.Generic;

public class Script_12_10
{
```

```csharp
[MenuItem("UI/Tools/Export")]
static void ExportUI()
{
    //Prefab 顶层目录
    string prefabPathRoot = "Assets/UI/UIPrefab";

    //base 类顶层目录
    string scriptBaseRoot = "Assets/UI/UIBase";


    if(Selection.activeTransform) {
        //如果选择了一个 UI 界面，那么开始导出
        if(Selection.activeGameObject.tag == "UIView") {
            //生成 Prefab
            string prefaPath = string.Format("{0}/{1}{2}", prefabPathRoot,
                Selection.activeGameObject.name,".prefab");
            GameObject prefab = AssetDatabase.LoadAssetAtPath<GameObject>
                (prefaPath);
            if(!prefab) {
                PrefabUtility.CreatePrefab(prefaPath, Selection.activeGameObject,
                    ReplacePrefabOptions.ConnectToPrefab);
            } else {
                PrefabUtility.ReplacePrefab(Selection.activeGameObject, prefab,
                    ReplacePrefabOptions.ConnectToPrefab);
            }
            //生成代码
            string basePath = string.Format("{0}/{1}{2}{3}", scriptBaseRoot,
                "Base", Selection.activeGameObject.name,".cs");
            //读取代码模板
            string script = AssetDatabase.LoadAssetAtPath<TextAsset>("Assets/
                UI/UIBase.cs.txt").text;


            StringBuilder property = new StringBuilder();
            StringBuilder get = new StringBuilder();
            //确保变量唯一
            Dictionary<string,int> uniqueName = new Dictionary<string, int>();
            //确保路径唯一
            HashSet<string> uniquePath = new HashSet<string>();

            foreach(var transform in Selection.activeGameObject.
                GetComponentsInChildren<Transform>(true)) {
                if(transform.tag == "UIProperty") {
                    Component component = transform.GetComponent<Button>();
                    if(component == null)
                        component = transform.GetComponent<Image>();
                    if(component == null)
                        component = transform.GetComponent<Text>();
                    if(component == null)
                        component = transform.GetComponent<RectTransform>();
                    if(component != null) {
                        string propertyName = component.name.ToLower();
                        int uniqueValue;
                        if(uniqueName.TryGetValue(propertyName, out uniqueValue)) {
```

```
                    uniqueValue = uniqueName [propertyName] += 1;
                } else {
                    uniqueValue = uniqueName [propertyName] = 0;
                }
                if(uniqueValue > 0) {
                    propertyName = string.Format("{0}_{1}",
                        propertyName, uniqueValue);
                }
                property.AppendFormat("\tprotected {0} m_{1};",
                    component.GetType().ToString(), propertyName).
                    AppendLine();

                string findPath = AnimationUtility.
                    CalculateTransformPath(component.transform,
                    Selection.activeTransform);
                if(!uniquePath.Contains(findPath)) {
                    uniquePath.Add(findPath);
                } else {
                    Debug.LogErrorFormat("读取路径发生重复：{0}",
                        findPath);
                }
                get.AppendFormat("\t\tm_{0}=transform.Find(\"{1}\").
                    GetComponent<{2}>();", propertyName, findPath,
                    component.GetType().ToString()).AppendLine();
            }
        }
    }

    script = script.Replace("#Name#", Path.GetFileNameWithoutExtension
        (basePath));
    script = script.Replace("#Property#", property.ToString ());
    script = script.Replace("#Get#", get.ToString());
    File.WriteAllText(basePath, script);

    AssetDatabase.Refresh();
        }
    }
}
```

导出界面时，还需要考虑变量的命名。游戏对象的名字很容易重复，可以记录名字出现的次数。如果遇到相同的，直接递增即可。

Hierarchy 树结构是允许同名的。transform.Find()用于查找是否存在路径相同的情况。为了保证代码的可读性，我是在导出界面的时候做一个错误的日志输出。如果一定需要保证唯一，可以调用 transform.GetChild(0).GetChild(0).GetComponent<UnityEngine.UI.Button>() 方法，不过这读起来就没有路径直观了。

最后，导出代码其实是很灵活的功能。这里我只做了最简单的对象导出，其实还可以衍生很多导出功能，比如自动监听按钮组件的事件等，读者可以自行摸索。

12.2.6 导出模型

例如,程序员和美术人员约定,所有模型都必须要有 effect_0 和 effect_1 这两个挂载点。我们可以做工具来帮助美术人员自动生成 Prefab,同时可以检查资源是不是没有这两个挂载点。如图 12-12 所示,如果美术人员没有提供挂载点或者挂载点的名字写错了,那么 Prefab 就不会被导出了,就不会影响到程序。

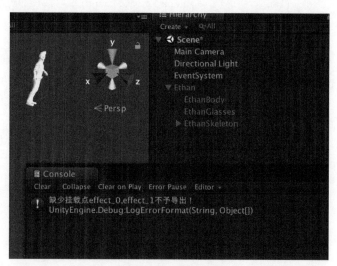

图 12-12 导出模型

如代码清单 12-12 所示,导出模型会自动遍历所有子节点,看它们是否有预先约定的挂载点。

代码清单 12-12 Script_12_11.cs 文件

```
using UnityEngine;
using UnityEditor;
using System.Text;
using UnityEngine.UI;
using System.IO;
using System.Collections.Generic;

public class Script_12_11
{
    [MenuItem("UI/Tools/Export")]
    static void ExportModel()
    {
        //Prefab模型顶层目录
        string prefabPathRoot = "Assets/Model";

        //待检查的挂载点
        List<string> checkPotint = new List<string>()
        {
            "effect_0",
```

```
            "effect_1",
        };

        if(Selection.activeTransform) {
            //检查模型的挂载点是否合法
            foreach(var transform in Selection.activeTransform.GetComponentsInChildren
                <Transform>(true)) {
                for(int i = 0; i < checkPotint.Count; i++) {
                    if(transform.name == checkPotint[i]) {
                        checkPotint.RemoveAt(i);
                    }
                }
            }
            if(checkPotint.Count > 0) {
                Debug.LogErrorFormat("缺少挂载点{0}不予导出！",string.Join(",",
                    checkPotint.ToArray()));
            } else {
                string prefaPath = Path.ChangeExtension(Path.Combine(prefabPathRoot,
                    Selection.activeGameObject.name), "prefab");
                //有 Prefab 的话更新，没有就创建
                GameObject prefab = AssetDatabase.LoadAssetAtPath<GameObject>(prefaPath);
                if(!prefab) {
                    PrefabUtility.CreatePrefab(prefaPath, Selection.activeGameObject,
                        ReplacePrefabOptions.ConnectToPrefab);
                } else {
                    PrefabUtility.ReplacePrefab(Selection.activeGameObject, prefab,
                        ReplacePrefabOptions.ConnectToPrefab);
                }
            }
        }
    }
}
```

12.2.7 不参与保存对象

当美术人员做好场景以后，可能需要放进去一个摄像机以及一些模型来作为参照物，这样他们可以检查场景做得是否正确。如果一不小心这些参照物的游戏对象也被保存在场景上，那么程序运行后再打开这个场景，参照物就显示出来了。如图 12-13 所示，可以将这些参照物对象标记成 EditorOnly，这样游戏最终打包后就看不见它们了，但是编辑模式下还是可以看到的。

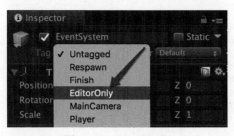

图 12-13　EditorOnly

12.2.8 导出场景

我们可以和美术人员约定一个节点。这里导出一份新场景，其中只保留 Map 节点下的元素，别的节点通通删除，如图 12-14 所示。另外，Map 节点下还有 static 节点，表示自动设置 Static 属性。为了避免美术人员忘记勾选，如果还有别的定制性需求，也可以通过导出的功能完成。

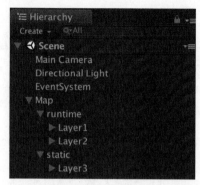

图 12-14　Map 节点

如图 12-15 所示，导出场景后，将自动删除无用节点并且保留 Map 节点，static 节点下的游戏对象将标记 Static 属性。另外，复制的对象会保留原场景的烘焙以及寻路信息，无须执行复制它们的操作。

图 12-15　导出场景

如代码清单 12-13 所示，首先复制一份新的场景出来，接着根据导出的规则动态修改这个场景，从而达到定制性需求。

代码清单 12-13　Script_12_12cs 文件

```
using UnityEngine;
using UnityEditor;
using System.Text;
using UnityEngine.UI;
using System.IO;
using System.Collections.Generic;
```

```csharp
using UnityEditor.SceneManagement;
using UnityEngine.SceneManagement;

public class Script_12_12
{

    [MenuItem("Tools/Map/ExportScene")]
    static void ExportSceneMap()
    {
        string scenePath = EditorSceneManager.GetActiveScene().path;
        string sceneName = System.IO.Path.GetFileName(scenePath);
        string generateScenePath = Path.Combine("Assets/NewScene/", sceneName);
        //复制到新场景中
        if(AssetDatabase.CopyAsset(scenePath, generateScenePath)) {

            EditorSceneManager.SaveOpenScenes();
            //打开新场景
            Scene baseScene = EditorSceneManager.OpenScene(generateScenePath);

            if(GameObject.Find("Map") == null) {
                Debug.LogError("场景缺失 Map 节点,不予导出");
                return;
            }

            //删除不是 Map 节点下的游戏对象
            foreach(var gameObject in GetRootGameObjects()) {
                if(gameObject.name != "Map") {
                    GameObject.DestroyImmediate(gameObject);
                }
            }

            //强制设置选择 Static
            GameObject staticGameObject = GameObject.Find("Map/static");
            if(staticGameObject) {
                foreach(var transform in staticGameObject.GetComponentsInChildren
                    <Transform>(true)) {
                    transform.gameObject.isStatic = true;
                }
            }
            //保存场景
            EditorSceneManager.SaveScene(baseScene);
            //打开保存前的场景
            EditorSceneManager.OpenScene(scenePath);
        }
    }
    ///<summary>
    ///获取顶层游戏对象
    ///</summary>
    ///<returns>The root game objects.</returns>
    static private List<GameObject> GetRootGameObjects()
    {
        GameObject[] pAllObjects = (GameObject[])Resources.FindObjectsOfTypeAll
            (typeof(GameObject));
        List<GameObject> list = new List<GameObject>();
```

```
        foreach(GameObject pObject in pAllObjects) {
            if(pObject.transform.parent != null)
                continue;
            if(pObject.hideFlags == HideFlags.NotEditable || pObject.hideFlags ==
                HideFlags.HideAndDontSave)
                continue;
            if(Application.isEditor) {
                string sAssetPath = AssetDatabase.GetAssetPath(pObject.transform.
                    root.gameObject);
                if(!string.IsNullOrEmpty(sAssetPath))
                    continue;
            }
            if(pObject.transform.parent == null && GameObject.Find(pObject.name)) {
                if(!list.Contains(pObject)) {
                    list.Add(pObject);
                }
            }
        }
        return list;
    }
}
```

12.2.9 过滤无用场景

只要场景被添加到 Build Settings 面板中,打包就会带上这个场景。由于场景的制作是阶段性的,美术人员会陆陆续续地制作,所以不要把开发中的场景也添加在 Build Settings 中。另外,很多场景并没有被策划人员投放。策划人员一般会在数据表中配置场景,打包之前可以读取这个配置,提取出来需要用到的场景,过滤掉那些无用场景。如代码清单 12-14 所示,EditorBuildSettings.scenes 用来动态设置最终需要参与打包的游戏场景。

代码清单 12-14　Script_12_13.cs 文件

```
using UnityEngine;
using UnityEditor;

public class Script_12_13
{
    [MenuItem("Tools/Map/SetScene")]
    static void SetScene()
    {
        //在脚本中添加关联场景
        UnityEditor.EditorBuildSettings.scenes = new EditorBuildSettingsScene[]
        {
            new EditorBuildSettingsScene("Assets/Scene.unity",true),
            new EditorBuildSettingsScene("Assets/Scene1.unity",true),
        };
    }
}
```

12.3 自动打包

Unity 编辑器提供了手动打包的功能，但还是很麻烦。每次策划人员需要游戏包了，都要过来找程序员打包。为了彻底解放程序员的双手，可以考虑让策划人员来打包，或者每天凌晨自动打包。但是打包也不是那么容易的，通常需要在打包前后自动执行一些特殊的代码，例如 Logo、图标、编译选项、复制和第三方库的依赖等。

12.3.1 打包前后事件

打包时，我们需要监听一些事件，比如打包前设置版本号、游戏名和图标等，打包后可能需要复制或者压缩游戏包等。如图 12-16 所示，打开 Build Settings 面板，在左边选择一个需要打包的平台，在右边单击 Build 或者 Build And Run 按钮即可。

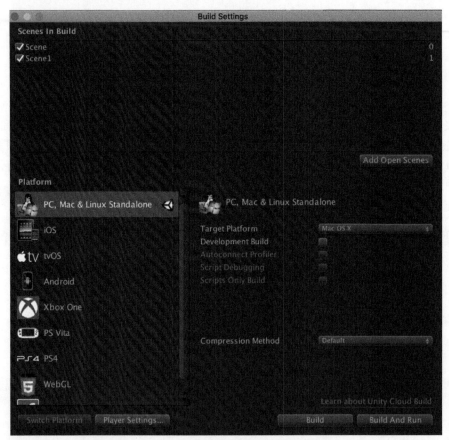

图 12-16　打包面板

如代码清单 12-15 所示，我们可以监听打包前和打包后的事件。

代码清单 12-15　Script_12_14.cs 文件

```
using UnityEngine;
using UnityEditor;
using UnityEditor.Build;

public class Script_12_14 : IPreprocessBuild, IPostprocessBuild
{

    int IOrderedCallback.callbackOrder {
        get {
            return 0;
        }
    }

    //打包前事件
    void IPreprocessBuild.OnPreprocessBuild(BuildTarget target, string path)
    {
        //设置版本号和游戏名
        PlayerSettings.bundleVersion = "2.0.0";
        PlayerSettings.productName = "雨松 momo";
    }

    //打包后事件
    void IPostprocessBuild.OnPostprocessBuild(BuildTarget target, string path)
    {
        Debug.LogFormat("游戏包生成路径:{0}", path);
    }

}
```

12.3.2　打包后自动压缩

在 macOS 操作系统中，最常用来操作文件的就是 shell 脚本了，Unity 也免不了调用 shell 脚本的需求。在 Mac 上，是可以打 Windows 游戏包的。打包后，会多生成一个文件夹，这样管理起来就很混乱了，所以希望打包后自动压缩成一个 ZIP 文件。本例中，打完包后，会自动打开终端，执行 shell 脚本，将它压缩成一个 ZIP 文件，相关代码如代码清单 12-16 所示。如图 12-17 所示，打包后，会自动将 pc.exe 和 pc_Data 压缩成 project.zip。

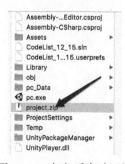

图 12-17　打包后自动压缩

代码清单12-16   Script_12_15.cs 文件

```csharp
using UnityEngine;
using UnityEditor;
using UnityEditor.Build;
using System.IO;

public class Script_12_15 : IPreprocessBuild, IPostprocessBuild
{
    //打包前后的事件
    void IPreprocessBuild.OnPreprocessBuild(BuildTarget target, string path)
    {
        //清空上次打包的残留
        string data = Path.Combine(Path.GetDirectoryName(path),
            Path.GetFileNameWithoutExtension(path)+"_Data");

        if(File.Exists(path))
            File.Delete(path);
        if(File.Exists(data))
            File.Delete(data);

    }

    int IOrderedCallback.callbackOrder {
        get {
            return 0;
        }
    }

    //打包后的事件
    void IPostprocessBuild.OnPostprocessBuild(BuildTarget target, string path)
    {
        string data = Path.Combine(Path.GetDirectoryName(path),
            Path.GetFileNameWithoutExtension(path)+"_Data");
        string zip = Path.Combine(Path.GetDirectoryName(path), "project.zip");

        //清空上次打包的ZIP文件
        if(File.Exists(zip))
            File.Delete(zip);

        //压缩资源包
        System.Diagnostics.Process process = new System.Diagnostics.Process();
        process.StartInfo.FileName = "osascript";
        process.StartInfo.Arguments = string.Format("-e 'tell application \"Terminal\"
            to do script \"zip -r {0} {1} {2}\"'",zip,path,data);
        process.Start();
        process.WaitForExit();
        process.Close();
    }
}
```

上述代码中，当EXE包构建出来后，自动将它压缩成ZIP格式。

12.3.3 用 C# 调用 shell 脚本

首先，写 shell 脚本，它实现的功能很简单，就是复制文件，将参数 1 复制到参数 2 中，如代码清单 12-17 所示。

代码清单 12-17　copy.sh 文件

```sh
#!/bin/sh

from=$1
to=$2

#复制文件
cp -vf "${from}" "${to}"
```

如代码清单 12-18 所示，在 C# 中调用这个 shell 脚本，并且将参数传递给它，其调用原理就是 `Process.Start()` 方法。

代码清单 12-18　Script_12_16.cs 文件

```csharp
using UnityEngine;
using UnityEditor;
using UnityEditor.Build;
using System.IO;

public class Script_12_16
{
    [MenuItem("Tool/CopyShell")]
    static void CopyShell()
    {
        string shell = Path.Combine(Application.dataPath, "copy.sh");
        string arg1 = Path.Combine(Application.dataPath, "Cube.prefab");
        string arg2 = Path.Combine(Application.dataPath, "Cube1.prefab");
        string argss =  shell +" "+ arg1 +" " + arg2;
        System.Diagnostics.Process.Start("/bin/bash", argss);
    }
}
```

12.3.4 等待 shell 结束

有时候，我们需要通过 shell 脚本来执行一些压缩指令。由于它并不能在一帧内完成，所以需要等待 shell 脚本彻底执行完后再执行后面的 C# 代码。首先，在 shell 脚本中模拟一个耗时操作，此时就会用 `for` 循环，具体如代码清单 12-19 所示。

代码清单 12-19　copy.sh 文件

```sh
#!/bin/sh

count=$1

#耗时操作
for((i=1;i<=$count;i++));do
```

```
        echo $(expr $i \* 4);
done
```

接着，在 C# 中传入 count=10000 来阻塞主线程。如代码清单 12-20 所示，在执行 shell 脚本前后，都输出一下当前的时间。

代码清单 12-20　Script_12_17.cs 文件

```csharp
using UnityEngine;
using UnityEditor;
using UnityEditor.Build;
using System.IO;
using System;

public class Script_12_17
{
    [MenuItem("Tool/CopyShell")]
    static void CopyShell()
    {
        string shell = Path.Combine(Application.dataPath, "copy.sh");
        int arg1 = 10000;
        string argss =  shell +" "+ arg1;
        Debug.LogFormat("begin : {0}",DateTime.Now.ToString("hh:mm:ss").ToString());
        System.Diagnostics.Process.Start("/bin/bash", argss).WaitForExit();
        Debug.LogFormat("end : {0}",DateTime.Now.ToString("hh:mm:ss").ToString());
    }
}
```

在 WaitForExit() 方法后，就表示 shell 执行完毕了。如图 12-18 所示，可以看出这段代码阻塞的时间。

图 12-18　结束事件

12.3.5　shell 脚本调用 C#

如果 shell 脚本要调用 C#，那么必须自动打开 Unity 项目。首先，需要在工程中写一个静态方法来接收 shell 的调用，具体如代码清单 12-21 所示。

代码清单 12-21　Script_12_18.cs 文件

```csharp
using UnityEngine;
using UnityEditor;
using UnityEditor.Build;
using System.IO;
```

```csharp
using System;
using System.Collections.Generic;

public class Script_12_18
{
    static void Build()
    {
        Dictionary<string, string> args = GetArgs("Script_12_18.Build");
        foreach(var item in args) {
            Debug.LogFormat("key = {0} value = {1}", item.Key, item.Value);
        }
    }


    ///<summary>
    ///提取参数
    ///</summary>
    static Dictionary<string, string> GetArgs(string methodName)
    {
        Dictionary<string, string> args = new Dictionary<string, string>();
        bool isArg = false;
        foreach(string arg in System.Environment.GetCommandLineArgs()) {
            if(isArg) {
                if(arg.StartsWith("--")) {
                    int splitIndex = arg.IndexOf("=");
                    if(splitIndex > 0) {
                        args.Add(arg.Substring(2, splitIndex - 2), arg.Substring
                            (splitIndex + 1));
                    } else {
                        args.Add(arg.Substring(2), "true");
                    }
                }
            } else if(arg == methodName) {
                isArg = true;
            }
        }
        return args;
    }
}
```

调用 shell 脚本后，Unity 会自动打开这个工程，并且调用 Script_12_18.Build() 静态方法。如图 12-19 所示，工程会打开，并且将 shell 传递的参数输出。

图 12-19 调用方法

如代码清单 12-22 所示,在 shell 脚本中,需要指定 Unity 的安装目录、项目的目录和传递的参数。

代码清单 12-22 build.sh 文件

```sh
#!/bin/sh

#Unity 在本机的安装目录
BUILD_UNITY="/Applications/Unity2017/Unity.app/Contents/MacOS/Unity"

#项目在本地的目录
BUILD_UNITY_PROJECT="/Users/senumatsu/Unity3D 游戏开发第二版/第十二章/CodeList_12_19"

#打开 Unity 并且执行 Script_12_18.Build 方法
${BUILD_UNITY} -projectPath ${BUILD_UNITY_PROJECT} -executeMethod Script_12_19.Build --version="2.0.0" --name="雨松 momo"
```

12.3.6 shell 脚本自动打包

通过上面的调用方法,其实就可以将脚本自动打包了,但是我们需要的是自动打开 Unity 工程,打完包后自动关闭掉 Unity,以及输出打包过程中的日志。所以,下面来拓展一下打包脚本,具体如代码清单 12-23 所示。

代码清单 12-23 build.sh 文件

```sh
#!/bin/sh

#Unity 在本机的安装目录
```

```
BUILD_UNITY="/Applications/Unity2017/Unity.app/Contents/MacOS/Unity"
#项目在本地的目录
BUILD_UNITY_PROJECT="/Users/senumatsu/Unity3D游戏开发第二版/第十二章/CodeList_12_19"
#日志路径
BUILD_LOG="${BUILD_UNITY_PROJECT}/build.log"
#打包平台
BUILD_PLATFORM="Win64"
#最终EXE文件生成的位置
BUILD_OUT="${BUILD_UNITY_PROJECT}/Out/pc.exe"

#打开Unity并且执行Script_12_24.Build方法
${BUILD_UNITY} -quit -batchmode -nographics -projectPath ${BUILD_UNITY_PROJECT}
-logFile "${BUILD_LOG}" -buildTarget "${BUILD_PLATFORM}" -executeMethod
Script_12_23.Build --version="2.0.0" --name="雨松momo" --out="${BUILD_OUT}"
```

在上述代码中,我们通过shell脚本自动调用Script_12_24.cs类中的Build()方法。如代码清单12-24所示,在C#代码中继续处理自动打包的逻辑。

代码清单12-24   Script_12_19.cs文件

```csharp
using UnityEngine;
using UnityEditor;
using UnityEditor.Build;
using System.IO;
using System;
using System.Collections.Generic;

public class Script_12_19
{
    static void Build()
    {
        Dictionary<string, string> args = GetArgs("Script_12_19.Build");

        PlayerSettings.bundleVersion = args ["version"];
        PlayerSettings.productName = args ["name"];

        //执行打包,最终输出在args["out"]路径下
        BuildPipeline.BuildPlayer(GetBuildScenes(),args["out"],BuildTarget.
            StandaloneWindows64,BuildOptions.Development);
    }

    ///<summary>
    ///获取打包场景
    ///</summary>
    static string[] GetBuildScenes()
    {
        List<string> names = new List<string>();

        foreach(EditorBuildSettingsScene e in EditorBuildSettings.scenes) {
            if(e == null)
                continue;

            if(e.enabled)
                names.Add(e.path);
```

```csharp
        }
        return names.ToArray();
    }

    ///<summary>
    ///提取参数
    ///</summary>
    static Dictionary<string, string> GetArgs(string methodName)
    {
        Dictionary<string, string> args = new Dictionary<string, string>();
        bool isArg = false;
        foreach(string arg in System.Environment.GetCommandLineArgs()) {
            if(isArg) {
                if(arg.StartsWith("--")) {
                    int splitIndex = arg.IndexOf("=");
                    if(splitIndex > 0) {
                        args.Add(arg.Substring(2, splitIndex - 2), arg.Substring
                            (splitIndex + 1));
                    } else {
                        args.Add(arg.Substring(2), "true");
                    }
                }
            } else if(arg == methodName) {
                isArg = true;
            }
        }
        return args;
    }
}
```

在 shell 脚本中，可以传递参数到 C# 中，调用 `System.Environment.GetCommandLineArgs()` 方法就可以获取传递的参数。

如图 12-20 所示，最终的 PC 包已经在指定目录下了。这里我们只演示了 Windows 平台下自动打包的例子，其实别的平台也可以自动打包。只要读者掌握了本章内容，可以自行拓展打包脚本。

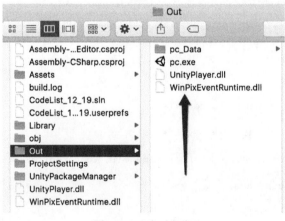

图 12-20　自动打包

12.3.7 彻底解放程序员的双手

上一节中，我们学习了如何使用脚本来完成自动化打包，但是脚本还需要程序员来执行，这样还会耽误时间，所以要想真正解放程序员的双手，还需要三样东西。

- ❑ 公共打包机一台。
- ❑ 在打包机上配置 Jenkins 服务。
- ❑ 在打包机上配置下载服务。

Jenkins 就是持续集成，将它配置在打包机上后，可以通过网页的形式访问。其打包原理其实就是执行上面我们写的 shell 脚本。Jenkins 提供了丰富的接口，比如打包前 SVN 自动更新、自动 revert，每天定时构建任务等。如图 12-21 所示，打开 Jenkins 网页后，单击 Build with Parameters 按钮，还可以配置打包参数。

Jenkins 打包后，可以设置复制的目录，可以给这个目录配置下载服务。如图 12-22 所示，这样内网用户可以从网页上直接下载打好的游戏包了。这就彻底解放程序员的双手了，打包工作完全可以交给策划人员或者美术人员来做了。

图 12-21 Jenkins 操作页面

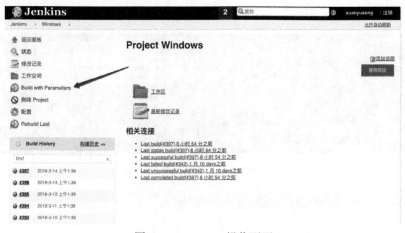

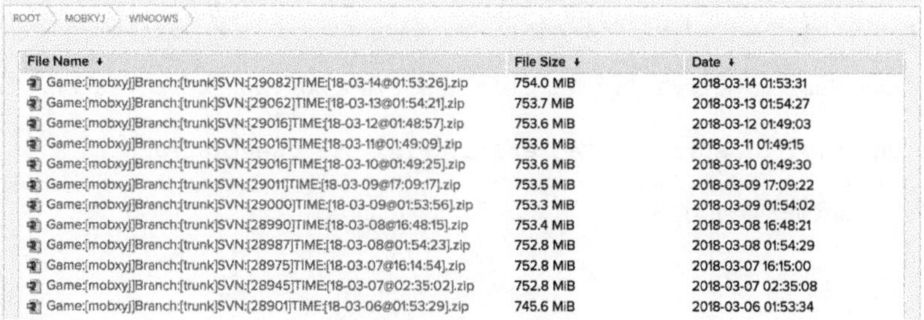

File Name	File Size	Date
Game:[mobxyj]Branch:[trunk]SVN:[29082]TIME:[18-03-14@01:53:26].zip	754.0 MiB	2018-03-14 01:53:31
Game:[mobxyj]Branch:[trunk]SVN:[29062]TIME:[18-03-13@01:54:21].zip	753.7 MiB	2018-03-13 01:54:27
Game:[mobxyj]Branch:[trunk]SVN:[29016]TIME:[18-03-12@01:48:57].zip	753.6 MiB	2018-03-12 01:49:03
Game:[mobxyj]Branch:[trunk]SVN:[29016]TIME:[18-03-11@01:49:09].zip	753.6 MiB	2018-03-11 01:49:15
Game:[mobxyj]Branch:[trunk]SVN:[29016]TIME:[18-03-10@01:49:25].zip	753.6 MiB	2018-03-10 01:49:30
Game:[mobxyj]Branch:[trunk]SVN:[29011]TIME:[18-03-09@17:09:17].zip	753.5 MiB	2018-03-09 17:09:22
Game:[mobxyj]Branch:[trunk]SVN:[29000]TIME:[18-03-09@01:53:56].zip	753.3 MiB	2018-03-09 01:54:02
Game:[mobxyj]Branch:[trunk]SVN:[28990]TIME:[18-03-08@16:48:15].zip	753.4 MiB	2018-03-08 16:48:21
Game:[mobxyj]Branch:[trunk]SVN:[28987]TIME:[18-03-08@01:54:23].zip	752.8 MiB	2018-03-08 01:54:29
Game:[mobxyj]Branch:[trunk]SVN:[28975]TIME:[18-03-07@16:14:54].zip	752.8 MiB	2018-03-07 16:15:00
Game:[mobxyj]Branch:[trunk]SVN:[28945]TIME:[18-03-07@02:35:02].zip	752.8 MiB	2018-03-07 02:35:08
Game:[mobxyj]Branch:[trunk]SVN:[28901]TIME:[18-03-06@01:53:29].zip	745.6 MiB	2018-03-06 01:53:34

图 12-22 下载服务

12.4 自动构建图集与压缩

为了降低 DrawCall，我们需要将多个图片构建在图集上。构建图集还有个好处，那就是可以自动将图片补齐 2 的幂次方或者正方形图，这样就可以进行 ETC 和 PVRTC 的压缩了。

12.4.1 UI 图集的压缩

如果是 UI 图片，设置了相同 packingTag 的 Sprite 会合并在一个图集，但是游戏中有很多本身就很大的图片，比如背景图或者大头像图。由于这些图片不一定是 2 的幂次方或正方形图，所以我们可以自动给它们设置成一个唯一的 packingTag，这样它们也可以进行压缩了。

需要注意的是，如果有些大图需要设置成了 RGBA32 或者 RGBA16，并且自身并不是 2 的幂次方或者正方形图的话，在去加 pakcingTag 就不太合适了，这样只会让图片变得更大。所以，在自动构建图集的时候可以做个判断，当图片出现这类情况时不再构建图集。

如图 12-23 所示，在 Sprite Packer 中，可以自定义图集的构建策略，这样就可以进一步灵活地控制图集的构建方式，可以调节图集的压缩类型以及大小等。本示例的完整代码详见代码清单 12-25。

图 12-23 自定义图集构建策略

代码清单 12-25 CustomPackerPolicySample.cs 文件

```
using System;
using System.Linq;
using UnityEngine;
using UnityEditor;
using System.Collections.Generic;

public class CustomPackerPolicySample : UnityEditor.Sprites.IPackerPolicy
{
    bool UnityEditor.Sprites.IPackerPolicy.AllowSequentialPacking {
        get {
            return true;
        }
    }

    protected class Entry
    {
```

```csharp
    public Sprite             sprite;
    public UnityEditor.Sprites.AtlasSettings settings;
    public string             atlasName;
    public SpritePackingMode  packingMode;
    public int                anisoLevel;
}

private const uint kDefaultPaddingPower = 3;

public virtual int GetVersion() { return 1; }
protected   string TagPrefix { get { return "[RECT]"; } }
protected   bool AllowTightWhenTagged { get { return true; } }
protected   bool AllowRotationFlipping { get { return true; } }

//判断图片是否为压缩格式
public static bool IsCompressedFormat(TextureFormat fmt)
{
    if(fmt >= TextureFormat.DXT1 && fmt <= TextureFormat.DXT5)
        return true;
    if(fmt >= TextureFormat.DXT1Crunched && fmt <= TextureFormat.DXT5Crunched)
        return true;
    if(fmt >= TextureFormat.PVRTC_RGB2 && fmt <= TextureFormat.PVRTC_RGBA4)
        return true;
    if(fmt == TextureFormat.ETC_RGB4)
        return true;
    if(fmt >= TextureFormat.ETC_RGB4 && fmt <= TextureFormat.ETC2_RGBA8)
        return true;
    if(fmt >= TextureFormat.EAC_R && fmt <= TextureFormat.EAC_RG_SIGNED)
        return true;
    if(fmt >= TextureFormat.ETC2_RGB && fmt <= TextureFormat.ETC2_RGBA8)
        return true;
    if(fmt >= TextureFormat.ASTC_RGB_4x4 && fmt <= TextureFormat.ASTC_RGBA_12x12)
        return true;
    if(fmt >= TextureFormat.DXT1Crunched && fmt <= TextureFormat.DXT5Crunched)
        return true;
    return false;
}

public void OnGroupAtlases(BuildTarget target, UnityEditor.Sprites.PackerJob
    job, int[] textureImporterInstanceIDs)
{
    List<Entry> entries = new List<Entry>();

    foreach(int instanceID in textureImporterInstanceIDs)
    {
        TextureImporter ti = EditorUtility.InstanceIDToObject(instanceID)
            as TextureImporter;

        TextureFormat desiredFormat;
        ColorSpace colorSpace;
        int compressionQuality;
        ti.ReadTextureImportInstructions(target, out desiredFormat, out colorSpace,
            out compressionQuality);
```

```csharp
            TextureImporterSettings tis = new TextureImporterSettings();
            ti.ReadTextureSettings(tis);

            Sprite[] sprites =
                AssetDatabase.LoadAllAssetRepresentationsAtPath(ti.assetPath)
                    .Select(x => x as Sprite)
                    .Where(x => x != null)
                    .ToArray();

            foreach(Sprite sprite in sprites)
            {
                //设置图片的压缩参数
                Entry entry = new Entry();
                entry.sprite = sprite;
                entry.settings.format = desiredFormat;
                entry.settings.colorSpace = colorSpace;
                entry.settings.compressionQuality = IsCompressedFormat(desiredFormat)
                    ? compressionQuality : 0;
                entry.settings.filterMode = Enum.IsDefined(typeof(FilterMode),
                    ti.filterMode)
                    ? ti.filterMode
                    : FilterMode.Bilinear;
                entry.settings.maxWidth = 2048;
                entry.settings.maxHeight = 2048;
                entry.settings.generateMipMaps = false;
                entry.settings.enableRotation = AllowRotationFlipping;
                entry.settings.paddingPower = (uint)EditorSettings.
                    spritePackerPaddingPower;
                entry.settings.allowsAlphaSplitting = ti.GetAllowsAlphaSplitting();
                entry.atlasName = ParseAtlasName(ti.spritePackingTag);
                entry.packingMode = GetPackingMode(ti.spritePackingTag,
                    tis.spriteMeshType);
                entry.anisoLevel = ti.anisoLevel;

                entries.Add(entry);
            }
            Resources.UnloadAsset(ti);
        }

        //将相同图集的名称划分成不同图集
        var atlasGroups =
            from e in entries
            group e by e.atlasName;
        foreach(var atlasGroup in atlasGroups)
        {
            int page = 0;
            //相同的图集中压缩格式不同的图划分成不同的组
            var settingsGroups =
                from t in atlasGroup
                group t by t.settings;
            foreach(var settingsGroup in settingsGroups)
            {
                //已经设置了非压缩的图，这里不做成图集，避免图片变大
                bool skipAtlas = (settingsGroup.Count() == 1 && !IsCompressedFormat
```

```
                    (settingsGroup.ElementAt(0).settings.format));
                if(skipAtlas) {
                    continue;
                }

                string atlasName = atlasGroup.Key;

                if(settingsGroups.Count() > 1)
                    atlasName += string.Format("(Group {0})", page);

                UnityEditor.Sprites.AtlasSettings settings = settingsGroup.Key;
                settings.anisoLevel = 1;
                if(settings.generateMipMaps)
                    foreach(Entry entry in settingsGroup)
                        if(entry.anisoLevel > settings.anisoLevel)
                            settings.anisoLevel = entry.anisoLevel;

                job.AddAtlas(atlasName, settings);
                foreach(Entry entry in settingsGroup)
                {
                    job.AssignToAtlas(atlasName, entry.sprite, entry.packingMode,
                        SpritePackingRotation.None);
                }

                ++page;
            }
        }
    }

    protected bool IsTagPrefixed(string packingTag)
    {
        packingTag = packingTag.Trim();
        if(packingTag.Length < TagPrefix.Length)
            return false;
        return (packingTag.Substring(0, TagPrefix.Length) == TagPrefix);
    }

    private string ParseAtlasName(string packingTag)
    {
        string name = packingTag.Trim();
        if(IsTagPrefixed(name))
            name = name.Substring(TagPrefix.Length).Trim();
        return(name.Length == 0) ? "(unnamed)" : name;
    }

    private SpritePackingMode GetPackingMode(string packingTag, SpriteMeshType
        meshType)
    {
        if(meshType == SpriteMeshType.Tight)
            if(IsTagPrefixed(packingTag) == AllowTightWhenTagged)
                return SpritePackingMode.Tight;
        return SpritePackingMode.Rectangle;
    }
}
```

如果打包前需要自动指定策略,可以强制设置本地打包使用什么图集策略,然后将策略以及打包平台传入即可:

```
static private void BuildAtlas(BuildTarget buildTarget)
{
    Packer.SelectedPolicy=typeof(CustomPackerPolicySample).Name;
    Packer.RebuildAtlasCacheIfNeeded(buildTarget);
}
```

如图 12-24 所示,我们在 Packing Tag 中设置了特殊的名字。由于它的压缩格式是 RGBA32,并且 Packing Tag 中没有别的图片与它相同,那么此图片将不会被打成图集,这样就可以优化一些图片占用的内存了。

图 12-24　过滤特殊图集

12.4.2　非 UI 图集压缩

角色、特效和场景等图片并不属于 UI 图片,如果美术人员并没有提供 2 的幂次方或者正方形图,就无法压缩了。如图 12-25 所示,在 Advanced 下拉面板中选择合适的 Non Power of 2,可以强制设置图片成 2 的幂次方。其中 ToNearest、ToLarger 和 ToSmaller 表示自动将图片拉伸成对应的规则大小,这样图片就可以进行压缩了。

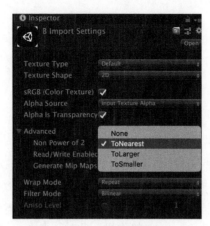

图 12-25 设置 2 的幂次方

12.5 小结

本章中，我们学习了 Unity 的自动化，监听资源的导入/导出事件。自动化利于自动修改资源，减少手动犯错的成本。例如，和美术人员约定好目录的含义，程序自动设置资源的格式。策划人员和美术人员很容易犯错，可以通过自动化的方式避免他们犯错。程序员不直接使用美术人员提交的资源，而是通过工具来自动生成，同时可以做检查操作，避免资源做的格式不正确带来的隐患。最后，我们学习了自动化打包，C# 与 shell 之间的相互调用。这里彻底解放程序员的双手，让策划人员或者美术人员来打包。

第 13 章

3D 游戏开发

本章是本书的最后一章，是将前面的知识融会贯通的一章。从场景来说，Unity 支持多场景编辑功能。我们可以在游戏中同时打开多个场景，但是场景之间需要相互配合。在渲染方面，需要两个摄像机：一个 3D 主相机和一个 UI 摄像机。此时，点击事件需要区分是点击在 UI 上还是点击在场景上了。如果 UI 需要显示 3D 模型，还可以使用 RenderTexture 将相机渲染在纹理上。物理引擎与碰撞检测需要配合使用 Rigidbody 和 Collider 组件，这和第 6 章介绍的 2D 碰撞类似。动画系统则需要配合使用 Animation 和 AnimationController。首先，我们来学习 Shader 在渲染管线中的应用。

13.1 Shader

Shader（着色器）是一段渲染管线运行的小程序，它负责告诉 GPU 如何渲染图形。在游戏开发中，CPU 一直都是瓶颈，所以有些计算完全可以通过 Shader 放在 GPU 中来做。例如，UV 动画，如果在 CPU 上做，需要每一帧向 GPU 传递移动的范围；如果直接在 GPU 中处理，就不需要传递数据这部分开销了，直接在 Shader 中通过时间函数来移动即可。类似的需求其实还有很多，比如水流、岩浆、天气、雾和溶解等效果都可以在 Shader 中完成。

如图 13-1 所示，在 Unity 中可创建 4 种不同类型的 Shader。除了 Compute Shader 以外，其余的 Shader 都属于可编程渲染管线以及 Shader Variant Collection（着色器变体采集）。

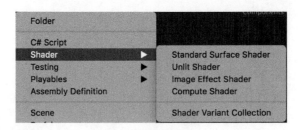

图 13-1　创建 Shader

13.1.1　固定渲染管线

固定渲染管线（Fixed Function Shader）是 OpenGL ES 1.0 所使用的渲染管线，属于最古老的一种渲染管线,无法自由灵活地控制渲染的每个片段。它只提供了一些渲染功能的开关项,这就好比工厂生成零件部分,相同模具生成的零件部分是完全一样的,只能选择是否生产这部分零件。例如,控制渲染管线打开漫反射功能,或者关闭环境光功能等。从 OpenGL ES 2.0 开始,Unity 全面支持可编程渲染管线(Vertex and Fragment Shader),已经不建议使用固定渲染管线了,但是我们需要对它有一个简单的了解。如代码清单 13-1 所示,我们来写一个简单的固定渲染管线 Shader。

代码清单 13-1　FixedFunctionShader.shader 文件

```
Shader "Unlit/FixedFunctionShader"
{
    Properties
    {
        //主纹理
        _MainTex("Texture", 2D) = "white" {}
    }

    SubShader
    {
        Pass {
            //设置主纹理
            SetTexture [_MainTex] {
                combine texture
            }
        }
    }
}
```

下面简要介绍一下上述代码。

- `Shader "Unlit/FixedFunctionShader"{}`:表示 Shader 的显示目录,以及 Shader 代码语块。
- `Properties{}`:表示 Shader 的属性部分,这里配置渲染管线需要用到的参数,比如图片类型或者值类型等。
- `SubShader{}`:表示子着色器。Shader 可以包含多个 `SubShader`,Unity 会自动找到当前设备硬件支持的 `SubShader` 执行。
- `Pass{}`:表示渲染通道。在 SubShader 中,可以添加多个 `Pass`,但是每个 `Pass` 就是一个 DrawCall,所以尽量只使用一个 `Pass`。
- `SetTexture [_MainTex]{}`:设置纹理。其中 `combine texture` 表示显示合并纹理。

将这个 Shader 绑定在待显示的 Mesh Renderer 组件的材质上,即可立即看到效果,我们来给它加上光照信息。如代码清单 13-2 所示,在 `Pass` 代码块内需要开启光照信息,在 `Material` 代码块内开启需要启动的光照类型。

代码清单 13-2　FixedFunctionShader_Plus.shader 文件

```
Shader "Unlit/FixedFunctionShader_Plus"
{
    Properties
```

```
    {
        //主纹理
        _MainTex("Texture", 2D) = "white" {}
        //颜色
        _Color("Main Color", Color) = (1,1,1,0)
    }
    SubShader
    {
        Pass {
            Lighting On
            Material {
                Diffuse [_Color]//接收漫反射光
                Ambient [_Color]//接收环境光
             }

            //设置主纹理
            SetTexture [_MainTex] {
                combine previous * texture
            }
        }
    }
}
```

如图 13-2 所示,右边的立方体收到灯光的影响,打开了漫反射以及环境光。固定渲染管线还可以设置镜面高光反射(Specular)、镜面高光反射强度(Shininess)、自发光(Emission)等。在 Material 代码块内,也就是开启某项光照功能的开关,打开或者关闭它们,会应用模型的全部。假如只希望模型身体的一部分接收漫反射而另一部分不接收,这种情况下固定渲染管线是无法做到的,所以真正最灵活的方式应该是可编程渲染管线。

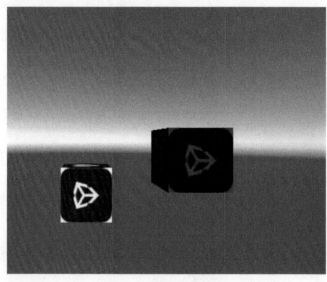

图 13-2　固定渲染管线

13.1.2 可编程渲染管线

在可编程渲染管线中,程序代码可以对每一个片段进行着色。就像工厂中生产零件,可以委培很多工人对一个零件进行特殊处理,这样生产出来的零件就各式各样了。如果需要对每一个片段像素点做特殊着色,那么在 Shader 中首先需要获取几何图形对应的顶点以及 UV 信息,然后通过 UV 以及贴图拿到当前片段的像素信息,最终就可以自定义着色了。如代码清单 13-3 所示,我们先来学习一个简单的可编程渲染管线。

代码清单 13-3　VertexandFragmentShader.shader 文件

```
Shader "Unlit/VertexandFragmentShader"
{
    Properties
    {
        _MainTex("Texture", 2D) = "white" {}
    }
    SubShader
    {
        Pass
        {
            //标记CG程序块的起始位置
            CGPROGRAM
            //执行顶点着色方法
            #pragma vertex vert
            //执行片段着色方法
            #pragma fragment frag
            #include "UnityCG.cginc"

            struct v2f
            {
                float2 uv : TEXCOORD0;
                float4 vertex : SV_POSITION;
            };

            sampler2D _MainTex;
            float4 _MainTex_ST;

            v2f vert(appdata_base v)
            {
                //输出几何图形顶点以及UV信息
                v2f o;
                o.vertex = UnityObjectToClipPos(v.vertex);
                o.uv = TRANSFORM_TEX(v.texcoord, _MainTex);
                return o;
            }

            fixed4 frag(v2f i) : SV_Target
            {
                //输入自定义着色信息
                fixed4 col = tex2D(_MainTex, i.uv);
                return col;
```

```
        }
        //标记CG程序块的结束
        ENDCG
    }
}
```

使用 CGPROGRAM 和 ENDCG 来标记 CG 程序块,这是标准的 C++ 语法。#pragma vertex vert 表示执行顶点着色方法,调用下方的 vert()函数输出几何图形顶点以及 UV 信息。接着执行#pragma fragment frag,它表示片段着色方法,将调用下方 frag()函数。v2f 就是刚刚顶点着色方法返回的数据。最终,将每个点的颜色返回去交给 GPU 渲染。如代码清单 13-4 所示,我们在可编程渲染管线中实现漫反射环境光的光照。

代码清单 13-4　VertexandFragmentShader_Plus.shader 文件

```
Shader "Unlit/VertexandFragmentShader_Plus"
{
    Properties
    {
        _MainTex("Texture", 2D) = "white" {}
    }
    SubShader
    {
        //设置光照类型为前向渲染
        Tags {"LightMode"="ForwardBase"}

        Pass
        {
            //标记CG程序块的起始位置
            CGPROGRAM
            //执行顶点着色方法
            #pragma vertex vert
            //执行片段着色方法
            #pragma fragment frag
            #include "UnityCG.cginc"
            #include "UnityLightingCommon.cginc"

            struct v2f
            {
                float2 uv : TEXCOORD0;
                float4 vertex : SV_POSITION;
                fixed4 diff : COLOR0; //漫反射光颜色
                fixed3 ambient : COLOR1;//环境光颜色
            };

            sampler2D _MainTex;
            float4 _MainTex_ST;

            v2f vert(appdata_base v)
            {
                //输出几何图形的顶点以及UV信息
```

```
            v2f o;
            o.vertex = UnityObjectToClipPos(v.vertex);
            o.uv = TRANSFORM_TEX(v.texcoord, _MainTex);
            //计算漫反射环境光的颜色
            half3 worldNormal = UnityObjectToWorldNormal(v.normal);
            half nl = max(0, dot(worldNormal, _WorldSpaceLightPos0.xyz));
            o.diff = nl * _LightColor0;
            o.ambient = ShadeSH9(half4(worldNormal,1));
            return o;
        }

        fixed4 frag(v2f i) : SV_Target
        {
            //输入自定义着色信息
            fixed4 col = tex2D(_MainTex, i.uv);
            //添加漫反射以及环境光
            fixed3 lighting = i.diff + i.ambient;
            col.rgb *= lighting;
            return col;
        }
        //标记CG程序块的结束位置
        ENDCG
    }
  }
}
```

如图13-3所示，漫反射和环境光已经添加到立方体对象身上了。除了漫反射和环境光以外，光照模型还有很多。光照模型的算法是一致的，然而都需要在vertex方法中实现。为了方便，我们还可以使用可编程渲染管线的表面着色器。

图 13-3　可编程渲染管线

13.1.3　可编程渲染管线的表面着色器

可编程渲染管线的表面着色器（Surface Shader）可以省略编写`#pragma vertex vert`方法，并且Shader中不需要写Pass代码块。如代码清单13-5所示，`#pragma surface surf Lambert`表示执行光照模型，`SurfaceOutput`表示`vertex`输出的结构对象。

代码清单13-5　SurfaceShader.shader文件

```
Shader "Unlit/SurfaceShader"
{
    Properties
    {
        _MainTex("Texture", 2D) = "white" {}
    }
    SubShader
    {
        //标记CG程序块的起始位置
        CGPROGRAM
        #pragma surface surf Lambert addshadow

        sampler2D _MainTex;

        struct Input {
            float2 uv_MainTex;
        };
        void surf(Input IN, inout SurfaceOutput o) {
            half4 c = tex2D(_MainTex, IN.uv_MainTex);
            o.Albedo = c.rgb;
            o.Alpha = c.a;
        }
        //标记CG程序块的结束位置
        ENDCG
    }
}
```

在Unity安装目录下，我们可以找到内置的CG代码，它位于CGIncludes文件夹下。打开Lighting.cginc文件后，就可以看到`SurfaceOutput`的结构定义：

```
struct SurfaceOutput
{
    fixed3 Albedo;
    fixed3 Normal;
    fixed3 Emission;
    half Specular;
    fixed Gloss;
    fixed Alpha;
};
```

这些数据并不需要我们主动赋值，Surface Shader会返回给我们，最后只需要处理片段着色即可。

13.1.4 深度排序

模型之间是有遮挡关系的，所以需要设置模型之间的渲染顺序，这可以在 Shader 中标明，例如 `Tags{"RenderType"="Opaque"}`。一共分为 5 个类型，它们代表的数值也不一样。

- `Background`：代表 1000，天空盒或者背景，其他元素都需要盖在它前面。
- `Geometry`：代表 2000，几何体，地形、地上的房子、树木等并不需要带透明通道的模型。
- `AlphaTest`：代表 2450，透明测试。
- `Transparent`：代表 3000，透明或半透明的模型。
- `Overlay`：代表 4000，渲染在最前面，比如 UI 一类的。

数值越小，越先渲染，所以数值大的会挡住数值小的。此外，也可以直接加减对应的数值，例如，`Tags{"RenderType"=" Geometry +1"}`。如图 13-4 所示，选择一个材质后，既可以对 Render Queue 进行二次编辑，也可以自定义一个新的数值。

图 13-4 Render Queue

如图 13-5 所示，可以在 Frame Debug 窗口中依次查看当前的渲染顺序。这里提供一个小技巧，在游戏中地形之上，还会绘制很多建筑一类的元素，如果先绘制地形再绘制建筑的话，那么重合的像素点就需要画多遍，所以我们可以将地形的 `RenderType` 值设置得比地上建筑大，这样就会先绘制建筑，然后再绘制地形。

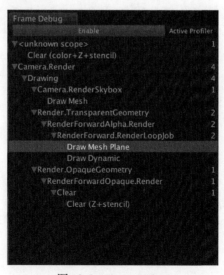

图 13-5 Frame Debug

在 Shader 中，通常可以看到 ZTest LEqual、ZWrite On 的字样，其中 ZTest 表示开启深度测试。比如，前面的建筑遮挡住了地形，CPU 到底是渲染建筑还是地形呢？通过深度测试，能判断出摄像机相对它的距离，然后就知道谁先谁后了。此时如果需要覆盖填充像素，就需要开启 ZWrite 了。这两个参数在 Shader 中默认是开启的。在游戏中，角色在场景中移动可能会被建筑挡住。如图 13-6 所示，为了凸显被挡住的部分，将其标记成另外一种颜色，实现这样的效果就需要用到 ZTest。

图 13-6　遮挡

在 Shader 中，需要两个 Pass 来分别绘制挡住与被挡住的部分。首先是下半身，需要开启 ZTest Greater 和 ZWrite Off，表示突出显示被挡住的地方。上半身则设置 ZTest Lequal，表示正常遮挡，完整代码如代码清单 13-6 所示。

代码清单 13-6　ZTestShader.shader 文件

```
Shader "Unlit/ZTestShader"
{
    Properties {
        _NotVisibleColor("NotVisibleColor(RGB)", Color) = (0.3,0.3,0.3,1)
        _MainTex("Base(RGB)", 2D) = "white" {}
    }
    SubShader {
        Tags { "Queue" = "Geometry+500" "RenderType"="Opaque" }
        LOD 200
        
        Pass {
            ZTest Greater
            ZWrite Off
            Blend SrcAlpha OneMinusSrcAlpha
            SetTexture [_MainTex] { ConstantColor [_NotVisibleColor] combine constant
                * texture }
        }
        
        Pass {
```

```
        ZTest LEqual
        SetTexture [_MainTex] { combine texture }
    }
}
FallBack "Diffuse"
}
```

13.1.5 透明

在 Shader 中要实现透明效果,可以使用 Alpha Test 和 Alpha Blend 这两种方式。如图 13-7 所示,我们在两个不同面片的后方添加一个背景图。可以看到,Alpha Blend 透明部分会和背景混合,而 Alpha Test 则不会,只会出现透明和不透明两种结果。

Alpha Test 无法做混合。而且由于移动平台下不支持 Early-Z,它的效率会比 Alpha Blend 慢。游戏中有时候还需要用到它,比如类似自身溶解的效果。Alpha Blend 的使用场景就非常多了,比如粒子特效、角色身体、翅膀等发光效果。

另外,还需要注意游戏中应当尽量减少使用透明通道。由于透明会出现混合的现象,这样渲染队列必须是从后向前渲染,此时就会出现大量的过度绘制(overdraw)的现象。如果是不透明的话,可以将它设置到 Geometry 上,这样渲染顺序就会从前向后渲染。由于后面的像素挡住了前面的像素,所以将大量降低过度绘制。或者也可以向前面介绍的方法主动修改渲染的顺序,从而降低过度绘制的现象。所以,在设计阶段产品经理需要多和美术人员沟通,能不用透明的地方就不要使用透明。

图 13-7 透明

最后，大家在写 Shader 的时候，尤其需要注意复杂的数学函数。在片段着色器中，每个点都需要做运算，可想而知它的效率。强烈推荐使用 Mobile 下的 Shader，其运行效率更高，但是可选参数很少。大部分情况下，Mobile 下的 Shader 可以满足需求。如果有更复杂的需求，就再拓展一个特殊的 Shader 在这种情况下用。总之，要保证 Shader 中的代码简单。

13.1.6 裁切

Shader 中还提供了 `Stencil`（模板），它和深度测试比较像，测试是否能写入像素，测试成功后即可写入像素。它还有个重要的功能，那就是可以做裁切，即设置裁切区域显示其中一部分，UGUI 中的裁切其实也是这个原理。首先，在裁切与被裁切的 `Properties` 代码块中添加一个唯一标识的 `ID`：

```
Properties
{
    _MainTex("Texture", 2D) = "white" {}
    _ID("Mask ID", Int) = 1
}
```

在需要裁切的模型上写入 `Stencil`，其中 `Ref [_ID]` 表示唯一标识符，`Comp equal` 用来和裁切区域比较是否显示这个像素：

```
Stencil
{
    Ref [_ID]
    Comp equal
}
```

此外，还需要设置一个裁切区域，其中 `Ref [_ID]` 和裁切模型匹配，`Comp always` 和 `Pass replace` 表示在这个区域内的像素永远显示，否则将被裁切掉：

```
Stencil
{
    Ref [_ID]
    Comp always
    Pass replace
}
```

如图 13-8 和图 13-9 所示，通过拉伸裁切区域，可以直接裁切后面的模型，相关代码见 CodeList_13-06。

 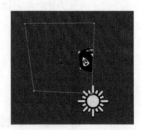

图 13-8　裁切　　　　　　　　　　图 13-9　裁切

13.1.7 着色器变体采集

Unity 内置了很多 Shader，但是比较通用，效率上可能就不是最优的了。通常，我们会在它的基础上修改，去掉一些没用的东西。这样，本地工程里就会有一些 Shader。那么它们应该放在哪里才不影响打包或者 AssetBundle 包呢？通常，Shader 可以直接放在 Resources 目录下。运行时，可以这样读取：

```
Shader a = Resources.Load<Shader>("shader name");
Shader b = Shader.Find("shader name");
```

这样加载出来的 Shader 第一次赋值给材质时，会进行解析，但这会带来一点卡顿。为了避免卡顿，可以将 Shader 放在 Shader Variant Collection 中提前进行预热。如图 13-10 所示，在 Graphics Settings 窗口中，拉到最下方，点击 Save to asset...按钮，即可创建 Shader 并将其包含进 Shader Variant Collection 中，后续也可以灵活地进行修改。

图 13-10　创建着色器变体采集器

然后在初始化的地方进行预热就可以了：

```
Resources.Load<ShaderVariantCollection>("NewShaderVariants").WarmUp();
```

此外，Unity 还提供了一种将 Shader 预制在包体中的功能。具体操作方法是，如图 13-11 所示，在 Graphics Settings 窗口中的 Always Included Shaders 处，将需要的 Shader 拖曳进来即可。这样做确实非常省心，但是会有一个巨大的隐患，那就是变体（Variant）。

如果 Shader 预制在 Always Included Shaders 中，那么所有的变体组合都会进行打包，这会大幅度增加包体，并且在加载时会带来额外的内存开销。如果 Shader 不放在 Always Included Shaders 中，当需要使用它时，就会进行条件判断，找到符合的结果，反之则不用。

图 13-11　预制 Shader

因此，要先确定 Shader 的变体到底有多少，再决定将它放在哪里。如图 13-12 所示，选择一个 Shader 后，点击 Compile and show code，即可看到它的变体数量。如果数量比较多的话，千万不要放进 Always Included Shaders 中。

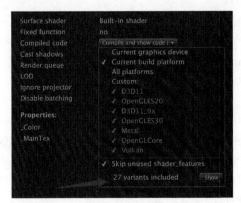

图 13-12　变体

13.2　摄像机

摄像机组件前面提到过，但是没有专门讲过它。其实摄像机是最重要的 3D 组件之一，它就像人的眼睛一样看着游戏世界。但是一双眼睛有时候是不够的，这时就需要多个摄像机了，这就涉及摄像机之间的管理、排序、3D 坐标与平面坐标转换等内容了。

13.2.1　3D 坐标转换屏幕坐标

Unity 的屏幕坐标系规定左下角是原点，X 轴向右为正，Y 轴向上为正。如图 13-13 所示，用鼠标点击面片，获取一个 3D 世界坐标，接着将它转成屏幕坐标，用 GUI 绘制出来。

图 13-13　3D 坐标转屏幕坐标

由于 GUI 的坐标系规定左上角是原点，X 轴向右为正，Y 轴向下为正，所以需要用屏幕高度减去鼠标点击屏幕的 Y 轴坐标，相关代码如代码清单 13-7 所示。

代码清单 13-7　Script_13_07.cs 文件

```csharp
using UnityEngine;

public class Script_13_07 : MonoBehaviour
{
    private Vector2 m_ScreenPotin= Vector3.zero;
    void Update()
    {
        if(Input.GetMouseButtonDown(0))
        {
            Ray ray = Camera.main.ScreenPointToRay(Input.mousePosition);
            RaycastHit hit;
            if(Physics.Raycast(ray, out hit))
            {
                //获取屏幕坐标
                m_ScreenPotin = Camera.main.WorldToScreenPoint(hit.point);
            }
        }
    }

    void OnGUI()
    {
        GUI.color = Color.red;
        GUI.Label(new Rect(m_ScreenPotin.x, Screen.height - m_ScreenPotin.y, 200,
            40),string.Format("鼠标{0}",m_ScreenPotin));
    }
}
```

在上述代码中，首先向鼠标点击的位置发射一条射线，得到点击模型的 3D 坐标，接着通过 Camera.main.WorldToScreenPoin() 方法将 3D 坐标换算成屏幕坐标。

13.2.2　3D 坐标转换 UI 坐标

UI 也有一个 3D 正交摄像机，3D 坐标转 UI 坐标其实就是先转成屏幕坐标，再由屏幕坐标转成 UI 相机的 3D 坐标。如图 13-14 所示，在 3D 世界中任意移动立方体的位置，UI 血条会同步跟随移动，相关代码如代码清单 13-8 所示。

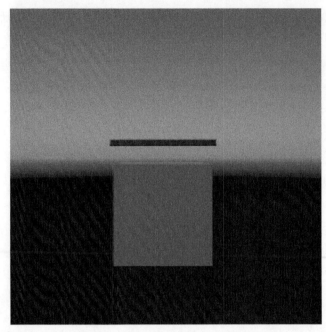

图 13-14　3D 坐标转 UI 坐标

代码清单 13-8　Script_13_08.cs 文件

```
using UnityEngine;

public class Script_13_08 : MonoBehaviour
{
    //英雄头顶
    public Transform heroTransform;
    //血条
    public RectTransform hpTransform;
    //UI 摄像机
    public Camera UICamera;
    //Canvas
    public RectTransform canvasTransform;

    void Update()
    {
        //先将3D坐标转成屏幕坐标
        Vector2 screenPoint = RectTransformUtility.WorldToScreenPoint(Camera.main,
            heroTransform.position);
        //再将屏幕坐标转换UI坐标
        Vector2 localPoint;
        if(RectTransformUtility.ScreenPointToLocalPointInRectangle(canvasTransform,
            screenPoint,UICamera, out localPoint)){
            hpTransform.anchoredPosition = localPoint;
        }
    }
}
```

在上述代码中，先将 3D 坐标转换成屏幕坐标。此时还没有完，由于 UGUI 并不是屏幕坐标，所以需要使用 `RectTransformUtility.ScreenPointToLocalPointInRectangle()` 方法再次将其转换成 UI 坐标。

13.2.3 主摄像机

主摄像机也是 MainCamera。如图 13-15 所示，只要给摄像机设置 MainCamera 标识即可。在代码中，可以通过 `Camera.main` 直接访问摄像机对象。但是这样访问其实未必安全。如果项目中同时有多个摄像机对象被标记了 MainCamera，那么就无法区分出正确的主摄像机了。

图 13-15　主摄像机

13.2.4 在 UI 上显示模型

UI 上是不能直接显示模型的，我们可以添加一个模型层，再创建一个新的摄像机来单独看这个模型层。如图 13-16 所示，将模型摄像机的深度设置到最高，那么它将显示在场景和 UI 的最上面。

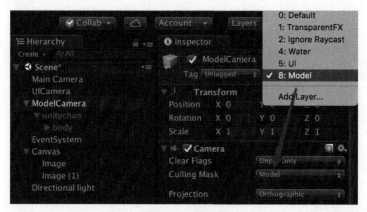

图 13-16　模型层

如图 13-17 所示，由于 Canvas 的 ScreenMatchMode 选择的是 Expand 模式来自适应屏幕，那么模型需要在不同的分辨率下进行缩放。总之，需要将模型填充在一个规定的 UI 区域内。

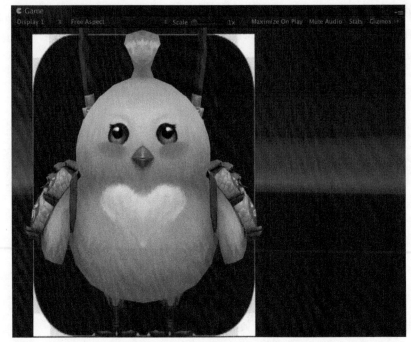

图 13-17 显示模型

将代码清单 13-9 挂在 ModelCamera 上,随便拉伸屏幕模型,依然可以保持相对比例自适应在背景的 UI 中。

代码清单 13-9　ModelCamera.cs 文件

```
using System.Collections;
using System.Collections.Generic;
using UnityEngine;
using UnityEngine.UI;
[RequireComponent(typeof(Camera))]
public class ModelCamera : MonoBehaviour {

    //UI
    public RectTransform uiArea;
    //模型
    public Transform model;

    void Update()
    {
        model.transform.position = uiArea.transform.position;
        //自适应分辨率
        float designWidth = 1136;//开发时分辨率的宽度
        float designHeight = 640;//开发时分辨率的高度
        float designScale =   designWidth/designHeight;
        float scaleRate =    (float)Screen.width/(float)Screen.height;
        float scaleFactor = scaleRate / designScale;
```

```
        if(scaleRate<designScale)
        {
            model.localScale = Vector3.one * scaleFactor;
        }else{
            model.localScale = Vector3.one;
        }
    }
}
```

在上述代码中，通过计算当前分辨率与开发分辨率的比例计算模型的缩放，以确保任何分辨率下模型都显示在 UI 中。

13.2.5 Render Texture

添加新摄像机来显示模型是有缺陷的。因为摄像机的深度决定了模型的渲染顺序，它要么显示在 UI 下面，要么显示在 UI 上面，如果想叠层显示在两个 UI 之间，就很麻烦了。我们可以使用 Render Texture，它可以将某个摄像机看到的内容渲染到纹理上，最后再将这个纹理显示在 UI 的 RawImage 上。如图 13-18 所示，将创建的 Render Texture 文件绑定在摄像机的 Target Texture 上即可。

图 13-18　Render Texture

> 注意：开发中尽可能不要在 Project 中创建 Render Texture 文件，因为我们并不知道游戏中到底会用到多少个，总不能预先创建一堆放在这里吧。Render Texture 是可以在内存中新建的，这样本地就不需要保存这个文件了。

13.2.6 在 Render Texture 上显示带特效的模型

粒子特效大都在使用 Alpha Blending，它需要和后面的物体来做融合。如果放在 Render Texture 上，由于后面没有东西和它混合，显示就不正确了。UI 的 Sprite 用在 Sprite Renderer 上，可以在特效的后面放一张图，让它与特效进行混合。如图 13-19 所示，特效与背景图混合后，再渲染到 Render Texture 上，最终 RawImage 也就正确显示了。

背景图用 Sprite Renderer 的好处是，它可以和 UI 混合使用。在代码中，我们可以根据 RawImage 的大小来自动计算背景图的区域，保证它可以挡住特效。由于 Render Texture 可能会显示得不太清楚，因此可以将它的大小设置成是 RawImage 的 1.5 倍。

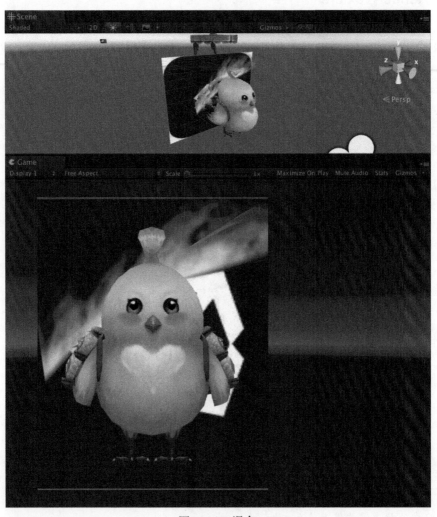

图 13-19　混合

首先，给背景的 Sprite Renderer 挂一个脚本来自适应模型的摄像机，相关代码如代码清单 13-10 所示。

代码清单 13-10　SpriteFull.cs 文件

```csharp
using System.Collections;
using System.Collections.Generic;
using UnityEngine;
using UnityEngine.UI;
public class SpriteFull : MonoBehaviour {

    //设置 Sprite Renderer 和摄像机的距离
    private float distance = 3f;
    //模型摄像机
    public Camera modleCamera;

    private SpriteRenderer spriteRenderer = null;
    void Start()
    {
        spriteRenderer = GetComponent<SpriteRenderer>();
        spriteRenderer.material.renderQueue =2980; //这段代码非常重要！！！大家务必要加上，
                                                   //不然透明的渲染层级会出错
    }

    void Update()
    {
        Camera camera = modleCamera;
        camera.transform.rotation = Quaternion.Euler(Vector3.zero);

        float width = spriteRenderer.sprite.bounds.size.x;
        float height = spriteRenderer.sprite.bounds.size.y;

        float worldScreenHeight,worldScreenWidth;
        //这里分别处理正交摄像机和非正交摄像机
        if(camera.orthographic) {
            worldScreenHeight = camera.orthographicSize * 2.0f;
            worldScreenWidth = worldScreenHeight / Screen.height * Screen.width;
        } else {
            worldScreenHeight = 2.0f * distance * Mathf.Tan(camera.fieldOfView * 0.5f
                * Mathf.Deg2Rad);
            worldScreenWidth = worldScreenHeight * camera.aspect;
        }
        transform.localPosition = new Vector3(camera.transform.position.x, camera.
            transform.position.y, distance);
        transform.localScale = new Vector3(worldScreenWidth / width, worldScreenHeight
            / height, 0f);
    }
}
```

接着，给模型摄像机挂脚本来自动创建 Render Texture 组件，并将其设置到 RawImage 上，相关代码如代码清单 13-11 所示。

代码清单 13-11　ModelCamera.cs 文件

```
using System.Collections;
using System.Collections.Generic;
using UnityEngine;
using UnityEngine.UI;

[RequireComponent(typeof(Camera))]
public class ModelCamera : MonoBehaviour {

    //RawImage
    public RawImage rawImage;

    void Start()
    {
        //为了避免Render Texture 显示不清楚，所以乘以显示区域的1.5倍
        float rate = 1.5f;
        int width = (int)(rawImage.rectTransform.sizeDelta.x * rate);
        int height = (int)(rawImage.rectTransform.sizeDelta.y * rate);
        RenderTexture renderTexture = RenderTexture.GetTemporary(width, height, 16,
            RenderTextureFormat.ARGB32);
        GetComponent<Camera>().targetTexture = renderTexture;
        rawImage.texture = renderTexture;
    }

}
```

使用 Render Texture 比直接使用摄像机看的效果差很多，所以可以提高 Render Texture 的尺寸大小，最终输出在 RawImage 上即可。

13.2.7　LOD Group

在场景中，元素多了是很耗渲染的。摄像机能看到的内容是有限的，如果一个离摄像机很远的物体也正常渲染，那效率就太低了。应对这个问题，Unity 提供了 LOD Group 组件。同样的模型可以让美术人员做两份，一个是高精度的，一个是低精度的。最后，根据摄像机的远近来动态更换模型，或者剔除部分渲染。

如图 13-20 所示，首先需要在 Quality Settings 面板中设置。

- Lod Bias：表示 Lod 的偏移值，它会影响剔除的距离。
- Maximum LOD Level：表示 Lod 的最大等级。运行时修改它，可以优化低端机器。

图 13-20　设置参数

它的原理就是根据物体在摄像机内的百分比来调节显示的级别，那么百分比是怎么来的呢？首先，会给物体添加平面的包围盒，摄像机发生移动后，即可计算与这个包围盒的百分比。如图 13-21 所示，可以添加任意数量的 LOD 并且设置每个等级显示或不显示什么。

图 13-21　Lod Group

总体来说，LOD Group 技术就是用内存来换时间，预先加载了模型的好几个显示状态，然后根据摄像机与它的距离来切换不同的状态以保证渲染效率。

13.3　场景管理

场景中可以放置任意游戏对象，场景之间可以任意切换。Unity 还提供了多场景管理，可以同时打开或者切换多个场景。

13.3.1　切换场景

Unity 提供了同步切换场景以及异步切换场景的方法。场景切换的原则是删除当前场景，切换新场景：

```
//同步切换场景
SceneManager.LoadScene("sceneName");
//异步切换场景
SceneManager.LoadSceneAsync("sceneName");
```

13.3.2　**DontDestroyOnLoad**

切换新场景会自动卸载老场景，但我们并不希望切换场景时删除某些游戏对象，例如 UI 节点和主角等，此时可以给这些游戏对象标记 DontDestroyOnLoad，这样他们就不会被卸载掉了。

```
GameObject.DontDestroyOnLoad(gameObject);
```

13.3.3　异步加载场景以及进度

同步加载场景会引起卡顿，所以游戏中通常都会采用异步方式来加载场景。此时，就需要获取加载场景的进度以及结束的事件了。首先，封装一下加载的管理器类，相关代码如代码清单 13-12 所示。

代码清单 13-12　SceneLoadManager.cs 文件

```csharp
using System.Collections;
using System.Collections.Generic;
using UnityEngine;
using UnityEngine.Events;
using UnityEngine.SceneManagement;

public class SceneLoadManager : MonoBehaviour
{
    static AsyncOperation m_AsyncOperation;
    static UnityAction<float> m_Progress;

    ///<summary>
    ///加载场景
    ///</summary>
    ///<param name="name">场景名</param>
    ///<param name="progress">回调加载进度</param>
    ///<param name="finish">回调加载场景结束</param>
    static public void LoadScene(string name,UnityAction<float>progress,
        UnityAction finish)
    {
        new GameObject("#SceneLoadManager#").AddComponent<SceneLoadManager>();
        m_AsyncOperation  = SceneManager.LoadSceneAsync("New Scene", LoadSceneMode.
            Single);
        m_Progress = progress;

        //加载完毕后抛出事件
        m_AsyncOperation.completed += delegate(AsyncOperation obj) {
            finish();
            m_AsyncOperation = null;
        };
    }

    void Update()
    {
        if(m_AsyncOperation != null) {
            //抛出加载进度
            if(m_Progress != null) {
                m_Progress(m_AsyncOperation.progress);
                m_Progress = null;
            }
        }
    }
}
```

最后，需要监听加载进度，并在加载结束的地方调用它。拿到加载进度，就可以在 UI 上显示了，相关代码如代码清单 13-13 所示。

代码清单 13-13　Script_13_11.cs 文件

```csharp
using UnityEngine;
using UnityEngine.UI;
using UnityEngine.SceneManagement;
```

```csharp
public class Script_13_11 : MonoBehaviour
{
    void Start()
    {
        //禁止切换场景时卸载初始化的游戏对象
        GameObject[]InitGameObjects = GameObject.FindObjectsOfType<GameObject>();
        foreach(GameObject go in InitGameObjects){
            if(go.transform.parent == null)
            {
                GameObject.DontDestroyOnLoad(go.transform.root);
            }
        }

        //加载场景
        SceneLoadManager.LoadScene("New Scene", delegate(float progress) {
            Debug.LogFormat("加载进度:{0}",progress);
        }, delegate() {
            Debug.Log("加载结束");
        });
    }
}
```

13.3.4 多场景

Unity 支持多场景编辑，即可以把元素放在两个不同的场景中，这样就可以实现无缝拼接。如图 13-22 所示，我们可以同时在 Hierarchy 视图中打开两个场景，并且相互编辑互不影响。这样就可以把多个小场景拼接成一个非常庞大的场景，运行游戏时，可以动态打开以及卸载这些小场景。

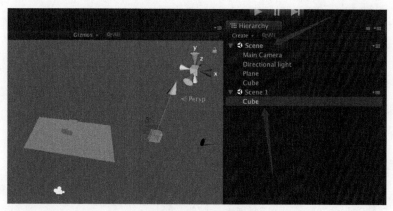

图 13-22 多场景

在代码中加载或者卸载多场景时，需要注意加载和卸载都提供了异步接口，可以按照上节方法来获取加载以及卸载的进度，相关代码如代码清单 13-14 所示。

代码清单 13-14　Script_13_12.cs 文件

```csharp
using UnityEngine;
using UnityEngine.SceneManagement;

public class Script_13_12 : MonoBehaviour {

    void OnGUI()
    {
        if(GUILayout.Button("<size=50>加载多场景</size>")) {
            SceneManager.LoadSceneAsync("Scene 1", LoadSceneMode.Additive);
        }

        if(GUILayout.Button("<size=50>卸载多场景</size>")) {
            SceneManager.UnloadSceneAsync("Scene 1");
        }
    }
}
```

在上述代码中，我们分别使用 `SceneManager.LoadSceneAsync()` 和 `SceneManager.UnloadSceneAsync()` 进行异步加载和卸载场景。

13.3.5　多场景游戏对象管理

因为同时存在多场景，所以游戏对象就需要考虑到底实例化在哪个场景中，并且需要判断游戏对象属于哪个场景，相关代码如代码清单 13-15 所示。

代码清单 13-15　Script_13_13.cs 文件

```csharp
using System.Collections;
using System.Collections.Generic;
using UnityEngine;
using UnityEngine.SceneManagement;

public class Script_13_13 : MonoBehaviour {

    void Awake()
    {
        //同时打开多个场景
        SceneManager.LoadScene("Scene 1",LoadSceneMode.Additive);
        SceneManager.LoadScene("Scene 2",LoadSceneMode.Additive);
    }
    void Start()
    {
        //获取场景对象
        Scene scene1 = SceneManager.GetSceneByName("Scene 1");
        Scene scene2 = SceneManager.GetSceneByName("Scene 2");

        GameObject g1 = new GameObject("g1");
        GameObject g2 = new GameObject("g2");

        //将游戏对象放入对应场景中
        SceneManager.MoveGameObjectToScene(g1, scene1);
        SceneManager.MoveGameObjectToScene(g2, scene2);
```

```
            //判断游戏对象属于哪个场景
            Debug.LogFormat("{0} 属于 Scene 1",g1.scene.name);
            Debug.LogFormat("{0} 属于 Scene 2",g1.scene.name);
        }
    }
```

13.3.6 场景切换事件

在多场景中，需要指定一个为激活场景，默认第一个为激活场景。如果添加新游戏对象时不指定场景，都将添加在激活场景中。另外，还提供了场景激活、场景加载、场景卸载的回调事件，这可以在代码中监听，具体如代码清单 13-16 所示。

代码清单 13-16　Script_13_14.cs 文件

```
using System.Collections;
using System.Collections.Generic;
using UnityEngine;
using UnityEngine.SceneManagement;

public class Script_13_16 : MonoBehaviour {

    void Awake()
    {
        //同时打开多个场景
        SceneManager.LoadScene("Scene 1",LoadSceneMode.Additive);
        SceneManager.LoadScene("Scene 2",LoadSceneMode.Additive);

        SceneManager.activeSceneChanged += delegate(Scene arg0, Scene arg1) {
            Debug.LogFormat("场景激活：{0} {1}",arg0.name,arg1.name);
        };

        SceneManager.sceneLoaded += delegate(Scene arg0, LoadSceneMode arg1) {
            Debug.LogFormat("场景加载：{0} 加载模式：{1}",arg0.name,arg1.ToString());
        };

        SceneManager.sceneUnloaded += delegate(Scene arg0) {
            Debug.LogFormat("场景卸载：{0} ",arg0.name);
        };
    }

    void Start()
    {
        //设置场景激活状态
        Scene scene2 = SceneManager.GetSceneByName("Scene 2");
        SceneManager.SetActiveScene(scene2);
        //如果不指定场景，对象将创建在当前激活的场景中
        new GameObject("MyNewGameObject");
    }
}
```

在上述代码中，我们监听了场景的激活、加载和卸载事件。

13.3.7 多场景烘焙

由于烘焙贴图是和场景绑定的,那么为了保证多场景没有烘焙接缝,可以使用脚本来烘焙。如图 3-23 所示,同时烘焙 3 个场景,会自动生成 3 个对应的 LightingData 文件。

图 13-23 多场景烘焙

如代码清单 13-17 所示,调用 Lightmapping.BakeMultipleScenes() 方法可以指定多场景烘焙。

代码清单 13-17　Script_13_15.cs 文件

```
using UnityEngine;
using UnityEditor;

public class Script_13_15 {

    [MenuItem("Tool/BakeMultipleScenes")]
    static void BakeMultipleScenes()
    {
        //指定烘焙场景
        Lightmapping.BakeMultipleScenes(new string[] {
            "Assets/Scene.unity",
            "Assets/Scene 1.unity",
            "Assets/Scene 2.unity",
        });
    }
}
```

13.4 输入事件

Unity 中的输入事件可分为两种:一种是全局触发的,需要在更新每一帧来判断;还有一种是监听式触发的,就是监听点击的事件回调从而处理后面的逻辑。

13.4.1 全局事件

全局输入事件需要使用 Input 这个类,它可以监听键盘、鼠标、手势以及移动设备上的 3D Touch 事件等。例如点击事件,它只能监听屏幕中的事件,并不能判断是点在了 3D 模型上还是点在了 UI 上。并且它没有提供触发事件，需要在 Update()方法中通过每一帧去判断，具体如代码清单 13-18 所示。

代码清单 13-18　Script_13_16.cs 文件
```
using UnityEngine;
public class Script_13_16 : MonoBehaviour
{
    void Update()
    {
        //按下空格键
        if(Input.GetKeyDown(KeyCode.Space)) {
        }
        //抬起空格键
        if(Input.GetKeyUp(KeyCode.Space)) {
        }
        //按下空格键
        if(Input.GetKey(KeyCode.Space)) {
        }
        //按下鼠标左键，手机上则是按下屏幕
        if(Input.GetMouseButton(0)) {
            Debug.LogFormat("点击屏幕坐标:{0}", Input.mousePosition);
        }

        //手指触摸屏幕中
        if(Input.touchCount > 0) {
            Touch touch = Input.GetTouch(0);
            //开始触摸
            if(touch.phase == TouchPhase.Began) {
            }
            //触摸移动
            if(touch.phase == TouchPhase.Moved) {
            }
            //触摸结束
            if(touch.phase == TouchPhase.Ended) {
            }
            //是否支持 3D Touch
            if(Input.touchPressureSupported) {
                Debug.LogFormat("3DTouch 的力度:{0}", touch.pressure);
            }
        }
    }
}
```

在上述代码中，我们通过 Input 类实现按键以及触摸的事件。

13.4.2　射线

射线就是由某一个点向一个方向发射的一条无尽的线，它通常用来做鼠标的 3D 拾取。以摄像机为原点向屏幕中的一点发射射线，当发生射线碰撞时，即可拾取鼠标点击在 3D 世界的坐标了。射线可能会碰到多个碰撞器（Collider），也可以全部拾取出来。如代码清单 13-19 所示，通过发射线来实现鼠标点选的位置。

代码清单 13-19　Script_13_17.cs 文件

```
using UnityEngine;

public class Script_13_17 : MonoBehaviour
{

    void Update()
    {

        if(Input.GetMouseButtonDown(0)) {
            Ray ray = Camera.main.ScreenPointToRay(Input.mousePosition);

            RaycastHit hit;
            if(Physics.Raycast(ray, out hit)) {
                Debug.LogFormat("Raycast: {0} 3D 坐标:{1}", hit.collider.name,
                    hit.point);
            }

            RaycastHit[] hits = Physics.RaycastAll(ray);
            foreach(var h in hits) {
                Debug.LogFormat("RaycastAll: {0} 3D 坐标:{1}", hit.collider.name,
                    hit.point);
            }
        }
    }
}
```

在上述代码中，`Physics.Raycast()` 表示只检测最先碰撞到的射线上的点，`Physics.RaycastAll()` 表示检测所有碰撞到射线上的点。

另外，Unity 还提供了一个层可以忽略射线。如图 13-24 所示，如果将游戏对象设置成 Ignore Raycast，此时这个游戏对象将不再接受射线碰撞。

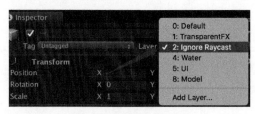

图 13-24　射线忽略层

13.4.3　点选 3D 模型

Unity 提供了 Event System 组件，它可以用来处理 UI 与 3D 对象的点击。它的原理也是发送射线，但是它封装得会更好一些。首先，需要给主摄像机绑定 Physics Raycaster 组件。如图 13-25

所示,让它来发送射线,其中 Event Mask 可以过滤掉某些不需要的层。使用 Physics Raycaster 组件还有个好处,如果 UI 一部分挡在 3D 模型上面,会优先响应 UI 事件。

图 13-25 物理射线组件

接着,将 Click3DEvent.cs 脚本(如代码清单 13-20 所示)挂在需要点击的 3D 模型上,在统一的地方来监听并处理它们的点击事件。如图 13-26 所示,点选后,会输出对应的游戏对象名。

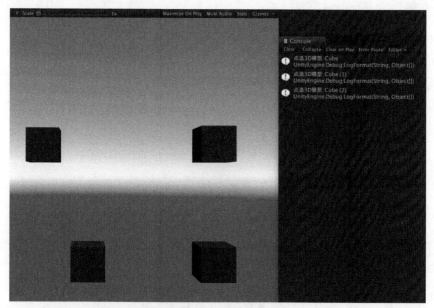

图 13-26 点选 3D 模型

代码清单 13-20 Click3DEvent.cs 文件

```
using System.Collections;
using System.Collections.Generic;
using UnityEngine;
using UnityEngine.EventSystems;
using UnityEngine.Events;

public class Click3DEvent : UnityEvent<GameObject,PointerEventData>{}

public class Click3D : MonoBehaviour,IPointerClickHandler {

    public static Click3DEvent click3DEvent = new Click3DEvent();
```

```
#region IPointerClickHandler implementation

public void OnPointerClick(PointerEventData eventData)
{
    click3DEvent.Invoke(gameObject, eventData);
}

#endregion
}
```

最后，统一处理点击事件，相关代码如代码清单 13-21 所示。

代码清单 13-21　Script_13_18.cs 文件
```
using UnityEngine;
using UnityEditor;
using UnityEngine.EventSystems;
public class Script_13_18: MonoBehaviour
{
    void Start()
    {
        Click3D.click3DEvent.AddListener(delegate(GameObject gameObject,PointerEventData arg1) {
            Debug.LogFormat("点选3D模型：{0}",gameObject.name);
        });
    }
}
```

13.4.4　通过点击控制人物移动

学习了鼠标拾取，就可以点击地面来控制角色移动了，如图 13-27 所示。由于移动是一个过程，可以使用 Vector3.MoveTowards 根据步长来移动模型。

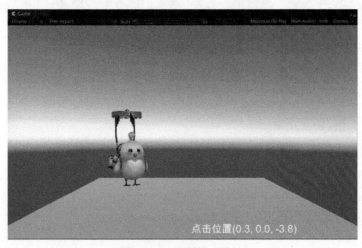

图 13-27　控制模型移动

如代码清单 13-22 所示，鼠标点选到地面的坐标后，通过 Vector3.MoveTowards() 将角色移动过去即可。

代码清单 13-22　Script_13_19.cs 文件

```
using UnityEngine;
using UnityEditor;
using UnityEngine.EventSystems;
public class Script_13_19 : MonoBehaviour
{
    //模型
    public Transform model;
    //3DTextMesh
    public TextMesh textMesh;
    //移动目的地
    private Vector3 m_MoveToPosition=Vector3.zero;

    void Update()
    {
        if(Input.GetMouseButtonDown(0)) {
            Ray ray = Camera.main.ScreenPointToRay(Input.mousePosition);


            RaycastHit hit;
            if(Physics.Raycast(ray, out hit)) {
                //面朝选择点
                m_MoveToPosition = new Vector3(hit.point.x,model.position.y,
                    hit.point.z);
                model.LookAt(m_MoveToPosition);

                textMesh.text = string.Format("点击位置{0}", hit.point);
                textMesh.transform.position = hit.point;
            }
        }
        if(model.position != m_MoveToPosition) {
            //步长
            float step = 5f * Time.deltaTime;
            model.position = Vector3.MoveTowards(model.position,
                m_MoveToPosition, step);
        }
    }
}
```

13.5　物理碰撞

在 Unity 中，3D 物理碰撞和 2D 碰撞的用法都差不多，使用的都是 Rigidbody 和 Collider 组件。有关这两个组件的用法，大家可参照第 6 章，这里将不再赘述。不过 3D 提供了一个角色控制器的组件，专门用来处理人物胶囊体碰撞。

13.5.1 碰撞器

如图 13-28 所示，3D 部分一共提供了六类碰撞器。

- Box Collider：最常用的立方体碰撞器。
- Sphere Collider：球形碰撞器。它的效率是最高的，因为球体的直径是相同的。
- Capsule Collider：胶囊体碰撞器，适用于控制的主角。
- Mesh Collider：网格碰撞器，根据美术人员的模型网格生成的碰撞器，其效率也是最低的。建议项目中不要使用它，或者可以让美术人员做一个专门用来检测碰撞的简易网格。
- Wheel Collider：轮胎碰撞器，就像汽车的 4 个轮胎一样在地面上跑。
- Terrain Collider：地形碰撞器，可以在地形编辑时减少网格的数量，这样就能优化效率。

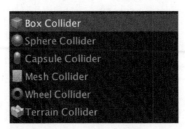

图 13-28　控制模型移动

13.5.2 角色控制器

角色控制器（Character Controller）是用来控制第三人称移动的，它并不会像刚体那样被另外的刚体对象击飞，更适合于人形控制器。如图 13-29 所示，绑定角色控制器后，即可出现胶囊体区域，可以再次编辑它的区域，并且设置爬坡的角度以及皮肤的宽度。如果不希望模型走上坡，可以将 Slope Limit 设置成 0 或者比较小的值。

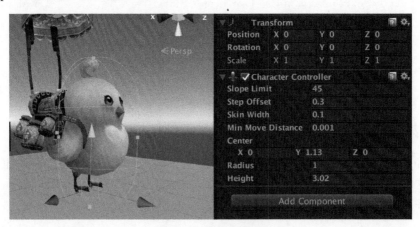

图 13-29　角色控制器

在代码中使用 Move() 方法，就可以控制移动了：

```
CollisionFlags flags = controller.Move(controller.transform.rotation *
    Vector3.forward * Time.deltaTime * 2f);
```

另外，还提供了 SimpleMove() 方法，它只能控制 X 轴和 Z 轴的移动。其中 Z 轴是依靠重力计算的，如果是不平的地面，可以掉下去。而 Move() 方法可以控制 X、Y 和 Z 这 3 个轴向的移动，并且忽略了重力，不会掉下去。

13.5.3　碰撞区域

控制胶囊体移动时，会和碰撞器发生碰撞，此时就有可能出现脚与地面碰撞，头与房顶碰撞，身子四周与墙发生碰撞。调用 Move() 后，即可判断是否发生了碰撞。如代码清单 13-23 所示，通过 CollisionFlags 可判断具体哪个方向发生了碰撞。

代码清单 13-23　Script_13_20.cs 文件

```csharp
using UnityEngine;
public class Script_13_20 : MonoBehaviour
{
    public CharacterController controller;
    void Start()
    {
        CollisionFlags flags = controller.Move(Vector3.forward);
        if(flags == CollisionFlags.None)
            print("没有发生碰撞");

        //可能出现身体的四周发生碰撞，并且脚与地面也发生碰撞的现象
        if((flags & CollisionFlags.Sides) != 0)
            print("身体的四周发生碰撞");

        if(flags == CollisionFlags.Sides)
            print("只有身体的四周发生碰撞");

        if((flags & CollisionFlags.Above) != 0)
            print("头部与上面发生碰撞");

        if(flags == CollisionFlags.Above)
            print("只有头部与上面发生碰撞");

        if((flags & CollisionFlags.Below) != 0)
            print("脚与地面发生碰撞");

        if(flags == CollisionFlags.Below)
            print("只有脚与地面发生碰撞");
    }
}
```

注意：Unity 只提供了上面、下面和周围这 3 个方面的碰撞，如果还需要检测身体的正面或者背面，需要根据自身旋转角度来计算。

13.5.4 碰撞检测

角色控制器只需要将脚本绑在自身，就可以监听碰撞物的信息了。我们来做一个例子：用鼠标控制模型移动，当发生碰撞后停止移动。如图 13-30 所示，控制模型移动，并且监听它与周围碰撞器碰撞的信息。

图 13-30　移动碰撞

将代码清单 13-24 所示的脚本挂在角色控制器对象身上即可。

代码清单 13-24　Script_13_21.cs 文件

```csharp
using UnityEngine;
[RequireComponent(typeof(CharacterController))]
public class Script_13_21: MonoBehaviour
{
    CharacterController controller;
    void Awake()
    {
        controller = GetComponent<CharacterController>();
    }

    private Vector3 m_MoveToPosition;
    void Update()
    {

        if(Input.GetMouseButtonDown(0) {
            Ray ray = Camera.main.ScreenPointToRay(Input.mousePosition);

            RaycastHit hit;
            if(Physics.Raycast(ray, out hit)) {
                //面朝选择点
```

```csharp
            m_MoveToPosition = new Vector3(hit.point.x,controller.transform.
                position.y,hit.point.z);
            controller.transform.LookAt(m_MoveToPosition);
        }
    }
    //只要角色与目标的距离小于0.1米,就移动
    if(Vector3.Distance(m_MoveToPosition, controller.transform.position) > 0.1f) {
        Vector3 back = controller.transform.position;
        CollisionFlags flags = controller.Move(controller.transform.rotation *
            Vector3.forward * Time.deltaTime * 2f);
        if((flags & CollisionFlags.Sides) != 0){
            //发生碰撞时停下来,并且还原坐标
            m_MoveToPosition = controller.transform.position = back;
        }
    }
}


void OnControllerColliderHit(ControllerColliderHit hit)
{
    //获取控制器
    CollisionFlags flags =  hit.controller.collisionFlags;
    //可能出现身体的四周发生碰撞,并且脚与地面也发生碰撞的现象
    if((flags & CollisionFlags.Sides) != 0)
        Debug.LogFormat("身体的四周与{0}发生碰撞",hit.collider.name);

    if(flags == CollisionFlags.Sides)
        Debug.LogFormat("只有身体的四周与{0}发生碰撞",hit.collider.name);

    if((flags & CollisionFlags.Above) != 0)
        Debug.LogFormat("头部与上面的{0}发生碰撞",hit.collider.name);

    if(flags == CollisionFlags.Above)
        Debug.LogFormat("只有头部与上面的{0}发生碰撞",hit.collider.name);

    if((flags & CollisionFlags.Below) != 0)
        Debug.LogFormat("脚与地面的{0}发生碰撞",hit.collider.name);

    if(flags == CollisionFlags.Below)
        Debug.LogFormat("只有脚与地面的{0}发生碰撞",hit.collider.name);
    }
}
```

13.5.5 碰撞触发器

在前面的例子,我们是通过控制主角来监听主动碰撞到的物体,它会被别的碰撞器挡住无法前进。例如,需要一个传送门一类的东西,并不希望它会挡住主角,而只需要监听触发的事件即可。如图13-31所示,找到需要监听的立方体对象,选中 Is Trigger 复选框,并且绑定 Trigger Listener 组件。

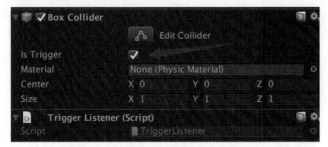

图 13-31　碰撞触发器

如代码清单 13-25 所示，通过 TriggerEvent 来监听触发器的事件

代码清单 13-25　TriggerEvent.cs 文件

```
using System.Collections;
using System.Collections.Generic;
using UnityEngine;
using UnityEngine.Events;

public class TriggerEvent : UnityEvent<Collider>{}

public class TriggerListener : MonoBehaviour {

    public static TriggerEvent triggerEventEnter = new TriggerEvent();

    void OnTriggerEnter(Collider collider)
    {
        //抛出事件
        triggerEventEnter.Invoke(collider);
    }
}
```

接着，在需要处理这个触发事件的地方添加监听即可：

```
TriggerListener.triggerEventEnter.AddListener(delegate(Collider collider) {
    Debug.LogFormat("触发到{0}",collider.name);
});
```

13.5.6　物理调试器

模型的显示和它的碰撞区域是两个不同的组件。例如，很多空气墙一类的碰撞器（Collider）组件，可能会分别设置显示和真实碰撞的参数。如果碰撞器多了以后，管理起来会非常麻烦，无法统一预览所有的碰撞器。在导航菜单栏中选择 Window→Physics Debug 菜单项，即可打开调试器窗口，如图 13-32 所示，这里可以批量隐藏/显示触发器和碰撞器等，还可以自定义颜色，更方便全局预览所有的物理元素。

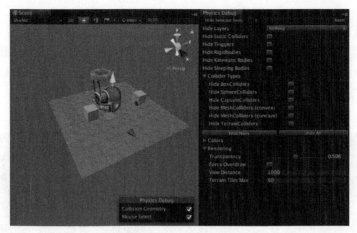

图 13-32 物理调试器

13.6 实战技巧

在 3D 游戏开发中,还有一些实战技巧,比如头顶显示文字、冒血数字、合并网格、合并贴图等,这一节就来盘点一下。

13.6.1 3D 头顶文字

前面我们介绍了通过 3D 人物的坐标来换算成 UI 坐标从而显示血条和文字,这种做法其实并不太好。因为只要人物移动或者摄像机移动,就需要一直换算坐标,这将带来一定开销。如图 13-33 所示,创建 Text Mesh 组件并将其直接挂在角色身上,这样头顶文字将自动跟着角色移动,并且自动根据摄像机的远近调整字号。

图 13-33 Text Mesh 组件

13.6.2 图文混排

如果显示文字的同时还需要显示图标，那么可以使用 RichText 的功能。如图 13-34 所示，将需要参与图文混排的图片放入一个新材质中，并且绑定在 Text Mesh 的 Materials 数组中，接着写入混排代码，其中 `<quad material=1 />` 表示显示数组中索引为 1 的图片：

```
<quad material=1 />图文混排啦<quad material=1 />
```

图 13-34　图文混排

RichText 的标签还包括 b、i、size 和 color，它们分别表示粗体、斜体、大小和颜色，可以相互嵌套使用。另外，RichText 的这些标签中除了 quad 以外，其他都可以在 UGUI 系统中使用。本例的详细工程见 CodeList_13_23。

13.6.3 图片数字

美术人员一般会设计出来一些图片数字，例如战斗时被击打产生的掉血数字。为了使用方便，需要实现一套将数字直接转换为图片数字的脚本。如图 13-35 所示，在 Text 文本区域中输入显示的数字，并且将它们转成图片数字显示在屏幕中。

图 13-35　图片数字

如代码清单 13-26 所示，首先将 Sprite 精灵拖入 numberRes 并且设置成相同的 atlasName，这样所有数字只会有一个 DrawCall。接着，通过输入的数字找到数字里对应保存的 Sprite，并且自动创建出对应的 Sprite Renderer 组件显示即可。

代码清单 13-26　UINumber.cs 文件

```
using UnityEngine;

public class UINumber : MonoBehaviour {

    //原始资源精灵
    [SerializeField]
    private Sprite[] numberRes;
    //输入区域
    [TextArea(2,4)][SerializeField]
    private string text;

    void Start()
    {
        Refresh();
    }


    void Refresh()
    {
        for(int i = 0; i < text.Length; i++)
        {
            int a;
            if(int.TryParse(text[i].ToString(), out a))
            {
                //如果缓存中没有，则创建新的
                SpriteRenderer spriteRenderer = new GameObject().AddComponent
                    <SpriteRenderer>();
                spriteRenderer.sprite = numberRes[a];
```

```
            spriteRenderer.gameObject.SetActive(true);
            spriteRenderer.gameObject.name = a.ToString();
            spriteRenderer.transform.SetParent(transform, false);
            spriteRenderer.transform.localPosition = new Vector3(i * 0.2f, 0f, 0f);
        }
    }
}
```

这段代码比较简单，实际游戏中会有更复杂的情况。由于数字可能是一直蹦出的，如果每次都创建新的 Sprite Renderer，岂不是很费效率，所以一定要考虑游戏中的缓存池。

13.6.4 游戏对象缓存池

单纯地创建游戏对象其实很快。但是由于游戏对象上会绑定脚本，这就会造成一些卡顿，原因就是每个脚本上的 `Awake()` 和 `OnEnable()` 方法会同步执行。如果有大量复用性强的游戏对象，一定要做好缓存机制。

缓存池的原理其实很简单，只需要对外提供两个接口——获取和回收。获取只需查看池子里有没有，没有则创建。回收就是将游戏对象放回池子，下次使用时就不用再创建新的了。由于缓存池可能有多种类型的，所以将它做成成员对象。相关代码如代码清单13-27所示。

代码清单13-27 GameObjectPool.cs 文件

```csharp
using System.Collections.Generic;
using UnityEngine;

public class GameObjectPool
{
    private GameObject m_ObjectContainer;
    private readonly Dictionary<string, Stack<GameObject>> m_AssetObjects;

    public GameObjectPool(string name)
    {
        m_AssetObjects = new Dictionary<string, Stack<GameObject>>();
        m_ObjectContainer = new GameObject("#POOL#" + name);
        m_ObjectContainer.SetActive(false);
    }

    //获取游戏对象
    public GameObject GetObject(string assetPath)
    {
        GameObject go = GetObjectFromPool(assetPath);
        if(go == null) {
            go = Object.Instantiate<GameObject>(Resources.Load<GameObject>(assetPath));
            go.AddComponent<PoolObject>().value = assetPath;
        }
        return go;
    }
```

```csharp
//释放游戏对象
public void ReleaseObject(GameObject go)
{
    var comp = go.GetComponent<PoolObject>();
    if(comp != null && comp.value != null)
    {
        Stack<GameObject> objects;
        if(m_AssetObjects.TryGetValue(comp.value, out objects))
        {
            objects.Push(go);
            go.transform.SetParent(m_ObjectContainer.transform, false);
            return;
        }
    }
    Object.Destroy(go);
}

private GameObject GetObjectFromPool(string assetPath)
{
    Stack<GameObject> objects;
    if(!m_AssetObjects.TryGetValue(assetPath, out objects)) {
        objects = new Stack<GameObject>();
        m_AssetObjects[assetPath] = objects;
    }
    return objects.Count > 0 ? objects.Pop() : null;
}

class PoolObject : MonoBehaviour
{
    public string value;
}
}
```

使用的时候，需要创建一个自己的对象池，如代码清单 13-28 所示。

代码清单 13-28　Script_13_25.cs 文件

```csharp
using System.Collections.Generic;
using UnityEngine;

public class Script_13_25 : MonoBehaviour {

    private GameObjectPool m_Pool = null;
    private Stack<GameObject> m_Objects = null;

    private void Awake()
    {
        m_Pool = new GameObjectPool("Cube");
        m_Objects = new Stack<GameObject>();
    }

    private void OnGUI()
    {
```

```
if(GUILayout.Button("<size=50>Get</size>")){
    GameObject go= m_Pool.GetObject("Cube");
    go.transform.SetParent(transform, false);
    m_Objects.Push(go);
}
if(GUILayout.Button("<size=50>Release</size>"))
{
    if(m_Objects.Count > 0)
    {
        m_Pool.ReleaseObject(m_Objects.Pop());
    }
}
        }
    }
```

如图 13-36 所示，点击 Get 按钮即可从缓存池中获取对象，如果没有，则创建新的。点击 Release 按钮，则将创建过的对象放回至缓存池中，并且设置它为隐藏状态。

图 13-36　缓存池

13.6.5　游戏资源缓存池

Unity 内部其实已经做了一个缓存池。例如，立方体对象引用了一张贴图资源，当它首次被实例化进游戏中时，贴图资源会自动读取。如果此时立方体对象被释放掉，贴图则被 Unity 临时缓存着，下次如果又用到了这个贴图，就不会再重新读取了，这将加快资源的访问速度。另外，这个贴图并不会被 Unity 永远临时缓存着，它会在下一次垃圾回收或者切换场景时被系统回收掉。对于比较重要的游戏资源，也需要做资源缓存池，其代码的实现思路和游戏对象缓存池类似，这里就不再赘述了。

13.6.6　运行时合并网格

由于换装游戏的身体部分可以被拆开，组合在一起后 DrawCall 数量就上去了。为了避免这种情况，我们可以在运行时合并网格和贴图。Unity 引擎计算 DrawCall 的方法也比较简单、粗暴，即同 Mesh、同材质、同 Shader、同参数、同贴图就是一个 DrawCall。模型换装其实是不影响骨骼和动画的，也就是说只需要合并 Mesh，就可以让骨骼动画复用，最终将合并后的 Mesh 重新赋给 Skinned Mesh Renderer 即可。

如图 13-37 所示，身体和头部是两个 Mesh，此时就算它们共用了相同的材质和贴图，也会进行多次 DrawCall，所以需要将各部位合并成一个 Mesh。如果还需要合并贴图以及骨骼动画，那么还需要保证 UV、骨骼权重和骨骼绑定姿势正确合并。

图 13-37　合并网格

如代码清单 13-29 所示，通过代码将不同模型的部位合并在一起，首先将需要合并的 Mesh 放入 CombineInstance 数组中并且设置它的位置，最终使用 CombineMeshes() 方法合并它们。

代码清单 13-29　Script_13_26.cs 文件

```
using UnityEngine;

public class Script_13_26 : MonoBehaviour
{
    public SkinnedMeshRenderer[] combinMeshs;
    public SkinnedMeshRenderer skinnedMeshRenderer;

    private void Awake()
    {
        CombineInstance[] combines = new CombineInstance[combinMeshs.Length];

        for(int i = 0; i < combinMeshs.Length; i++)
        {
            combines[i].mesh = combinMeshs[i].sharedMesh;
```

```
            combines[i].transform = combinMeshs[i].transform.localToWorldMatrix;
        }
        var mesh = new Mesh();
        mesh.name = "combine";
        mesh.CombineMeshes(combines);
        skinnedMeshRenderer.sharedMesh = mesh;
    }
}
```

如图 13-38 所示，最终将各部位合并在一起。之前的章节中，我们讲过 Unity 还提供了静态合并 Mesh 的方法，它也可以在运行时合并，但是合并后子节点下的 Mesh 是无法发生位移变化的。静态合并比较适合不需要移动的物体，而这里的动态合并适合可以发生移动的物体。

图 13-38　合并结果

游戏中的换装系统其实也不一定非得使用动态合并 Mesh 的方法。如果换的部件比较少，完全可以用多个 Skinned Mesh Renderer 的方式来做。如果换的部件比较多，就一定要动态合并了，不然 DrawCall 就太多了。

13.6.7　运行时合并贴图

和美术人员可以约定好贴图每一部分的含义，比如图片左半部分是身体，右半部分是头、武器等，运行时根据实际情况将贴图合并在一起，重新组成一张贴图。如果是 RGBA32 格式的贴图，那么每个像素是由 32 位来表示的，分别代表 R（红）、G（绿）、B（蓝）和 A（透明），其中每个颜色通道的取值范围是 0~255。一个像素占用 4 字节大小。如图 13-39 所示，一张完整的贴图是由一维数组表示的，并且排列顺序是从左到右，从下到上。

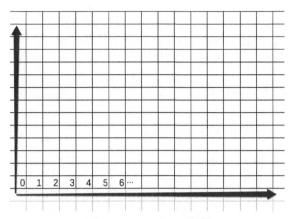

图 13-39　RGBA32 像素

但是其实游戏中很少使用 RGBA32，大多数情况都使用压缩贴图，比如 DXT5、ETC2_RGBA8 和 ASTC_RGBA_4x4。如图 13-40 所示，它们的像素排列格式是按照 4×4（每 16 个）为一个单位进行排列，总的排列顺序依然是从左到右，从下到上。

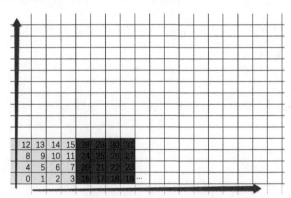

图 13-40　压缩格式像素

如图 13-41 所示，我们将一张 256×256 压缩贴图动态合并到一张 1024×1024 贴图的左下角处，并且保持原有的压缩格式。

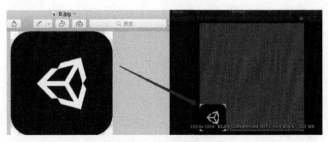

图 13-41　合并贴图

如代码清单 13-30 所示，由于这几个压缩格式中每个像素由 8 位组成，也就是一个字节来表示一个像素，所以 1024 大小的贴图就有 1024×1024 个像素组成。由于它们都是由 4×4 个 block 组成，所以每 16 个像素表示一个 block，依次复制它们即可。

代码清单 13-30　Script_13_27.cs 文件

```csharp
using System;
using UnityEngine;
using UnityEngine.UI;

public class Script_13_27 : MonoBehaviour
{
    public Texture2D texture;

    private void Awake()
    {
        if(texture.format ==  TextureFormat.DXT5 ||
            texture.format == TextureFormat.ETC2_RGBA8||
            texture.format == TextureFormat.ASTC_RGBA_4x4)
        {
            byte[] data = new byte[1024 * 1024];
            int blcokBytes = 16;
            //从小图的(0,0)位置开始，复制至大图的(0,0)，并且保持图片的宽和高分别是 256 和 256
            CombineBlocks(texture.GetRawTextureData(), 0, 0, data, 0, 0, 256, 256, 4,
                blcokBytes, 1024);
            var combinedTex = new Texture2D(1024, 1024, texture.format, false);
            combinedTex.LoadRawTextureData(data);
            combinedTex.Apply();
            GetComponent<RawImage>().texture = combinedTex;
        }
    }

    void CombineBlocks(byte[] src, int srcx, int srcy, byte[] dst, int dstx, int dsty,
        int width, int height, int block, int bytes,int maxWidth)
    {
        var srcbx = srcx / block;
        var srcby = srcy / block;

        var dstbx = dstx / block;
        var dstby = dsty / block;

        for(int i = 0; i < height / block; i++)
        {
            int dstindex = (dstbx + (dstby + i) * (maxWidth / block)) * bytes;
            int srcindex = (srcbx + (srcby + i) * (width / block)) * bytes;
            Buffer.BlockCopy(src, srcindex, dst, dstindex, width / block * bytes);
        }
    }
}
```

在上述代码中，我们首先通过 GetRawTextureData() 获取到贴图的压缩原始数据，接着再创建一个 1024×1024 的贴图，将之前贴图的原始数据复制进去。

> **注意**：运行时合并贴图需要读取图片的原始数据，所以必须开启贴图的 Read Enable/Write Enable 属性，这样内存就会占用双倍。平常开发中，如果没有这样大量换装的需求，最好不要开启它，或者可以考虑打包的时候将贴图的原始数据写在本地，不再使用贴图资源。运行时根据原始数据，再合并成一个新的贴图，这样就可以不设置 Read Enable/Write Enable 属性了。

13.7 小结

本章学习了 3D 游戏开发的基本元素，包括 Shader 渲染管线、多摄像机的管理、平面坐标与 3D 坐标/2D 坐标的转换。摄像机可以将显示渲染在 Render Texture 上，从而绑定在 UI 的 RawImage 上。但是如果模型带了粒子特效，渲染出来的就会有问题，此时可以在 UI 下面放一个 Spite Renderer 组件，这样粒子特效就可以和它进行混合了，最终就显示正确了。接着，我们还介绍了多场景管理，游戏对象创建在不同场景之上，场景的获取等。最后，介绍了角色控制器的使用，它与刚体的区别，以及模型之间的碰撞以及触发器事件、常用实战技巧等。

此刻，本书已经结束，祝大家学习愉快！

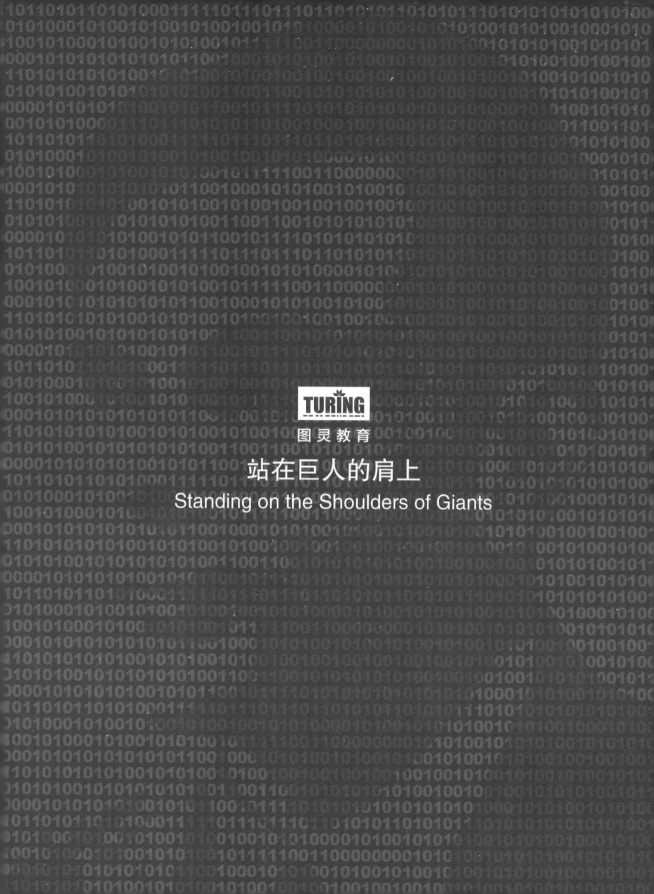